T0314205

Mathematics for the Liberal Arts

Mathematics for the Liberal Arts

Donald Bindner
Department of Mathematics and Computer Science
Truman State University
Kirksville, MO

Martin J. Erickson
Department of Mathematics and Computer Science
Truman State University
Kirksville, MO

Joe Hemmeter
Farmington, MI

A JOHN WILEY & SONS, INC., PUBLICATION

Cover Credit: @ usetrick/iStockphoto

Copyright © 2013 by John Wiley & Sons, Inc. All rights reserved

Published by John Wiley & Sons, Inc., Hoboken, New Jersey

Published simultaneously in Canada

No part of this publication may be reproduced, stored in a retrieval system, or transmitted in any form or by any means, electronic, mechanical, photocopying, recording, scanning, or otherwise, except as permitted under Section 107 or 108 of the 1976 United States Copyright Act, without either the prior written permission of the Publisher, or authorization through payment of the appropriate per-copy fee to the Copyright Clearance Center, Inc., 222 Rosewood Drive, Danvers, MA 01923, (978) 750-8400, fax (978) 750-4470, or on the web at www.copyright.com. Requests to the Publisher for permission should be addressed to the Permissions Department, John Wiley & Sons, Inc., 111 River Street, Hoboken, NJ 07030, (201) 748-6011, fax (201) 748-6008, or online at http://www.wiley.com/go/permission.

Limit of Liability/Disclaimer of Warranty: While the publisher and author have used their best efforts in preparing this book, they make no representations or warranties with respect to the accuracy or completeness of the contents of this book and specifically disclaim any implied warranties of merchantability or fitness for a particular purpose. No warranty may be created or extended by sales representatives or written sales materials. The advice and strategies contained herein may not be suitable for your situation. You should consult with a professional where appropriate. Neither the publisher nor author shall be liable for any loss of profit or any other commercial damages, including but not limited to special, incidental, consequential, or other damages.

For general information on our other products and services or for technical support, please contact our Customer Care Department within the United States at (800) 762-2974, outside the United States at (317) 572-3993 or fax (317) 572-4002.

Wiley also publishes its books in a variety of electronic formats. Some content that appears in print may not be available in electronic formats. For more information about Wiley products, visit our web site at www.wiley.com.

Library of Congress Cataloging-in-Publication Data:

Bindner, Donald, author.
  Mathematics for the liberal arts / Donald Bindner, Department of Mathematics and Computer Science, Truman State University, Kirksville, MO, Martin J. Erickson, Department of Mathematics and Computer Science, Truman State University, Kirksville, MO, Joe Hemmeter, Farmington, MI.
    pages cm
  Includes bibliographical references and index.
  ISBN 978-1-118-35291-5 (hardback)
  1. Mathematics—History—Textbooks. I. Erickson, Martin J., 1963– author. II. Hemmeter, Joe, 1950– author. III. Title.
  QA21.B56 2013
  510.9—dc23   2012023745

Printed in the United States of America.

10 9 8 7 6 5 4 3 2 1

To Linda, Christine, and Debbie

CONTENTS

PREFACE

This book is an introduction to mathematics history and mathematical concepts for liberal arts students. Students majoring in all fields can understand and appreciate mathematics, and exposure to mathematics can enhance and invigorate students' thinking.

The book can be used as the basis for introductory courses on mathematical thinking. These courses may have titles such as "Introduction to Mathematical Thinking." We describe the history of mathematical discoveries in the context of the unfolding story of human thought. We explain why mathematical principles are true and how the mathematics works. The emphasis is on learning about mathematical ideas and applying mathematics to real-world settings. Summaries of historical background and mini-biographies of mathematicians are interspersed throughout the mathematical discussions.

What mathematical knowledge should students have to read this book? An understanding of basic arithmetic, algebra, and geometry is necessary. This material is often taught in high school or beginning college-level courses. Beyond this background, the book is self-contained. Students should be willing to read the text and work through the examples and exercises. In mathematics, the best way (perhaps the only way) to learn is by doing.

Part I, comprising the first three chapters, gives an overview of the history of mathematics. We start with mathematics of the ancient world, move on to the Middle

Ages, and then discuss the Renaissance and some of the developments of modern mathematics. Part II gives detailed coverage of two major areas of mathematics: calculus and number theory. These areas loom large in the world of mathematics, and they have many applications. The text is rounded out by appendices giving solutions to selected exercises and recommendations for further reading.

A variety of courses can be constructed from the text, depending on the aims of the instructor and the needs of the students. A one-semester course would likely focus on selected chapters, while a two-semester course sequence could cover all five chapters.

We hope that by working through the book, readers will attain a deeper appreciation of mathematics and a greater facility for using mathematics.

Thanks to the people who gave us valuable feedback about our writing: Linda Bindner, Robert Dobrow, Suren Fernando, David Garth, Amy Hemmeter, Mary Hemmeter, Daniel Jordan, Kenneth Price, Phil Ryan, Frank Sottile, Anthony Vazzana, and Dana Vazzana.

Thanks also to the Wiley staff for their assistance in publishing our book: Liz Belmont, Sari Friedman, Danielle LaCourciere, Jacqueline Palmieri, Susanne Steitz-Filler, and Stephen Quigley.

PART I

MATHEMATICS IN HISTORY

CHAPTER 1

THE ANCIENT ROOTS OF MATHEMATICS

Mathematics—the unshaken Foundation of Sciences, and the plentiful Fountain of Advantage to human affairs.

ISAAC BARROW (1630–1677)

1.1 Introduction

Mathematics is a human enterprise, which means that it is part of history. It has been shaped by that history, and in turn has helped to shape it. In this chapter we will trace these connections.

Many societies have contributed to mathematics, but a main historical thread is discernible, one that has led directly to today's mathematics. That thread began in the ancient Mediterranean world, swelled mightily in ancient Greece, dwindled at the time of the Roman empire, was kept alive and augmented in the Muslim world, re-entered Western Europe in the Renaissance, developed in Europe for several centuries, then spread throughout the world in the 20th century. We will spend most of our time on this thread, in part because so much is known about it, with a few excursions into other cultures.

Mathematics for the Liberal Arts.
By Donald Bindner, Martin Erickson, Joe Hemmeter Copyright © 2012 John Wiley & Sons, Inc.

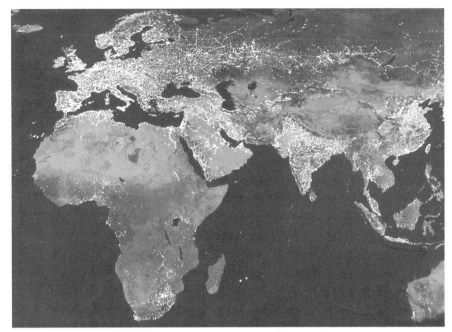

Eurasia and Africa.[1]

Fingers, Knots, and Tally Sticks

Experiments have shown that humans, and other animals, are born with innate mathematical abilities. They regularly distinguish between, say, one tree and two trees. The next logical step is counting, that is, establishing a one-to-one correspondence between sets of objects. This is no doubt also an ancient ability.

Once we can count objects, how do we communicate numbers to others? Most of what follows in this chapter is based on the historical, i.e., written record. But writing is a fairly recent invention. Before the written word, people used a variety of methods to represent numbers. Surely one of the first, and still important, methods was the use of various parts of the body. Some quite elaborate systems have been developed. The Torres Strait islanders, an indigenous Australian people, used fingers, toes, elbows, shoulders, knees, hips, wrists, and sternum to represent different numbers. Many languages preserve the remnants of such systems: the word for "five," for example, is "hand" in Persian, Russian, and Sanskrit. And it is no coincidence that our number system is based on ten, the number of fingers.

Perhaps the most popular numbering system used notches on sticks or bones, so-called *tally sticks*, from the French word *tailler*, to cut. These date back at least

---

[1]From "Earth at Night." C. Mayhew and R. Simmon (NASA/GSFC), NOAA/NGDC, DMSP Digital Archive.

35,000 years, and must rank as one of the most successful technologies ever. As recently as 1826, tally sticks were used in official English tax records.

Another popular counting device was the stone. Our word "calculation" derives from the Latin *calculus*, which is a small stone. Early versions of the abacus were stones on the ground; "abacus" likely derives from the Hebrew *abhaq*, dust.

Knotted strings were a popular accounting tool throughout the world. The most notable examples of these were the amazing Incan *quipu*, which consisted of multiple knotted cords (up to 2000 of them) joined together.

A leading theory of the origin of writing in Mesopotamia, proposed by Denise Schmandt-Besserat, relates to a different method of recording numbers. It starts with the use of small clay tokens, found in archaeological sites, beginning circa 8000 BCE[1]. These tokens, in various standard shapes, were used for accounting: one shape might represent one sheep, for example, another one goat, or ten sheep. Imagine you are a merchant, and have hired someone to deliver a herd of 27 sheep to a neighboring city. The buyer needs to have some way to verify that the number of sheep that arrive is the same number sent. The solution was to encase tokens representing 27 sheep in a clay "envelope," a hollow ball. The ball could be broken open at the destination, and the number of sheep verified.

Now imagine that the sheep's journey has two legs; person A delivers them to person B, who in turn delivers them to the buyer. If B breaks open the ball to verify the count, what is the buyer to do? The solution found was to make impressions on the ball, using the tokens, before they were placed inside. After the clay ball hardened, these impressions could serve as a record as well as the tokens. Eventually, it was realized that the tokens were unnecessary. The "writing" on the ball sufficed.

Agriculture and Civilizations

Some time around 10,000 years ago, humans began developing agriculture, inaugurating the Neolithic, the "new stone age." The first important crops were grains—large-seeded grasses—including wheat, sorghum, millet, and rice. Gradually, various animals were domesticated, notably cattle, sheep, oxen, pigs, and goats. This whole set of developments dramatically changed the way people lived. Instead of living in relatively small nomadic bands of "hunter-gatherers," they started settling into villages. This allowed a larger population density.

In some areas of the world, usually in the flood plains of great river valleys, the agricultural settlements developed civilizations. Among these areas were Mesopotamia, the Nile in Egypt, the Yellow River in China, the Indus River in Pakistan, and the Ganges in India. Civilizations were characterized by more central organization, often including irrigation, granaries to store surplus grain, and cities.

The civilizations were based on the existence of agricultural surplus, which freed people to work on other things. This led to the development of many new technolo-

---

[1]Circa (abbreviation c.), from the Latin, means "around." We will use it for approximate dates. BCE (Before Current Era) is becoming standard for dates before the year 0, what used to be written B.C. CE is used for dates after the year 0, in place of A.D.

gies. Among these were the plow, wheeled vehicles, and, most notably, writing and metallurgy.

Civilizations required different, more sophisticated, types of mathematics. Geometry was needed for surveying land, building canals, dikes, and ditches, and constructing larger buildings like granaries and palaces. Administering the new city-states, apportioning taxes, and paying workers made increasing demands on arithmetic and algebra, as did the expanded commercial activity.

With the rise of civilization came new class structures. Most people were farmers, but some became blacksmiths, leather workers, engineers, architects, merchants, priests, scribes, surveyors, and of course kings. Some of the new, specialized professions (such as surveyors) nurtured their own mathematical techniques, handing them down through the generations. In some societies, small groups inside the new classes turned their collective attention to developing mathematics generally. Society provided practical inspiration for the new mathematics, but some mathematicians pursued knowledge for its own sake.

EXERCISES

1.1   What mathematics would a pre-agricultural (hunter-gatherer) society need?

1.2   What mathematics would an agricultural village need that a hunter-gatherer society would not?

1.3   What mathematics would a city need that a agricultural village would not?

1.4   Look up Incan quipus in your favorite Internet search engine. What did they look like? How were they used?

1.2 Ancient Mesopotamia and Egypt

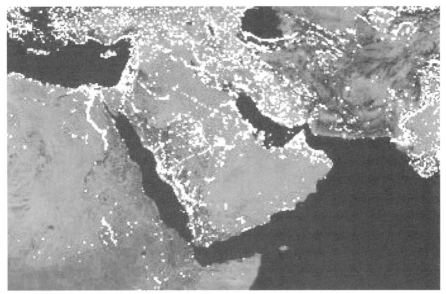

The Middle East.[1]

Two of the earliest civilizations arose in the Near East.

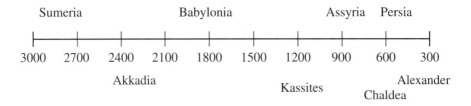

Ancient Mesopotamian history. All dates are BCE.

Mesopotamia (from the Greek, literally "between the rivers") is the plain between the Tigris and Euphrates rivers, about 600 miles long, in modern-day Iraq. Mesopotamia was home to the earliest known agriculture. The major crops were wheat and barley, but there were also fruits, including dates, grapes, figs, melons, and apples; vegetables, including eggplant, onions, radishes, beans, and lettuce; and sheep, cattle, goats, and pigs.

[1]From "Earth at Night." C. Mayhew and R. Simmon (NASA/GSFC), NOAA/NGDC, DMSP Digital Archive.

Figure 1.1   The Tigris and Euphrates rivers are above and to the left of center; the Nile is on the left.[1]

Between the Rivers

The farmers relied on the flooding of the Tigris and Euphrates. These floods, which could be violent and destructive, nonetheless left behind very fertile silt. Mesopotamia doesn't get much rain, so the other important ingredient to agriculture was irrigation. One of the critical functions of the government was the construction and maintenance of irrigation systems. The first Mesopotamian civilization was the Sumerian, named after the city-state of Sumer in southern Mesopotamia. It arose circa 3500–3000 BCE. Politically, the early Sumerians did not have an empire; empires came later. Instead, they were organized into city-states, ruled by priest-kings. These city-states built up bureaucracies to manage the irrigation systems and the surplus grains. They even had postal systems. The Sumerians are credited with the invention of plows, the potter's wheel, and wheeled carts. Their greatest invention was an improved writing system. Earlier writing systems had relied principally on *pictograms*, symbols which were meant to look like the thing represented. The Sumerians developed, over many hundreds of years, a system of standardized *ideograms*, symbols that represented ideas.

The Sumerians, and their successors in Mesopotamia, wrote by using a stylus, a reed cut at an angle, to make impressions in wet clay tablets. The tablets were then

[1] Map by Sémhur. Wikipedia Commons.

baked until hard. Their writing is called cuneiform, from the Latin *cuneus* (wedge) and *forma* (shape). It is from these tablets that we learn most of what we know about their history. Ironically, the preservation of this history was often assisted when the buildings housing the tablets were burned. This baked the tablets, making them more durable.

Mesopotamia is a crossroads. This allowed it to be a trading center and to profit from the ideas of other civilizations. It also was subject to regular raids and occasional full-scale invasions from its neighbors. So there was a succession of civilizations and empires through the years. The Sumerians were conquered by the Akkadians, whose most famous ruler was Sargon the Great, who lived around 2250 BCE. The Akkadians were replaced by the Babylonians.

The most important Babylonian king was Hammurabi, who ruled c. 1792–1750 BCE. He is famous for promulgating the first code of laws, a list of 282 short "laws." Here is one: "If a man puts out the eye of an equal, his eye shall be put out." Presumably he can put out the eye of an inferior with impunity. A "tooth for a tooth" is also here. Many of the laws end with "shall be put to death."

If you read about ancient Mesopotamia, you will often find it referred to as Babylon, even during those times when Babylon was not its most important city. Perhaps this reflects Babylonian cultural accomplishments. In particular, the high point of Mesopotamian mathematics was during this time. After the Babylonians, mathematics was mainly stagnant.

The Babylonians in their turn fell to the Kassites, who had a new weapon, horse-drawn chariots, the tanks of their day. In the 9th century BCE, the Assyrians ruled, relying on iron weapons. In the 7th century came the Chaldean empire, when Nebuchadnezzar built the hanging gardens of Babylon and sent many Hebrews into Babylonian exile. The Persians under Cyrus invaded in 538 BCE, and Alexander the Great took over in 330 BCE, bringing Greek culture with him.

Most of the invaders did not displace the local culture. Instead they adopted much of it. In particular, the bureaucracy essential to managing their conquests tended to stay in place. This bureaucracy included the scribes who were at the heart of mathematics. Early on, the Sumerians developed scribal schools, which taught writing and mathematics, among other subjects, to future bureaucrats. Most of these scribes came from wealthy families. They were the ruling elite of their day.

The schools for scribes became centers of culture, including mathematics. Their main emphasis, however, was business and administration. Irrigation systems had to be run, laws administered, lands apportioned, taxes levied. A very important responsibility, and one that had a profound influence on mathematics in Mesopotamia and elsewhere, was maintenance of the calendar, which required accurate measurements of the heavens.

Most of the tablets from which we learn about Mesopotamian mathematics were created at the schools, for the purpose of training scribes. They usually took the form of solving problems. The problems were stated in practical terms: measuring fields, apportioning inheritance, and so on. But the purpose of the tablets was to train students in mathematical methods rather than in practical problem-solving. In some ways, mathematical textbooks haven't changed.

Although the tablets reveal a strong mathematical tradition, they do not reveal a lot of theory, or even general methods. These methods are implied by the results, but apparently were restricted to an oral tradition. Sometimes historians have been able to infer the methods used, sometimes they have to guess them.

Numeration   One of the greatest accomplishments of the Mesopotamian culture was the development of the best number system of antiquity.

A problem that any sophisticated number system must address is how to group numbers. Small numbers may be expressed by simple ticks, but if we are to handle larger numbers, they must somehow be grouped together. The Sumerians were the first to establish a consistent grouping system. Unlike our number system, which groups by powers of 10 (1, 10, 100, ...), the Sumerians grouped by powers of 60 (1, 60, 3600, ...). This system is called *sexagesimal* (from the Latin *sexagesimus*, sixtieth), as opposed to our *decimal* system (from *decimus*, tenth).

No one knows exactly why they chose this system, but one of its useful features is that 60 has many divisors: 1, 2, 3, 4, 5, 6, 10, 12, 15, 20, 30, and 60. By contrast, 10 has only four divisors. More divisors make fractions easier to work with. Consider the multiplication $1/2 \times 3/5$. One way of doing this is to convert to a decimal representation: $.5 \times .6 = .3$. Thus, we can use our regular multiplication and not have to deal with fractions as ratios. It only works, however, because 2 and 5 are divisors of 10. Consider $2/3 \times 11/12$. This problem does not lend itself to an easy decimal representation in base 10. Ten doesn't have enough factors. We will see below how a sexagesimal system can handle this multiplication.

The legacy of the Mesopotamian sexagesimal system survives to this day: we divide hours into 60 minutes, minutes into 60 seconds, and we divide the circle into 360 degrees.

One of the most important advances in representing numbers was the idea of place-value notation, developed by the Babylonians. To understand this, let us look at how we represent numbers in our own place-value system. Consider the number 235:

$$235 = (2 \times 100) + (3 \times 10) + (5 \times 1) = (2 \times 10^2) + (3 \times 10^1) + (5 \times 10^0).$$

The meaning of each digit depends on its *place* in the number. So, for example, 235 is not the same as 253.

Since the Babylonians had a sexagesimal system, they would represent the number 235 in powers of 60. Thus

$$235 = (3 \times 60) + (55 \times 1) = (3 \times 60^1) + (55 \times 60^0),$$

so we might write this as 3,55, using a comma to separate powers of 60.

What number would 3,40,6 represent?

$$\begin{aligned} 3,40,6 &= (3 \times 60^2) + (40 \times 60^1) + (6 \times 60^0) \\ &= (3 \times 3600) + (40 \times 60) + (6 \times 1) \\ &= 10800 + 2400 + 6 \\ &= 13206 \end{aligned}$$

One of the advantages of a place-value system is the ability it gives us to express arbitrarily large numbers with a small set of symbols, ten symbols in the case of our decimal system. The number symbols used in Mesopotamia changed dramatically through the years. The Sumerians used hundreds of symbols, both pictorial and phonetic. Their successors, the Akkadians, developed a standardized system of number ideograms. These ideograms represented the *idea* of a number, divorcing it from concrete notions such as a "hand," for five.

The Babylonians reduced the number of symbols to two, one for 1 and one for 10. They repeated these symbols as necessary to get the numbers from 1 to 59, as in Figure 1.2. For numbers greater than 59, they used their place value system, as we do for numbers greater than 9. Figure 1.3 shows how they would write the number $2,34 = 2 \times 60 + 34$.

Figure 1.2   Babylonian symbols.

Figure 1.3   Babylonian 2,34.

This system could also handle numbers less than one, in the same way as our decimal system. We write $2/5$ as .4. If we use a semi-colon, instead of a decimal point, the Babylonians could use ; 24 for two-fifths, since 24 is two-fifths of 60. Since 60 has so many divisors, this was a convenient way of writing fractions. Another example: since 20 is one-third of 60, we would write $1/3 = ;20$. Here is an example of a mixed fraction.

$$70\frac{2}{15} = 1 \times 60 + 10 + 2 \times \frac{1}{15} = (1,10) + 2 \times (;4) = 1,10;8$$

The Babylonians did not use a semi-colon, or any indicator of where fractions started, so there was an ambiguity to their numerals. For example, they wrote all these numbers the same way:

$$2,5,0 = 2 \times 60^2 + 5 \times 60 = 7500$$

$$2,5 = 2 \times 60 + 5 = 125$$

$$2;5 = 2 + 5 \times 60^{-1} = 2\frac{1}{12}$$

$$;2,5 = 2 \times 60^{-1} + 5 \times 60^{-2} = \frac{5}{144}.$$

They would determine which number was meant by the context.

The other missing element in this system was the notion of zero. The Babylonians did not have a zero number, and would never write a number such as 2,5,0. They wrote 2,5 and interpreted it as 2,5,0 from context. They did have to develop some way to indicate skipped digits, such as what we mean when we write 0 in the middle of a number, as in 205. This gap was indicated in different ways, often with just a space.

Computation and Algebra   The Babylonian number system allowed for a sophisticated arithmetic. Like us, the Babylonians wrote down multiplication tables. They also had tables of squares and cubes of numbers. For division, they used tables of reciprocals. For example, consider the problem $32 \div 25$. If we had a table giving us .04 as the reciprocal of 25, we could translate the division problem into its equivalent multiplication problem, $32 \times .04$. This the Babylonians regularly did, aided by the fact that 60 has many divisors. Of course, this works well only when the reciprocal has a nice form; think of trying it with $32 \div 7$. The Babylonian reciprocal tables were usually restricted to the nicer reciprocals. For other divisions, approximation techniques were used.

The Babylonians were very good at calculating square roots. A tablet from around 1600 BCE gives the approximation $\sqrt{2} = 1;24,51,10$. In decimal terms, this is about 1.414213, while the correct value starts 1.414214 . . . . The approximation is within one-millionth of the correct value.

The method they used to obtain such accuracy is not known with certainty, but may be what later was called Heron's method, since Heron was the first to write it down, over 1500 years later. This method can be used to find the square root of any number N. You start with any guess for $\sqrt{N}$, say, x_1. You then generate a sequence of numbers x_2, x_3, x_4, . . . , as follows.

$$x_2 = \frac{1}{2}\left(x_1 + \frac{N}{x_1}\right)$$

$$x_3 = \frac{1}{2}\left(x_2 + \frac{N}{x_2}\right)$$

$$x_4 = \frac{1}{2}\left(x_3 + \frac{N}{x_3}\right)$$

You can continue this pattern as long as you like. The numbers x_1, x_2, x_3, x_4, . . . get closer and closer to $\sqrt{N}$.

As an example, we will approximate $\sqrt{2}$. We first guess $x_1 = 1.5$. (The initial guess doesn't have to be too accurate. Just pick something reasonable.) Then

$$x_2 = \frac{1}{2}\left(1.5 + \frac{2}{1.5}\right) \approx 1.41666666666667$$

$$x_3 = \frac{1}{2}\left(1.41666666666667 + \frac{2}{1.41666666666667}\right) \approx 1.41421568627451$$

$$x_4 = \frac{1}{2}\left(1.41421568627451 + \frac{2}{1.41421568627451}\right) \approx 1.41421356237309.$$

Already, all the digits given for x_4 are accurate.

Babylonian mathematicians had a good understanding of linear equations in one variable, $ax + b = c$, even though the scribes had no general notion of a variable. They could also solve systems of two linear equations in two unknowns.

Many of the tablets concern problems that involve solving quadratic equations, equations we would write as $ax^2 + bx + c = 0$. Here is an example: "I summed the areas of my two square-sides so that it was 0;21,40. A square-side exceeds the (other) square-side by 0;10."

Let us translate that into modern notation. The number 0;21,40 refers to $21/60 + 40/3600 = (21 \times 60 + 40)/3600 = 13/36$. Also, 0;10 is 10/60, or 1/6. So we have two squares, one of side x, say, and the other of side $x - 1/6$. Since the sum of their areas is $13/36$, the problem is to solve the equation

$$x^2 + \left(x - \frac{1}{6}\right)^2 = \frac{13}{36}.$$

The solution is then given step-by-step. It starts like this: "You break off half of 0;21,40 and you write down 0;10,50."

Note that the problem itself is stated in geometric form. This was typical; there was no clear distinction between algebra and geometry. The Babylonians did not have an algebraic notation. They also had no symbols for arithmetic operations, like $+, -, \times, \div$. The problems were stated in words.

The solution of the problem is given by a set of specific instructions on how to proceed. This also was typical. There was no notion of a general solution like our quadratic equation, even though the scribes clearly had methods for solving many such equations. The idea of a general theory had yet to be developed.

Given these restrictions, their accomplishments in algebra were impressive.

Geometry   The people of Mesopotamia dealt with many practical problems requiring geometric knowledge for their solutions. Surveyors had to measure distance and compute areas. Builders of large structures needed knowledge of distance, area, and volume.

Surviving tablets give us some insight into their geometry. The scribes had rules for computing the areas of triangles, rectangles, and various other plane figures such as pentagons, hexagons, and trapezoids. Rules for computing with circles usually have $\pi = 3$, although the better approximation of $3\frac{1}{8}$ was also used. The Babylonians had the correct formula for the volume of a truncated pyramid, which is a pyramid with its top cut off.

What we now call the Pythagorean Theorem, $a^2 + b^2 = c^2$, where a and b are the legs of a right triangle and c its hypotenuse, was known in Mesopotamia at least 1000 years before Pythagoras. A famous clay tablet from the Old Babylonian Period lists a number of Pythagorean triples, which are sets of three numbers obeying the Pythagorean equation, for example, 3–4–5 (since $3^2 + 4^2 = 5^2$). See Section 5.11 for more on these triples.

We must remind ourselves when dealing with cultures several thousand years in the past, that our knowledge is spotty. In the matter of geometry, it has been suggested that clay tablets were not the best medium for drawings. Perhaps the scribes did their best work drawing in sand, or some other medium which has been lost. Having said that, it is certain that their geometry never approached anything like the sophistication attained in ancient Greece. The existing tablets address particular problems, not general theory. There is no notion of proof.

Egypt before Alexander

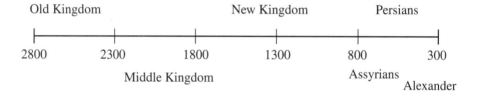

Ancient Egyptian history. All dates are BCE.

Gift of the Nile   Egypt consists of a desert cut by the Nile River. The upper Nile, about 600 miles from Aswan to Memphis, is a narrow valley, not more than 15 miles wide, bordered by cliffs. The lower Nile, about 150 miles from Memphis to the Mediterranean, is a fan-shaped marshy delta. Outside the river valley is only desert.

In ancient times, before modern flood control systems, the Nile flooded every year. The flood left behind rich soil. About 7000 years ago, the Egyptians began to farm this soil. As in Mesopotamia, there was little rain, so the agriculture that developed was reliant on irrigation. Egyptians raised wheat and barley and a variety of vegetables for food, and flax for clothing. The most important domesticated animals were cattle, but they also kept sheep, goats, and pigs. It was one of the most productive agricultural areas in the world, producing in good years a large surplus that could support a sophisticated civilization.

The Egyptians had a written language by about 3200 BCE, almost as early as in Mesopotamia. The *hieroglyphs*, from the Greek for "sacred carvings," were pictographs. A little later, the Egyptians developed the cursive *hieratic* script, and, in the first millennium BCE, an alphabetic system called *demotic*.

Hieroglyphs were painted or carved on monuments, while the hieratic and demotic systems were written using ink on papyrus. *Papyrus* was a type of paper made from a reed, *Cyperus papyrus*, found in the Nile Delta. Our word "paper" derives from papyrus. Papyrus sheets were cheaper than the clay tablets used in Mesopotamia, but they don't last as long. Thus much of what we know of Egyptian mathematics, with some exceptions noted below, comes from inscriptions on monuments.

One meaning of hieroglyphic in English is "hard to decipher," and Egyptian hieroglyphs were unreadable by modern scholars until the decipherment of the famous Rosetta Stone. This stele (inscribed stone slab) was found by Napoleon's armies in Egypt in 1799. It had a message written in Greek, hieroglyphic, and demotic, which allowed Jean Champollion, after much work, to decipher it in 1821.

Egypt was first brought together under a single ruler about 3100 BCE. In the Old Kingdom (c. 3000–2200 BCE), the Egyptians adopted much from the Sumerians, including irrigation systems, the plow, and metallurgy. They too had a class of scribes to assist in administration, and again it was from this class that most of their mathematics originated. It was during the Old Kingdom that the biggest pyramids were built, including the famous Great Pyramid of Giza (c. 2500 BCE).

Egypt was not organized into city-states like Mesopotamia, but instead was centrally ruled by the pharaoh, who was considered a god. Only under such a centralized system could monuments such as the pyramids be constructed. According to the Greek historian Herodotus (c. 484–425 BCE), who first called Egypt the "gift of the Nile," the Great Pyramid required the labor of 400,000 men at a time, for three months of the year, over twenty years. That was after the effort of ten years building the road needed to transport the materials. (Herodotus wrote 2000 years after the fact, so the precise numbers shouldn't be taken too seriously. Nonetheless, they are not far from modern estimates.)

A period of political unrest followed the Old Kingdom, until the arrival of the Middle Kingdom (c. 2100–1800 BCE). Egyptian culture flourished in this period. In fact, it was the high point of ancient Egyptian mathematics. The Middle Kingdom ended with the invasion of the Hyksos, from Syria-Palestine.

The Hyksos were expelled, starting the New Kingdom (c. 1600–1100 BCE). In this period, Egypt expanded its power into the Middle East (Palestine and Syria) and to the south (Nubia and the Sudan). The New Kingdom was followed by a period of weak kings and a number of invasions, including conquests by the Assyrians in 671 BCE, the Persians in 525 BCE, and finally, in 332 BCE, Alexander the Great.

Mathematics in ancient Egypt was applied to many of the same uses as in Mesopotamia: building irrigation systems and granaries, levying taxes, paying workers, and apportioning the surplus grain. Of particular note is surveying. Farms had to be marked off again after each yearly flood destroyed the previous year's boundaries. For this, surveyors needed *geometry*—from the Greek *geo*, Earth, and *metria*, measure.

As mentioned above, most of what we know about ancient Egypt is from the thousands of inscriptions on monuments they left behind. The most important sources for the later mathematics are, however, a dozen or so papyri. Two stand out. The *Moscow Mathematical Papyrus* (which is in the Moscow Museum of Fine Arts), dating from about 1850 BCE, contains a list of twenty-five problems. Eleven of these twenty-five concern ways of making different beers and breads. The Rhind Mathematical Papyrus (bought in the 19th century in Luxor, Egypt by a man named Rhind) is a scroll 13 inches wide and 18 feet long, which contains eighty-seven problems and tables. It dates from around 1650 BCE, but its author writes that he copied it from a work written 200 years before that.

Numeration and Arithmetic   The Egyptians had one system of numeration for each of their three writing systems: hieroglyphic, hieratic, and demotic. All of them were decimal, that is, they grouped numbers by powers of ten. The hieroglyphic system had symbols for 1, 10, 100, etc. Multiples of these were represented by repeating symbols. The hieratic and demotic systems added symbols for 2, 3, ..., 9 and 20, 30, ..., 90, and so on, which made writing large numbers much easier. Unlike the Mesopotamians, the Egyptians never developed a fully positional system.

Addition and subtraction in these systems was rather like in our own. For multiplication, they used a doubling system called *duplation*. It is similar to how modern computers multiply. Here is an example, computing $11 \cdot 17$. First we write powers of 2, with their corresponding multiples of 17. Each line is obtained by multiplying the previous line by 2.

$$
\begin{array}{cc}
1 & 17 \\
2 & 34 \\
4 & 68 \\
8 & 136 \\
\end{array}
$$

Why only four lines? Because we can write $11 = 1 + 2 + 8$. Therefore $11 \cdot 17 = (1 + 2 + 8) \cdot 17 = 1 \cdot 17 + 2 \cdot 17 + 8 \cdot 17$. To complete the multiplication, we need only add the entries in the right-hand column corresponding to 1, 2, and 8, to get $11 \cdot 17 = 17 + 34 + 136 = 187$. Division was handled using the same idea, although it was a bit more complicated due to remainders.

Egyptians didn't use fractions as we do. They only used *unit fractions* of the form $1/n$, for example, $1/2$, $1/3$, or $1/4$. The sole exception is their use of $2/3$. Other fractions were expressed as sums of unit fractions. A famous problem from the Rhind Papyrus asks how to divide six loaves of bread among ten men. The answer given was $1/2 + 1/10$. You can check that this equals $6/10$. In fact, dividing the loaves is easy using $1/2 + 1/10$. Cut five of the loaves in half, giving one-half to each man. Then cut the last loaf into tenths, giving each man one tenth. Each man ends up with one-half plus one-tenth.

The Egyptians did not prove that any fraction can be written as the sum of distinct unit fractions (they didn't have the notion of proof), but it can be proven. To do the actual computations, they used extensive tables. The Rhind Papyrus, for example, includes a table decomposing fractions of the form $2/n$ into sums of unit fractions.

Geometry   One area where Egyptian mathematics excelled was geometry. They knew how to compute areas of rectangles, triangles, and trapezoids, as well as volumes of rectangular boxes and various cylinders. They also used similar triangles.

Two triangles are similar if they have the same three interior angles. One triangle is a blown up version of the other. They are useful because the ratios of their corresponding sides are the same. As for circles, again we refer to the Rhind Papyrus, where the area of a circle is given as $(8d/9)^2$, where d is the diameter of the circle. This was equivalent to approximating π by $256/81$, about 3.16.

Given the importance of pyramids to them, it is no surprise that the Egyptians knew how to compute the volume of a pyramid. The *Moscow Papyrus* has a method for calculating the volume of a truncated pyramid. (You can see a drawing of a truncated pyramid on the back of a U.S. dollar bill, underneath an eye.) The method is equivalent to the formula

$$V = \frac{1}{3}h(a^2 + ab + b^2),$$

where a is the length of the lower base, b is the length of the upper base, and h is the height (see Figure 1.4).

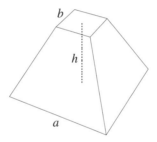

Figure 1.4   A truncated pyramid.

The Egyptians did not pursue theoretical geometry the way the Greeks later did. They were, however, unsurpassed in practical geometry. This can be seen in their monuments. The base of the Great Pyramid is an almost perfect square about 756 feet per side. The length of the sides differ by less than one foot—and this was done using 2.5 ton stone blocks (more than 2,300,000 of them). Later Egyptian construction never reached this level of precision.

Algebra   The Egyptians were able to solve linear equations, and some quadratic equations. They, as all ancients, were hampered by the lack of a good notation. For them, all problems were word problems.

One method the Egyptians scribes used is called the method of *false position*. This method was adopted by many other peoples, and used as late as the Middle Ages in Europe. As an example, the Rhind Papyrus offers this in Problem 26: "A quantity whose fourth part is added to it becomes 15. What is the quantity?" In modern notation, we want to solve the equation $x + x/4 = 15$ for x. The method of false position involves guessing a (probably incorrect) solution, then adjusting it using proportionality. In our problem, the scribe guessed that $x = 4$, to make the fraction $x/4$ easier. If $x = 4$, then $x + x/4 = 5$. Since we want $x + x/4 = 15$, we multiply by 3, because $15/5 = 3$. If we multiply $x = 4$ by 3, we get $x = 12$. You can check that $12 + 12/4 = 15$, so we have solved the problem.

Astronomy and the Calendar   In ancient Egypt, as in many places at many times, an important use of mathematics was in astronomy. The astronomers of Egypt were priests, which suggests that astronomy was not only a practical science. Astronomers kept track of the Sun, Moon, planets, and stars. One notion they used to track the seasons was that of a helical rising of a star, which meant that the star rose just before the Sun. The most important helical rising was that of Sothus (which we know as Sirius), the brightest star in the sky. This rising occurred in July, shortly before the onset of the Nile floods. Sothus was known as the Dog Star, and we still refer to this time of year as the dog days of summer.

The Egyptians developed the calendar on which ours is based. Their civil calendar, used for official record keeping (as opposed to the everyday lunar calendar), had 365 days, divided into twelve 30-day months, plus an extra five days at the end of the year. Actually, they knew that the year was about $365\frac{1}{4}$ days, but they never adjusted their calendar with leap years, as we do.

EXERCISES

1.5   Each of the numbers below is given in sexagesimal form. Translate each into our decimal form.

    a) 2

    b) 3,1,2

    c) 1,2;6

    d) ;1,40

1.6   Translate each of the decimal numbers into sexagesimal form.

    a) 2

    b) 122

    c) 7265

    d) .2

    e) $1\frac{1}{3}$

1.7   a) Write the fractions $2/5$ and $11/12$ in sexagesimal form.

    b) Use the results of (a) to write $2/5 + 11/12$ in sexagesimal form. (Hint: think of how adding decimals works.)

    c) Add $2/5 + 11/12$ in our usual way, and confirm that you get the same answer.

1.8   a) Write the fraction 1/15 in sexagesimal form.

    b) Use (a) to divide 7 by 15, expressing the result in in sexagesimal form.

1.9   a) Write the fraction 1/30 in sexagesimal form.

b) Use (a) to divide 43 by 30, expressing the result in in sexagesimal form.

1.10   Use Heron's method to approximate $\sqrt{3}$, accurate to eight decimal places. (A calculator may be necessary.) Check your answer by squaring it.

1.11   Solve the Babylonian problem given in the text: "I summed the areas of my two square-sides so that it was 0;21,40. A square-side exceeds the (other) square-side by 0;10."

1.12   a) Here is another problem from a Babylonian tablet, written around 2000 BCE: "I have added the area and two-thirds the side of my square and it is 0; 35. What is the side of my square?" Translate this into modern notation. The result should be a quadratic equation.

b) Solve the equation. Do you get the same answer as on the tablet? "You take 1. Two-thirds of 1 is 0; 40. Half of this, 0; 20, you multiply by 0; 20 and it, 0; 6, 40, you add to 0; 35 and the result 0; 41, 40 has 0; 50 as its square root. The 0; 20 which you have multiplied by itself, you subtract from 0; 50, and 0; 30 is the side of the square."

1.13   Use duplation to calculate 13 times 15.

1.14   Use duplation to calculate 15 times 22.

1.15   Show that the formula $(8d/9)^2$ given in the Rhind papyrus for the area of a circle is equivalent to approximating π by $256/81$.

1.16   a) Find the volume of a truncated pyramid with lower base 100 feet, upper base 30 feet, and height 69 feet.

b) No existing papyrus gives the volume of a whole (as opposed to truncated) pyramid, but this can easily be derived from the formula above. What is the formula for the volume of a whole pyramid? (Hint: what happens to b as a truncated pyramid gets closer to a whole pyramid?)

1.17   Around the year 1200 CE, the famous mathematician Fibonacci described a method for expressing any fraction as the sum of distinct unit fractions. The method was simple: find the largest unit fraction less than your number. Then subtract it from your number and repeat. For example, consider $41/42$. The largest unit fraction less than this is $1/2$. Subtracting, we get $41/42 = 1/2 + 20/42$. The largest unit fraction less that $20/42$ is $1/3$. Subtracting, $20/42 = 1/3 + 6/42$, so $41/42 = 1/2 + 1/3 + 6/42 = 1/2 + 1/3 + 1/7$, and we are done.

a) Express $7/10$ as the sum of distinct unit fractions.

b) Express $8/15$ are the sum of distinct unit fractions in two different ways.

1.18   Solve by the method of false position: "A quantity whose seventh part is added to it becomes 32. What is the quantity?"

1.19   Solve by the method of false position: "A quantity whose fifth part is subtracted from it becomes 6. What is the quantity?"

1.20 Suppose that a civilization had a system similar to the Babylonian, but based on 5 instead of 60. If they write the first five numbers as A, B, C, D, and E, how would they write these numbers?

 a) 23
 b) 72
 c) .2
 d) .24

1.3 Early Greek Mathematics: The First Theorists

> Mathematics, rightly viewed, possesses not only truth, but supreme beauty—a beauty cold and austere, like that of sculpture, without appeal to any part of our weaker nature, without the gorgeous trappings of painting or music, yet sublimely pure, and capable of a stern perfection such as only the greatest art can show. The true spirit of delight, the exaltation, the sense of being more than Man, which is the touchstone of the highest excellence, is to be found in mathematics as surely as poetry.
>
> BERTRAND RUSSELL (1872–1970)

Modern mathematics is distinguished not only by its techniques and results, but by its logical structure. A mathematician does not merely discover formulas, but proves theorems, starting from well-understood assumptions and definitions. This logical structure is the invention of the Greeks, surely one of the greatest inventions in human history. They also applied this new invention, especially in geometry, to produce some very sophisticated mathematics.

More than only mathematics, much of Western intellectual tradition dates to classical Greece. Mathematics was part of a larger philosophical movement in which the Greeks attempted to understand the world in rational, not mythical or religious, ways. This movement extended to the political and social spheres as well. The idea of democracy is usually dated to 5th century BCE Athens.

To understand the enormity of the Greek accomplishment, and appreciate that this advance was not inevitable, consider that most of recorded history occurred *before* classical Greek civilization.

Historians have learned a great deal about Greek mathematics. However, unlike the case in Egypt and especially Mesopotamia, none of this knowledge is first-hand. Papyrus did not last long in the moist climate of the Greek world, so what we have is copies of copies of Greek texts.

Greece before 600 BCE

The Middle East and the Mediterranean Sea.[1]

Unlike Mesopotamia and Egypt, Greece was not a great agricultural center. The land was too mountainous. Greece did have the sea, however. Very little of the mainland is far from the water, and there are many Greek islands in the Aegean Sea. Hence the Greeks were a seafaring people. By the 11th century BCE (1100–1000), they had spread across the Aegean to Ionia, on the shores of what is now Turkey.

By the middle of the 8th century BCE, the Greeks entered a period of expansion, physically and culturally. This was the time of Homer, author of the *Iliad* and the *Odyssey*. In 750 BCE the Greeks established a colony near modern-day Naples. Over the next three centuries, they set up many more colonies around the Mediterranean, in southern Italy, Sicily, Spain, and northern Africa.

The Greeks were borrowers. During this period, they adopted papyrus from Egypt, and adapted the Phoenician alphabet. (Our word "alphabet" comes from the first two Greek letters, *alpha* and *beta*.) As we will see, the early mathematicians were also travelers, and learned much from Mesopotamia and Egypt.

The early Greeks did not have great empires. The basic political unit was the city-state, the *polis*, the origin of our word "politics." There were many forms, from democracies to monarchies, but they were all distinguished by a respect for law. Another notable feature of Greek public life was debate and argumentation. But public life was not shared by all; slavery was common.

[1]From "The Blue Marble." R. Stöckli, R. Simmon. (NASA/GSFC)
http://visibleearth.nasa.gov/.

Numeration

The Greeks had a variety of numeral systems. The best known, which appeared in the 6th century BCE and was standard by the 3rd century BCE, was the Ionic system. It had 27 symbols: the 24 letters of the alphabet plus three others. These symbols represented the numbers 1, 2, 3, ..., 9, 10, 20, 30, ..., 90, 100, 200, 300, ..., 900. For example, γ was 3 and μ was 40, so 43 would be written $\mu\gamma$.

There were various ways of writing larger numbers, usually involving adding an extra symbol on top of, or next to, the existing symbols. A similar system handled fractions. The fraction $1/3$, for example, would be written $\alpha\acute{\gamma}$ (since $\alpha = 1$).

As you can see, this system was closer to the Egyptian system than to the superior Mesopotamian system. The development of Greek mathematics was not held back by the limitations of this system, since Greek mathematics did not rely heavily on numerical calculations.

Ionia, Miletus

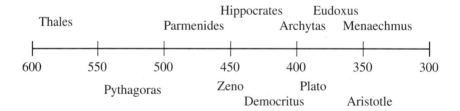

From Thales to the death of Alexander the Great. All dates are BCE.

Classical Greek civilization began in the 6th century BCE, in Ionia, the Greek colony located on the western shores of modern Anatolia in Turkey and some nearby islands. Unlike the Greek mainland, Ionia enjoyed good land for agriculture. Miletus, its greatest city, was an important trading center on the Meander River (which gave us our word "meander"). It was connected to Mesopotamia via overland trading routes. It was also a seaport whose ships traded with Egypt. Thus Miletus had access to the knowledge of these great civilizations.

Some time in the 6th century BCE, Western philosophy was born, and with it theoretical mathematics. In ancient times, philosophy was not distinct from science. The goal was to understand the world. The explanations that the new philosophers gave were natural, as opposed to supernatural. Religion was not abandoned; rather it was no longer considered adequate to explain the natural world by means of the actions of capricious gods in myths. In mathematics, it was not enough to empirically demonstrate results. One should argue why they were true. This was the beginning of formal deductive reasoning.

No one knows why this major intellectual development started in this place at this time. Certainly, it made a difference that Miletus had access to the major intellectual traditions of the Near East. Perhaps the new philosophy was an attempt to explain the contradictions in these traditions. It has also been suggested that the Greek habit of public debate fostered the notion that all assertions should be justified by careful argument.

Thales of Miletus (c. 625–547 BCE)

Thales was credited with beginning the new philosophy. Very little is known about his life. He was from Miletus, probably born into an aristocratic family. He was said to be a merchant, and to have traveled to both Egypt and Mesopotamia. He was famous as a statesman, astronomer, and engineer, as well as mathematician and philosopher.

Many stories have been told of Thales, but all of them date from well after his death, and most are no doubt apocryphal. There is a famous story, perhaps the first absent-minded professor story, of how, intent on studying the heavens, he fell into a ditch. On the other hand, there is another tale that when he was criticized for being impractical, he shrewdly cornered the market on olive presses, thereby making a fortune. Nothing he wrote has been preserved, so we have mostly the legends from later times.

It is said of Thales that, on a trip to Egypt, he impressed his hosts by demonstrating a method to determine the height of a pyramid by measuring shadows. Here is how he did it. Let A be the top of the pyramid, C the tip of the pyramid's shadow, and angle ABC be a right angle. (See Figure 1.5.) Suppose that a staff is held perpendicular to the ground. Let D be its tip, E its base, and F the tip of its shadow. Then the triangles ABC and DEF are similar, which means that the ratios of corresponding sides are equal. In particular, if the height of pyramid and staff are h_1 and h_2, respectively,

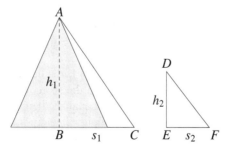

Figure 1.5   Measuring a pyramid's height by its shadow.

and the lengths of the shadows are s_1 and s_2, then $h_1/h_2 = s_1/s_2$. Since h_2, s_1, and s_2 can easily be measured, h_1 can be computed.

For example, if the staff is 6 feet, its shadow is 8 feet, and the pyramid's shadow is 640 feet, then $h_1/6 = 640/8$. Solving this, we get $h_1 = 480$ feet.

Eudemus of Rhodes wrote a book on early Greek geometers in the 4th century BCE. No copies of it remain, but a short bit of it was included in a book of the commentator Proclus (411–485 CE). Thales is credited with five theorems. They are fairly basic; one of them is that a circle is bisected by its diameter. Such a result would surely be known by anyone who had a practical need for it. Thales, however, was said to have been the first to seek a logical, not merely practical, basis for such theorems.

The most famous student of Thales was Anaximander. As with Thales, we know little of Anaximander. He was credited with introducing into Greece, from Mesopotamia, the gnomon, the center of the sundial, which casts the shadow. Anaximander also contributed to geography, drawing a circular map of the world.

In the last half of the 6th century BCE, due largely to the expansion of the Persian empire, Ionia declined as a cultural center. The center of the new Greek philosophy shifted west to Magna Graecia ("greater Greece" in Latin), the Greek colonies in southern Italy.

The Pythagoreans

In the 6th century BCE, an important group of thinkers emerged in Magna Graecia, centered around Pythagoras.

Pythagoras of Samos (c. 572–497 BCE)

Pythagoras was from the Ionian island of Samos. As with Thales, our knowledge of Pythagoras has been pieced together from reports written long after his death. He was said to have studied in Miletus, perhaps with Anaximander. He also traveled to Egypt, and reportedly spent seven years in Babylon, after which he returned to Samos. He was forced to leave Samos around 530 BCE, and settled in Crotona, a Greek seaport in southern Italy. It was there that he founded his society, known as the Pythagoreans, which also spread to neighboring cities, and was for a time very influential. Around 500 he was forced to move again, to the neighboring town of Metapontum, where he died.

Pythagoras left no writings, and his followers had a habit of attributing all of the group's discoveries to him. This practice, and the secrecy surrounding his organization, make it difficult to sort out his individual mathematical accomplishments. He certainly was a leading religious, philosophical, and political figure of his time, but perhaps his greatest accomplishment was founding the society that left such an important mark on our intellectual history.

The Pythagoreans were a society of a few hundred aristocrats. It is sometimes called a brotherhood, but there is one story that its original members included dozens of women. It was certainly selective and hierarchical. Members were divided into the *akousmatikoi*—listeners—who were expected to learn the master's teachings, and the *mathematikoi*, who could develop the teachings. The *mathematikoi* were among the first pure mathematicians. Our words "mathematics" and "mathematician" derive from *mathematikoi* (which in turn was based on *mathesis*, learning).

Religiously, the Pythagoreans practiced an asceticism, were probably vegetarian, and eschewed wine. They believed in the transmigration and reincarnation of souls, where the soul is reborn in another body after death.

Politically, the Pythagoreans were anti-democratic. In fact, democratic forces attacked them and burned their buildings in about 450. Their political influence waned thereafter, although the sect continued beyond that time and continued to produce important mathematics.

What distinguished the Pythagoreans from other mystery cults of the time was their philosophy that numbers, by which they meant counting numbers, were the foundation of the universe. An example of this was their discovery of the connection between musical harmonies and simple ratios. If one measures two strings on a lyre tuned an octave apart, their lengths are in the ratio 2 : 1. The musical interval of a fifth is associated with the ratio 3 : 2, a fourth with 4 : 3, and so on.

The Pythagoreans thought that these musical ratios were reflected in the heavens as well. They theorized that the planets, including the Sun and Moon and a couple of extra ones necessary to produce the number ten, traveled around on invisible spheres. These spheres were spaced according to the harmonic ratios they discovered, and produced sounds, the "music of the spheres."

The Pythagoreans dealt in numerology as well as number theory. The number 1 is associated with reason, 3 with harmony, odd numbers are masculine, even numbers are feminine. Ten, the sum of the first four numbers, was magical. They had the notion of "perfect" numbers, numbers with the property that the sum of their proper factors equals the number itself. The smallest such number is 6, as $6 = 1 + 2 + 3$. This may seem an odd notion to us, endowing numbers with human characteristics, but it inspired some lovely mathematics. Perfect numbers turn out to be associated with a family of prime numbers that is still studied. (See Section 5.7.)

Another area of interest to the Pythagoreans was the study of figurate, or polygonal, numbers, obtained from drawing regular figures with dots. Figure 1.6 shows the first few triangular (bowling pin) numbers. Note that the nth triangular number is the sum of the first n counting numbers. They attached special significance to the *tetractys*, the figure with $10 = 1 + 2 + 3 + 4$ dots. (See the prayer on p. 30.)

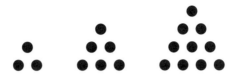

Figure 1.6   Triangular numbers.

A type of figurate numbers with which you may be familiar are the square numbers, which are numbers of the form n^2. As the nth triangular number is the sum of the first n counting numbers, the nth square is the sum of the first n odd numbers. For example, $4^2 = 1 + 3 + 5 + 7$. If you study Figure 1.7, you can see a geometric demonstration of this. The Pythagoreans also studied rectangular and pentagonal numbers. Figurate numbers have continued to fascinate mathematicians into modern times.

Figure 1.7   Square numbers.

Pythagoras is best known for the Pythagorean Theorem, that the area of the square on the hypotenuse of a right triangle is equal to the sum of the areas of the squares on the legs.

With the angles of the triangle labeled A, B, and C (the right angle), and the side lengths labeled a, b, and c, as in Figure 1.8, the Pythagorean Theorem says

$$c^2 = a^2 + b^2.$$

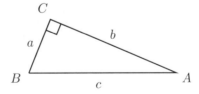

Figure 1.8   A labeled right triangle.

Figure 1.9 illustrates the Pythagorean Theorem. The area of the darker shaded square is equal to the sum of the areas of the two lighter shaded squares.

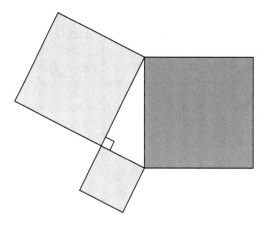

Figure 1.9   The Pythagorean Theorem (the area of the darker square equals the sum of the areas of the two lighter squares).

We have remarked that this theorem was known at least a thousand years before Pythagoras. But Pythagoras gets the credit for the theorem because his school was the first to provide a proof that covers all possible right triangles.

We offer the simplest proof of the Pythagorean Theorem that we know of. Let a right triangle be given. As in Figure 1.10, four copies of the right triangle are arranged inside a square whose side length is the sum of the two legs of the triangle. The shaded area is the area outside the four triangles and inside the large square. In the square on the left, the shaded area is the square on the hypotenuse. In the square on the right, the shaded area is the union of the squares on the legs of the triangle. Since the amount of shaded area doesn't change when we move the four triangles,

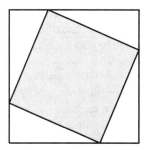 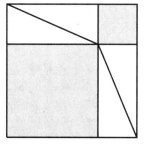

Figure 1.10   Proof of the Pythagorean Theorem.

the area of the square on the hypotenuse is equal to the sum of the areas of the squares on the two legs.

The Pythagoreans also studied Pythagorean triples (as did the Babylonians earlier), which are three positive integers a, b, and c such that $a^2 + b^2 = c^2$. They discovered an infinite family of such triples, namely,

$$a = 2n + 1$$
$$b = 2n^2 + 2n$$
$$c = 2n^2 + 2n + 1,$$

where n is any positive integer. For example, if $n = 2$, we get

$$a = 2 \cdot 2 + 1 = 5$$
$$b = 2 \cdot 2^2 + 2 \cdot 2 = 12$$
$$c = 2 \cdot 2^2 + 2 \cdot 2 + 1 = 13.$$

You can check that $5^2 + 12^2 = 13^2$. To see that these are always Pythagorean triples:

$$a^2 + b^2 = (2n + 1)^2 + (2n^2 + 2n)^2$$
$$= (4n^2 + 4n + 1) + (4n^4 + 8n^3 + 4n^2)$$
$$= 4n^4 + 8n^3 + 8n^2 + 4n + 1$$

and

$$c^2 = (2n^2 + 2n + 1)(2n^2 + 2n + 1)$$
$$= 4n^4 + 8n^3 + 8n^2 + 4n + 1.$$

Thus $a^2 + b^2 = c^2$.

The Pythagoreans proved that the sum of the angles in a triangle equals 180 degrees. Their proof is based on the equality of alternate angles, e.g., angles α and α' in Figure 1.11, where the two horizontal lines are parallel. This equality follows from the figure's symmetry; imagine rotating by $180°$, exchanging the parallel lines and leaving the diagonal line unchanged. This rotation also exchanges the angles α and α'.

Now let ABC be a triangle, and draw a line through B parallel to AC, as in Figure 1.12. Note that the angles α', β, and γ' together make a straight line, so $\alpha' + \beta + \gamma' = 180°$. Using our theorem on alternate angles, $\alpha = \alpha'$ and $\gamma = \gamma'$. If we substitute these, we obtain $\alpha + \beta + \gamma = 180°$. In other words, the sum of the angles in the triangle ABC is $180°$.

One of the most important Pythagorean discoveries was the existence of irrational numbers. A *rational number* is one that can be expressed as a ratio of integers. For example, $2/3$ is rational, as is $5 = 5/1$. It is a natural assumption that all numbers

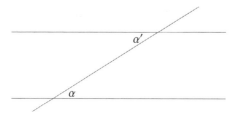

Figure 1.11   Equality of alternate angles.

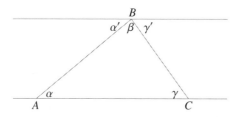

Figure 1.12   The sum of angles in a triangle is $180°$.

are rational, but this turns out not to be the case. For example, $\sqrt{2}$, the length of a diagonal from a square of side 1, turns out to be *irrational*, that is, not expressible as the ratio of two integers. This discovery, reportedly made by the Pythagorean Hippasus of Metapontum, certainly complicated theoretical mathematics. One can only guess at its effect on the Pythagoreans, whose philosophy was so dependent on whole numbers and their ratios. Legend has it that Hippasus made the discovery while at sea, and that the others, appalled by the idea, threw him overboard. More on this important theorem can be found in Section 5.3.

The Pythagoreans also had a profound effect on education. They identified four basic areas of education: arithmetic (which essentially meant the theory of numbers), geometry, music, and astronomy. These four were later extolled by Plato and Aristotle, and they loomed large into the Middle Ages, where they were known as the *quadrivium*.

Archytas of Tarentum (c. 438–347 BCE)

The Pythagorean Archytas was a politician and mathematician in southern Italy one hundred years after the death of Pythagoras. He was a number theorist and a geometer, devising a clever mechanical solution to the Delian problem (see "Three

Construction Problems" below). He is also credited with devising the educational system mentioned above, which became the quadrivium.

Archytas had two vastly influential students, Eudoxus and Plato.

Finally, lest we forget that these great mathematicians lived in a very different age, we end with one of their prayers.

> Bless us, divine number, thou who generated gods and men! O holy, holy Tetractys, thou that containest the root and source of the eternally flowing creation! For the divine number begins with the profound, pure unity until it comes to the holy four; then it begets the mother of all, the all-comprising, all-bounding, the first-born, the never-swerving, the never-tiring holy ten, the keyholder of all.
>
> PYTHAGOREAN PRAYER

Elea

Another important center of philosophy in Magna Graecia, not far from Crotona and Metapontum, was the city of Elea.

Parmenides (c. 515–450 BCE)

Parmenides, the founder of the Eleatic school, was born in Elea, into a wealthy family. We know little of his life. Philosophically, he was influenced by the poet Xenophanes, and was perhaps his pupil. Since Xenophanes was from Ionia, Parmenides was aware of the Ionian philosophers. It seems likely, given their proximity, that the Pythagoreans were also known to him. Plato writes that Parmenides visited Athens in 450, when he was an old man, and there met the young Socrates.

The only known work of Parmenides was a philosophical poem titled *On Nature*. Only fragments of it remain.

When Parmenides sang this poem (yes, he sang his philosophy), he entreated his listeners to ignore their senses, and instead follow pure reason. In particular, he insisted that movement and change were illusory; reality is unchanging, eternal.

The Eleatic school was important for its insistence on logic in philosophy. One did not merely assert beliefs, but needed to make formal, logically rigorous arguments in support of them. Members of the school were fond of the type of argument called *reductio ad absurdum*, in which a proposition is proved by demonstrating that its denial leads to a logical absurdity.

In mathematics, a reductio ad absurdum proof is often called a *proof by contradiction*. Here is a simple example, a proof that the number of integers is infinite. We begin by supposing the opposite, that the number of integers is finite. In this case, there must be a largest integer, say n. But then consider $n + 1$. It is clearly an integer, and it is larger than n, which gives us a contradiction (logical absurdity). Since this follows logically from assuming that the number of integers is finite, there must be an infinite number of integers.

Zeno (c. 490–425 BCE)

At this point, you will not be surprised to learn that our knowledge of Zeno's life is sketchy. Most of what we know is from Plato's book *Parmenides*. Zeno was born in Elea, and was a student of Parmenides. His importance rests on a book of paradoxes he wrote, in defense of Parmenides' philosophy. We do not even have copies of this book, only commentaries on parts of it written after his death.

Zeno's book was said to contain 40 paradoxes, of which nine have survived, although only as rephrased by other authors. Here are three of the most famous.

The Dichotomy: A runner is running toward a goal. In order to reach this goal, he must first reach the halfway point. He then must go halfway from that point to the goal, and so on. At every point, he must still traverse half the distance to the goal, so can never arrive.

Achilles and the Tortoise: Achilles, the fastest runner in the world, is chasing after a tortoise. In order to catch the tortoise, he must reach the point at which the tortoise started. (We assume that both are running in a straight line.) But when Achilles arrives at that point, the tortoise, slow as he may be, has moved on. So Achilles must then reach the new point at which the tortoise has arrived. This process continues; when Achilles reaches the point at which the tortoise is, the tortoise is no longer there. So Achilles can never catch the tortoise.

The Arrow: Consider a moving arrow at a particular instant of time. At that instant, the arrow occupies a particular place. But the place does not move, therefore the arrow is motionless. Thus motion is impossible.

A *paradox*, in English, can mean an argument or assertion that defies intuition. However, the sense in which Zeno's arguments are paradoxes is more specific. He presented arguments which ended in absurd conclusions, but the important point is that it was not clear exactly where the argument went wrong. The issue is not whether the conclusions are correct, but what exactly is wrong with the reasoning. (The word paradox comes from the Greek *para*, alongside or beyond, and *doxa*, opinion.)

Since we do not have Zeno's original words, we must infer his intent in presenting these paradoxes. One of his goals was probably to support his teacher's assertion that motion was an illusion. Another perhaps was to probe the nature of space and time. Are they infinitely divisible?

Zeno's paradoxes have inspired philosophers and mathematicians for millennia, because they force us to confront the thorny issues of infinity and continuity. These paradoxes are now considered solved, but the solution took the development of a sophisticated theory of limits, and was not completed until late in the 19th century, more than 2300 years after the paradoxes were first posed.

Democritus (c. 460–370 BCE)

Democritus was a native of Abdera, in Greece. He came from a wealthy family, was said to have traveled to many countries, and talked to many scholars before returning to Abdera.

Although not a resident of Elea, he was an academic grandson of Zeno, being a student of Leucippus who was in turn a student of Zeno. He wrote many works on mathematics, but none survive.

∼

Democritus is most famous for developing (with Leucippus) the atomic theory. The word "atom" comes from the Greek *an* (not) and *temnein* (to cut). This is the essence of the atomic theory, that the world consists of atoms that are indivisible. This theory was in part a response to Zeno's paradoxes.

Archimedes also gave Democritus credit for stating, but not proving, that the volume of a cone is one-third that of its related cylinder, and the volume of a pyramid is one-third that of its related prism. (See Figures 1.13 and 1.14.) The first proofs were given by Eudoxus, about 50 years after Democritus.

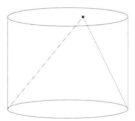

Figure 1.13   A cone and its cylinder.

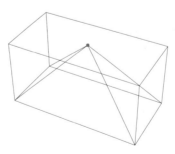

Figure 1.14   A pyramid and its prism.

There is also a tantalizing report by Plutarch (c. 100 CE) that Democritus studied thin sections cut from a cone by planes parallel to its base. This is an idea pursued fruitfully by Archimedes, and is an important part of the integral calculus developed in the 17th century.

Athens

By the middle of the 5th century, and until the late 4th century BCE, Athens was the most important center of Greek mathematics. It was the largest of the Greek city-states; we don't have good data, but estimates place its peak population at 300,000.

Athens was an important Mediterranean trading center. Its ships carried wine and olive oil, marble and silver. It had to import much of its grain. The economy of Athens was heavily dependent on slavery. Some scholars estimate that one-third of the population were slaves.

The political power of Athens waxed and waned in this period, but throughout it remained an important cultural center. It produced some of the most famous art, architecture, theater, science, and philosophy of antiquity. People still go to Athens to admire the Parthenon, and still stage productions of the plays of Aeschylus, Sophocles, Euripides, and Aristophanes. Its most famous philosophers were Socrates, Plato, and Aristotle.

In mathematics, Athenian scholars were responsible for refining the logical structure, overcoming the problem of irrational numbers, and developing theoretical geometry. They also established important institutions, where professional philosophers and mathematicians could flourish.

Three Construction Problems

Among the most important problems in ancient Greece, and beyond, were these three.

1. **Quadrature (or squaring) of the circle:** Construct a square whose area is the same as that of a given circle.

2. **Duplication of the cube (The Delian Problem):** Construct a cube whose volume is twice that of a given cube.

3. **Trisection of an angle:** Divide a given angle into three equal parts.

Obtaining the *quadrature* of a figure means constructing a square with the same area. The quadrature of a circle is in a class of problems that concern studying curved figures using simpler, straight-edged ones. Mathematicians had limited success with these problems until the advent of the calculus in the 17th century.

To duplicate a cube of side a, we need another cube of side $b = \sqrt[3]{2}a$. For then the volume of the cube of side b is $b^3 = (\sqrt[3]{2}a)^3 = 2a^3$, twice the volume of the cube of side a. So the difficulty is geometrically representing $\sqrt[3]{2}$.

What it means to "construct" a figure has been subject to different interpretations, the most famous being the use only of a straightedge and compass. The straightedge allows one to draw a straight line; the compass allows one to draw a circle of given center and radius.

Many mathematicians have worked on these problems, inventing important areas of mathematics in an attempt to solve them. The problems have provided inspiration far beyond ancient Greece. They were finally solved, for straightedge and compass, only in the 19th century, when it was shown that all three are impossible. (We should mention that Pappus, in the 4th century CE, asserted this impossibility, but without proof.)

Hippocrates of Chios (c. 470–410 BCE)

Not to be confused with the physician Hippocrates of Cos (of Hippocratic Oath fame), this Hippocrates was born on the Ionian island of Chios, not far from Pythagoras' birthplace of Samos. Early on, Hippocrates was a merchant, but, after setbacks in his business, he made his way to Athens where he became one of the foremost scholars.

Hippocrates was the leading geometer of his time and wrote an influential text on geometry, *Elements of Geometry*, which has been almost entirely lost. He taught mathematics, being one of the first to make his living that way.

$$\sim$$

Hippocrates worked on at least the first two of the three construction problems. His advances were typical in the history of difficult mathematical problems, reducing the unsolved problem to another that may be easier to solve. In the case of the quadrature of the circle, he studied a type of intersection of circular arcs called a lune. He showed that if one could always square the lune, then one could square the circle. He further showed how to square a particular type of lune. He was not able to complete this program, however, being unable to square an arbitrary lune.

Similarly, Hippocrates reduced the Delian problem (doubling the cube) to another problem in two dimensions, instead of three. Specifically, he reduced the problem of doubling the cube of side a to that of finding two *mean proportionals* between a and $2a$, that is, numbers x and y such that $a : x = x : y = y : 2a$, where $a : x$, for example, is the ratio of a to x.

Although Hippocrates' text has been lost, historians can make informed guesses about its content based on fragments of his writings included in later works. His work may have been the first to array geometric theorems in a logical sequence, from the simplest to the more advanced. The logic of his surviving proofs is not perfect, but it does demonstrate a sophistication well beyond that of the scholars of a century earlier. One area that had yet to be developed is a system of axioms, or assumptions, upon which to build a geometry.

Plato and His School

Plato (c. 429–347 BCE)

Plato was a son of Athenian aristocrats. When he was young, he became a student of Socrates, the stonemason turned philosopher. After Socrates was executed for irreverence in 399, Plato left Athens, reportedly traveled around Greece, to Egypt, and to Tarentum in Magna Graecia, where he studied with Archytas, the Pythagorean.

In 388 Plato returned to Athens, where he taught, wrote some of the most influential philosophical works ever, and founded his famous Academy. Except for two trips to Syracuse in Sicily, where the Pythagoreans were ensconced, Plato lived the rest of his life in Athens.

Plato was probably not much of a mathematician himself, being more interested in ethics. He had considerable influence on mathematics, however, in two important ways. The first was his philosophy. He believed that the world of the senses was imperfect, that what we experience is but a shadow of what he called "forms" or "ideas." For example, we can draw a circle but it is merely an imperfect representation of the Circle idea. Many mathematicians, though by no means all, subscribe to a form of Platonism that claims that mathematical objects have a real existence, independent of us. All that mathematicians do is study what is already out there. (Others reject this, believing that mathematics is an invention of humans, perhaps only a game we play.)

Plato's other major contribution to mathematics, and to learning in general, was his founding of his school, the Academy. The name, which gave us our word "academic," derived from the name of the site where the school was built. It was founded around 387 BCE.

Above the entrance gate of the Academy was inscribed "Let no man ignorant of geometry enter." The school's curriculum was based on the quadrivium—number theory, plane geometry, music, and astronomy—together with solid (3-dimensional) geometry. After the completion of these studies, the best students went on to study dialectics, a method of critical, persistent questioning, which Plato considered the way to arrive at truth.

Plato, very much the aristocrat, viewed the Academy as an institution to educate the ruling class. Its mathematics was therefore theoretical, not practical, with the exception of military applications. Interestingly, the word "school" derives from the Greek *skhole*, which means "leisure."

The Academy was more than a school; the best scholars from the Greek world came to Athens to study and teach at the school. The closest modern equivalent is the research university. Throughout most of the 4th century BCE, until the rise of the Museum in Alexandria, the Academy was the home of the cream of Greek scholarship. Even after that, it remained an important center of learning. It was finally closed, in 529 CE, by the Christian emperor Justinian, who viewed it as a pagan institution. It thus lasted more than 900 years. This is roughly the age of the oldest current European university, the University of Bologna, founded in 1088.

Eudoxus' Theory of Proportions

Eudoxus (c. 408–355 BCE)

Eudoxus was the most illustrious astronomer and mathematician of his time, and is generally considered to be the second best ancient mathematician, after Archimedes. He was born in the Ionian city of Cnidus. As a young man he traveled to Tarentum, in Sicily, where he studied both mathematics, with Archytas, and medicine. He then spent some months studying with Plato's circle in Athens. He was apparently too poor to live in Athens proper, so he lived in nearby Piraeus, and walked seven miles daily each way to the Academy. After Athens, he returned for a while to Cnidus.

Later, Eudoxus traveled to Egypt, where he worked primarily on astronomy. After Egypt, he returned to Ionia, specifically Cyzicus, where he founded a school and wrote his greatest astronomical works. He returned for a time to Athens, with some of his students, working there with Plato, Aristotle, and others. Finally, he returned to Cnidus, where he helped write a constitution for their new democracy, founded an observatory, taught, and practiced medicine and astronomy. He died at age 53.

Eudoxus' greatest mathematical contribution was his theory of proportions. To understand its importance, recall that the Pythagoreans had proved that the square root of 2 is irrational, that is, not representable as the ratio of two integers. The difficulty this presented to mathematicians was in making precise arguments about irrational numbers, and about geometric figures whose magnitudes might be irrational.

Modern mathematicians get around this difficulty by approximating an irrational number by a sequence of rational numbers, then taking a limit. Eudoxus did something similar, but a bit more complicated.

To start with, Eudoxus' theory did not deal with numbers as such, but with *magnitudes*, e.g., lengths or areas. The key to dealing with magnitudes was knowing how to compare them. Here is how he did it, as it appears in a definition from Euclid's *Elements*.

> Magnitudes are said to be in the same ratio, the first to the second and the third to the fourth, when, if any equal multiples whatever are taken of the first and third, and any equal multiples whatever of the second and fourth, the former multiples alike exceed, are alike equal to, or alike fall short of, the latter multiples respectively taken in corresponding order.

Here "multiple" means integer multiple. Let us translate this into more familiar terms. If we represent the ratio a to b as $a : b$, the question is when we have equality of two ratios, say $a : b = c : d$. The above definition says that we have equality provided the following three conditions are met, for every choice of positive integers m and n.

1. If $ma < nb$ then $mc < nd$.

2. If $ma = nb$ then $mc = nd$.

3. If $ma > nb$ then $mc > nd$.

It is understandable if this definition leaves you underwhelmed. It is difficult to fully appreciate its power unless you see it used in proofs, an exercise that is beyond this text. Mathematicians, however, were not able to replace this treatment of irrationals with anything of equal precision and usefulness until late in the 19th century.

Eudoxus also made rigorous a method called *exhaustion*, earlier invented by Antiphon, which is a limiting procedure to compute areas and volumes. He used the method of exhaustion to prove a result stated by Hippocrates, that the ratio of the areas of two circles is proportional to the square of the ratio of their diameters, or in modern terms, $A = kd^2$, where A is the area of a circle, d its diameter, and k some fixed constant of proportionality.

The method used successive approximations of a circle by polygons (see Figure 1.15). As the number of sides of the polygons grows, they "exhaust" the area of the circle. Combining this with Eudoxus' definition of equality of ratios, and some careful arguments, leads to the desired proof.

Figure 1.15   Approximating the area of a circle.

One effect of Eudoxus' subtle theory was to reinforce the Greek preference for geometry over algebra, a preference that influenced the course of mathematics for the next two thousand years. Greeks did solve some algebraic problems, usually by converting them first to geometric ones.

Finally, we note that Eudoxus introduced the use of spherical geometry in astronomy. His astronomical theory had stars and planets rotating on spheres centered at the Earth. Although this model wasn't accurate, it was sophisticated for its time and was immortalized by being included (in a modified form) in Aristotle's supremely influential works.

Logic

Aristotle (384–322 BCE)

Aristotle was from the Ionian colony of Stagira. His father was physician to the kings of neighboring Macedon. When he was seventeen or eighteen, Aristotle came to Athens to study at the Academy. He stayed on as a scholar until Plato's death in 347 BCE, after which he left Athens.

In 342 Aristotle became the tutor to the young prince Alexander of Macedon, later called Alexander the Great. He stayed as advisor until 335, when Macedonia took control of Athens. In that year, Aristotle returned to Athens and set up his own school, the Lyceum. He remained there until Alexander's death in 323, when he found it prudent to leave. He died the next year in Chalcis.

Aristotle is one of the most influential philosophers of all time. He wrote on a wide variety of topics and was the preeminent authority on the physical sciences for two thousand years. His importance to mathematics lies in his work on logic. He constructed a formal theory of logic, building on the ideas developed over the preceding 250 years or so of Greek philosophy.

Aristotle believed that the only way to certain knowledge was by the use of logic, deducing new knowledge based on old. One has to start somewhere, however. His starting point was a set of axioms and postulates, which were truths that needed no argument. An *axiom*, for Aristotle, was a truth that was not particular to any science, for example, "take equals from equals and equals remain." A *postulate* was a truth that concerned a particular area; for example, "through every two points a straight line may be drawn" is a geometric postulate. The philosopher should start with the minimum number of axioms and postulates needed, and some definitions, then proceed by logical argument to prove things.

This logical structure is still the basis for mathematics, although modern mathematicians do not distinguish between axioms and postulates, usually calling any initial assumption an axiom. We also tend not to use the word "truth" for our axioms, although axioms must be carefully constructed to be of use.

As an example of this aspect of logic, let us reconsider the earlier proof we gave that the number of integers is infinite.

We begin by supposing the opposite, that the number of integers is finite. In this case, there must be a largest integer, say n. But then consider $n + 1$. It is clearly an integer, and it is larger than n, which gives us a contradiction. Since this follows logically from assuming that the number of integers is finite, there must be an infinite number of integers.

This argument assumes some things, for example, that every finite set of integers has a largest element, and that for every integer n, there is an integer $n + 1$. These may seem obvious to you, but they are still logically required. So we may take them as axioms. The argument also assumes that we know what an integer is. Perhaps we might add a definition for integer. (We won't attempt such a definition here; it turns out to be a delicate matter.)

Thus, the revised argument would start with a definition of "integer" and two axioms: (1), that every finite set of integers has a largest element, and (2), that for every integer n, there is an integer $n + 1$. We would then proceed with the argument proper.

Perhaps you can think of other axioms or definitions that this argument needs. If so, you have begun to appreciate how difficult and subtle is the task of establishing a logical foundation for mathematics. The Greeks started this process, but as we shall see in later chapters, theirs was not the final word.

At the Lyceum, his school, Aristotle was famed for lecturing while walking the grounds with his students. As a result, teachers and students there were known as Peripatetics, from the Greek *peri* (around) and *patein* (to walk). They didn't spend all of their time walking, however; the Lyceum introduced written examinations into the educational system.

Conic Sections

Menaechmus (c. 380–320 BCE)

Biographical details of Menaechmus' life are sketchy. He and his brother studied at Eudoxus' school at Cyzicus, perhaps with Eudoxus himself. He was a friend of Plato, and perhaps a tutor to Alexander the Great. He later headed the school in Cyzicus, where he died.

Among his many contributions, Menaechmus studied, and probably discovered, conic sections. A *conic section* is a curve obtained by intersecting a double cone with a plane. (See Figure 1.16.) If the plane intersects both parts of the cone, we get a hyperbola. An ellipse, of which a circle is a special case, intersects only one part, and is finite. In between, if the plane is parallel to a side of the cone, we get a parabola, which, unlike the hyperbola, has only one branch and, unlike the ellipse, goes on to infinity.

Menaechmus used conic sections to solve the Delian problem: doubling the cube. Recall that Hippocrates had reduced the problem of doubling the cube of side a to that of finding two mean proportionals between a and $2a$, that is, numbers x and y such that $a : x = x : y = y : 2a$. We would write the equality of the ratios as

$$\frac{a}{x} = \frac{x}{y} = \frac{y}{2a}.$$

These are equivalent to $ay = x^2$ and $2ax = y^2$, the equations of two parabolas. So the problem of finding x and y is equivalent to the problem of finding a point (x, y) on both of the parabolas, i.e., finding an intersection of the parabolas. This is what

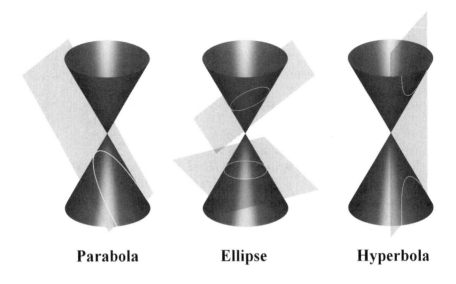

Parabola   **Ellipse**   **Hyperbola**

Figure 1.16   Conic sections.[1]

Menaechmus discovered. We don't know how he constructed his conic sections; they cannot be constructed via straightedge and compass, but the Greeks knew other techniques.

Menaechmus' solution to the Delian problem is an example, of which there are many in mathematics, of a discovery that turned out to be more important than its original inspiration would suggest. The Delian problem has faded into obscurity, but conic sections are very important. For instance, comets travel around the Sun in orbits that are conic sections.

EXERCISES

Below is a partial table of Ionic numerals.

Symbol	Number	Symbol	Number	Symbol	Number
α	1	ι	10	ρ	100
β	2	κ	20	σ	200
γ	3	λ	30	τ	300

1.21   What numbers would each of these represent?

[1] Diagram by Pbroks13. Wikipedia Commons.

a) κ

b) $\lambda\alpha$

c) $\sigma\iota\gamma$

d) $\rho\alpha$

1.22 Write each of these numbers in the Ionic system.

a) 21

b) 333

c) 220

1.23 Suppose that the shadow of a building measures 60 meters, at the same time that a 2 meter stick casts a 5 meter shadow. How tall is the building?

1.24 If the two legs of a right triangle have lengths 5 and 12, what is the length of the hypotenuse?

1.25 If the two legs of a right triangle both have length 1, what is the length of the hypotenuse?

1.26 If one of the legs of a right triangle has length 1 and the hypotenuse has length 2, what is the length of the other leg?

1.27 If the two legs of a right triangle have lengths $2n$ and $n^2 - 1$, where n is an integer greater than 1, what is the length of the hypotenuse?

1.28 A triangle with side lengths 8, 15, and 17 is inscribed in a circle. What is the diameter of the circle?

1.29 Find angles α, β, and γ.

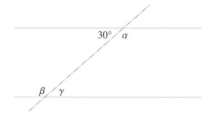

1.30 Another figurate number is the *oblong* number, which is a number of the form $n(n + 1)$. The first two oblong numbers are $2 = 1 \cdot 2$ and $6 = 2 \cdot 3$.

a) Find the first ten oblong numbers.

b) Recall that triangular numbers are of the form $1 + 2 + 3 + \cdots + n$, and square numbers are of the form $1 + 3 + 5 + \cdots + (2n - 1)$. Find a similar pattern for oblong numbers.

1.31 Use the formula for Pythagorean triples to find two other sets of triples, larger than $(5, 12, 13)$.

1.32   The first perfect number is 6. The next one is between 20 and 30. Find it.

1.33   Zeno's dichotomy paradox is related to a famous infinite sum. Find the sum $1/2 + 1/4 + 1/8 + \cdots$.

1.34   Eudoxus proved that a circle of diameter d has an area given by $A = kd^2$, for some constant k. What is k? (Hint: you know the area of a circle in terms of its radius r.)

1.35   How might the atomic theory of Leucippus and Democritus be used to explain Zeno's paradoxes?

1.36   Let $a = 1$, $x = \sqrt[3]{2} = 2^{1/3}$, and $y = 2^{2/3}$.
  a) Verify that x and y are mean proportionals between a and $2a$. (Hint: remember that $2^r/2^s = 2^{r-s}$.)
  b) Show that this solves the Delian problem for $a = 1$, i.e., that the cube of side x has twice the volume of the cube of side a.

1.37   Consider the proof of the equality of alternate angles given in the text. What axioms and definitions might be required to make this logically rigorous?

1.38   Look up the word *theorem*. What is its origin?

1.4   The Apex: Third Century Hellenistic Mathematics

There is no permanent place in the world for ugly mathematics.

G. H. HARDY (1877–1947)

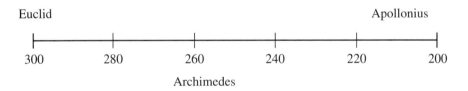

The Third Century BCE.

Greek mathematics, especially geometry, achieved its highest expression in the 3rd century BCE. The century started with the writing of the most famous math book ever, continued with the work of the greatest of the ancient mathematicians, and finished with the definitive Greek text on conic sections.

Alexander the Great

Macedonia was a small kingdom on the northern boundary of Greece. Under King Philip II, who ruled from 356 to 336 BCE, Macedonia gained effective control of most of Greece, and began a war against the Persian empire to the east. Shortly after the start of this war, Philip was assassinated and his army elected his son Alexander as his successor. Alexander, already a veteran general, was twenty years old.

The war against Persia continued. Alexander won a great battle at Issus, in modern-day Turkey, in 333. He continued his conquests, as far as modern-day Uzbekistan and Pakistan in Asia, and Egypt in Africa, creating the largest empire up to that time. He caught a fever and died in 323 BCE. He was 33 years old.

Among the many stories told of Alexander, perhaps the most famous is that of the Gordian knot. The legend has it that when Alexander entered the city of Gordium, he was shown a sacred knot tied around a pole. Supposedly, the man who could untie the knot was destined to become the king of Asia. Alexander's response was to take out his sword and slice the knot. The phrase "cutting the Gordian knot" is now used to indicate an audacious solution of a complicated problem. Perhaps it can also be considered a metaphor for Alexander's ruling philosophy. Another legend has it that as Alexander lay dying he was asked to whom would he leave his empire. His answer: "to the strongest."

In fact, the empire fell apart after Alexander's death. Major pieces were ruled by several of his generals, including Antigonus in Macedonia, Seleucus in the east, and Ptolemy in Egypt. What followed was a period of empire, much different than the time of the great Greek city-states. The successors of Seleucus, the Seleucids, ruled much of west Asia, gradually declining in power until their last holdings in Syria were conquered by the Romans in 64 BCE. The Ptolemys in Egypt ruled until the death of Cleopatra in 31 BCE, again falling to the Romans.

The civilization of this time, from Alexander until the rise of the Romans, is known as *Hellenistic*, distinguishing it from the earlier *Hellenic* period of Greek culture. Hellenistic culture spread across all of Alexander's empire; a form of Greek became the common language of trade and government. The rulers were educated in classical Greek culture.

Alexandria and Its Museum and Library

The greatest city of this time, both commercially and culturally, was Alexandria in Egypt. The city was founded by Alexander in 331, the first of seventeen cities of that name. It was a major port, located where the Nile empties into the Mediterranean. As such, it was situated to profit from Egypt's large surplus of grain, which was shipped to many Mediterranean ports.

Egypt was ruled by the Macedonian general Ptolemy I Soter from the time of Alexander's death in 323 until 283 BCE. His capital was Alexandria, and it was there that he built the Museum ("temple of the Muses") and Library. The Museum recruited the leading scholars of the Greek world, paying them a salary, providing free board and freedom from taxes. Originally, it was not a school; it has been compared to the modern Institute for Advanced Study in New Jersey, a place for scholars to discuss, and invent, ideas. Over time, students were attracted to the Museum, to learn from the experts.

The Library at Alexandria was the largest in the ancient world, eventually housing over 500,000 manuscripts. Ships sailing from Alexandria were instructed to gather any manuscripts they could, to add to the Library. One story has it that the Library borrowed manuscripts, copied them, then returned the copies, retaining the originals.

Alexandria quickly eclipsed Athens as the chief center of Greek learning. There were other centers, including Syracuse where Archimedes worked, but Alexandria was the greatest. The city and its Museum remained influential even after the rise of Rome.

Euclid's *Elements*

The *Elements* by Euclid is the most successful textbook in history. Written around 300 BCE, it is second only to the Bible in number of editions, certainly over one thousand. It was *the* mathematical textbook into at least the 19th century. It inspired many students from Abraham Lincoln to Albert Einstein. Edna St. Vincent Millay wrote a sonnet entitled *Euclid Alone Has Looked on Beauty Bare*.

The *Elements* is the culmination of the early period of Greek mathematics. In it, Euclid summarized much of the mathematics developed in the preceding three centuries. But he did more than that; he put this mathematics into a rigorous and consistent logical framework, starting with unproved assumptions and carefully, step-by-step, deducing more advanced theorems from them. It is this structure that gives the book its special character, and is no doubt the reason why high school geometry courses to this day often are students' first introduction to formal mathematical reasoning. It is also far from the modern multicolored, image-laden, mathematical textbook. In fact, the modern reader will find it dry as dust. It remains, however, one of the great intellectual milestones in history.

Euclid (c. 330–270 BCE)

Even less is known of Euclid than of many of his predecessors. He worked at the Museum in Alexandria, probably arriving some time around 300 BCE. It is reasonable to think that he had studied in Athens, perhaps at the Academy, because he was certainly familiar with the works of Eudoxus and other Athenians.

Euclid's date and place of birth are unknown, as is the date of his death. He wrote on many mathematical subjects, including astronomy and optics, perhaps a dozen books in all, but only the *Elements* has survived. Of course, tales are told of him, but they are all from many hundreds of years later. One of the most often repeated is that the king Ptolemy asked him if there were a shorter way to learn geometry than through the *Elements*. He was said to reply that there "is no royal road to geometry." He probably did not add that kings at least can afford tutors like Euclid.

The *Elements* is divided into thirteen "books." Book I starts with five postulates and five "common notions," both of which we would call axioms today. They are given below.

Postulates

1. To draw a straight line from any point to any point.

2. To produce a finite straight line continuously in a straight line.

3. To describe a circle with any center and radius.

4. That all right angles equal one another.

5. That, if a straight line falling on two straight lines makes the interior angles on the same side less than two right angles, the two straight lines, if produced indefinitely, meet on that side on which are the angles less than the two right angles.

Common Notions

1. Things which equal the same thing also equal one another.

2. If equals are added to equals, then the wholes are equal.

3. If equals are subtracted from equals, then the remainders are equal.

4. Things which coincide with one another equal one another.

5. The whole is greater than the part.

Some of this language may need translation. For example, Postulate 1 means that, given any two points, we can draw a straight line joining them. Common Notion 1 can be stated algebraically: If $a = b$ and $c = b$ then $a = c$.

Postulate 5 sticks out like a sore thumb. In modern language, it might be stated thus: if a line intersects lines 1 and 2, as in Figure 1.17, and the angles α and β sum to less than 180 degrees, then lines 1 and 2 must eventually intersect, on the same side of the third line as α and β. Postulate 5 is also called the *parallel postulate*, because Euclid used this postulate to prove theorems about parallel lines.

If you think about this postulate a bit, you may be able to convince yourself of its truth. Postulates and common notions, however, are supposed to be self-evident. This one seems a bit too involved. Over the centuries many people have attempted to show that it could be derived from the other postulates and axioms. In the 19th century, mathematicians developed Non-Euclidean geometries, in which the other axioms hold but the parallel postulate is false. Euclid was vindicated.

The rest of Book I is a careful, step-by-step, argument, culminating in the Pythagorean Theorem.

Book II contains a number of results which we would think of as algebraic, but in geometric form. In general, the Greeks preferred geometry to algebra, and would often cast problems into geometric form that we now would handle algebraically. As an example, consider the following.

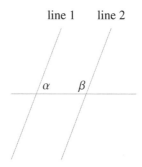

Figure 1.17   Parallel postulate.

Proposition II-4 If a straight line is cut at random, then the square on the whole equals the sum of the squares on the segments plus twice the rectangle contained by the segments.

We can picture this as in Figure 1.18, where the line at the left is cut into pieces of lengths a and b. The proposition states that the area of the big square equals the sum of the areas of the two smaller squares and the two rectangles. Of course, drawn as it is in the diagram, this is rather evident, but let's translate this into equations. The area of the larger square is $(a + b)^2$. The smaller squares have areas a^2 and b^2, and each of the rectangles has area ab. So the proposition states that

$$(a + b)^2 = a^2 + b^2 + 2ab.$$

This is our familiar way of squaring a binomial.

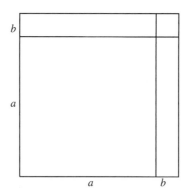

Figure 1.18   Proposition II-4.

Book II contains other results of this nature, including a geometric form of the quadratic formula, which we now write as

$$x = \frac{-b \pm \sqrt{b^2 - 4ac}}{2a},$$

giving solutions of the equation $ax^2 + bx + c = 0$.

Book III contains thirty-seven propositions on circles, starting from basic definitions. Here is one that you may have seen before.

Proposition III-20 In a circle the angle at the center is double the angle at the circumference when the angles have the same circumference as base.

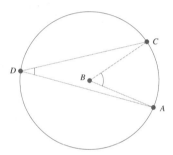

Figure 1.19   Proposition III-20.

Using the notation of Figure 1.19, angle ABC is twice angle ADC.

Book IV has results about inscribing polygons in circles and circles in polygons, and gives constructions of some regular polygons. The regular polygons of three and four sides, i.e., equilateral triangles and squares, are relatively easy to construct. The book ends with the construction of a regular pentagon (5-gon) and 15-gon.

Book V presents some basic results on magnitudes, which we would think of as lengths or areas. Here is an example.

Proposition V-1 If any number of magnitudes are each the same multiple of the same number of other magnitudes, then the sum is that multiple of the sum.

An algebraic version of this is, for a positive integer n and any magnitudes $a_1, a_2, \ldots, a_k$, we have $na_1 + na_2 + \cdots + na_k = n(a_1 + a_2 + \cdots + a_k)$. For us this follows from the associative law of numbers. Euclid didn't consider magnitudes to be the same as numbers, however. Magnitudes and numbers were different things for him.

Book V also presents Eudoxus' theory of proportions, and uses it to prove other results on magnitudes. Many of these results are used in Book VI, which is about similarity, defined as follows (see Figure 1.20).

Definition VI-1 Similar rectilinear figures are such as have their angles severally equal and the sides about the equal angles proportional.

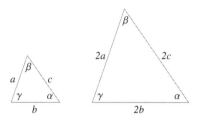

Figure 1.20 An example of similar triangles.

In Proposition VI-19 it is proved that the ratio of the areas of two similar triangles is the square of the ratio of their corresponding sides. As an illustration, the larger triangle in Figure 1.20 has four ($= 2^2$) times the area of the smaller.

Books VII, VIII, and IX cover number theory. These books do not rely on the results of the first six books, because they are about numbers, not magnitudes, which Euclid viewed as entirely different entities. Thus, Euclid proved again theorems such as the distributive law, even though he had a similar theorem about magnitudes. The number theory chapters are based on Pythagorean work, although this work had been later logically reorganized, perhaps by Theaetetus.

We mention four important results from these three books. The first has to do with greatest common divisors. Recall that the greatest common divisor of two integers is the largest integer dividing them both. Book VII contains the *Euclidean Algorithm*, a method of determining greatest common divisors that is still important. Details on this algorithm can be found in Section 5.4.

Two fundamental results on prime numbers are in these books. The first is a proof that the number of primes is infinite. We give this proof in Section 5.5. The second is the Fundamental Theorem of Arithmetic, that every positive integer can be expressed as the product of primes, and in only one way. This theorem is the subject of Section 5.8.

The culmination of Euclid's number theory chapters is a study of perfect numbers. Recall that a positive integer n is a perfect number if it is the sum of all its divisors, excluding n itself. Euclid was able to connect perfect numbers to certain types of primes. This topic is explored further in Section 5.7.

Book X undertakes a study of incommensurable magnitudes. This book starts with the following definition.

Definition X-1 Those magnitudes are said to be commensurable which are measured by the same measure, and those incommensurable which cannot have any common measure.

Here is the idea. Suppose that we have two lines, of lengths a and b. Then a and b (thought of as magnitudes, not numbers) are commensurable if there is another line of length c that fits into both a and b a whole number of times. For example, 6 and 10 are commensurable because 2 fits into both, i.e., $6 = 3 \cdot 2$ and $10 = 5 \cdot 2$. In general, if a and b are commensurable, say by c, then $a = mc$ and $b = nc$, for some integers m and n. But then $a/b = m/n$. In other words, the ratio of a to b is a *rational* number. Studying incommensurable magnitudes is tantamount to studying irrational numbers. It should therefore come as no surprise that Euclid relies on Eudoxus' theory of proportions in this chapter.

Books XI–XIII concern solid (3-dimensional) geometry. Books XI and XII prove some extensions of earlier theorems of plane geometry, and obtain results on volumes of a number of solids. For example, proofs are included for the theorems that the volume of a cone is one-third that of the related cylinder, and the volume of a pyramid is one-third that of its related prism. The following important result is also proven.

Proposition XII-18 Spheres are to one another in triplicate ratio of their respective diameters.

"Triplicate ratio" means cube, so the proposition is that the volume of a sphere is proportional to the cube of its diameter. This is equivalent to writing $V = kr^3$, where V is the volume, r the radius, and k some constant of proportionality. We now know that $k = 4\pi/3$, but this was not known to Euclid.

Many of the results in Books XI and XII are proved by the method of exhaustion, that is, by exhausting the area or volume of shape being studied by figures of known properties.

Finally, Book XIII is a study of convex regular polyhedra, the *Platonic solids*. These are 3-dimensional equivalents of regular polygons, solids whose faces are all congruent regular polygons, arranged the same way around each vertex. See Figure 1.21. These polyhedra make good dice and are used in various role-playing games.

| Cube | Tetrahedron | Octahedron | Dodecahedron | Icosahedron |

Figure 1.21   The Platonic solids.

Book XIII shows how to construct the Platonic solids, and gives some of their properties. It culminates in the proof that, unlike the regular polygons, there are only a finite number of convex regular polyhedra. In fact, the five in Figure 1.21 are the only ones.

The *Elements* contains some mistakes, and Euclid made some implicit assumptions without stating them. Still, Euclid skillfully surveyed the whole of Greek mathematics of his time. Most of the content of his book was originally due to others, but he put it together so well that previous summaries have not even been preserved.

Archimedes and Higher Geometry

Euclid was not the last word in Greek geometry. The 3rd century BCE was the high water mark for all of ancient geometry. The two most notable figures were Archimedes and Apollonius.

Archimedes (c. 287–212 BCE)

Widely considered to be the greatest mathematician of antiquity, Archimedes was born and raised in the Greek settlement of Syracuse in Sicily. His father was an astronomer, and no doubt his first mathematics teacher. Later, Archimedes traveled to Egypt, probably studying at the Alexandria Museum with students of Euclid. He then returned to Syracuse to work.

In addition to mathematics, Archimedes made important discoveries in physics, especially hydrostatics (the physics of water) and was an engineer of note. The most famous, probably apocryphal, story told of Archimedes concerns his solution of a problem presented to him by King Hiero. The king had recently acquired a golden crown, and wanted to verify that it was indeed made entirely of gold, without in any way modifying the shape. Archimedes was pondering the problem while in his bath, and it occurred to him that he could use the displacement of water when the crown was lowered into a bath to determine its density, hence its composition. He was so excited that he jumped out of the bath and ran naked through the streets yelling "Eureka! Eureka!" Eureka means "I have found it."

Among his many mechanical inventions was the Archimedean screw, a mechanical device for lifting water. Although it has now been mostly replaced for that purpose by other sorts of pumps, it is still used in a variety of applications from sewage treatment plants to fish hatcheries.

Archimedes was also noted for his prowess in designing and building war machines. They were a factor in allowing Syracuse to hold off a Roman siege for three years during the second Punic War. In the end, however, Syracuse was overrun, and Archimedes was slain by a Roman soldier.

Archimedes wrote many works on a wide variety of topics. Unlike Euclid, he did not write textbooks, but rather original research monographs. As with the other ancients, none of the originals have survived, and many of his works have been lost.

He was a pioneer of mathematical physics, the creation of mathematical models for physical situations. He proved the law of the lever, a result that had been previously known but not rigorously studied. This law states that the two sides of the lever are in balance when the product of the weight and the distance from the fulcrum on one side of the lever equals the similar product from the other side. Archimedes used this law in many of his mechanical inventions. He is said to have claimed: "Give me a lever long enough and a fulcrum on which to place it, and I shall move the world."

Archimedes also studied centers of gravity of various shapes, a topic that combined his interest in physics and geometry. The center of gravity of a body is the point from which it can (at least conceptually) be suspended and be at rest.

Hydrostatics is yet another area of science where Archimedes excelled. He discovered the *Archimedes principle*, that a body immersed in water will displace a volume of fluid that weighs as much as the body would weigh in air. (This was the theoretical principle behind the solution of the golden crown problem.)

Among his many mathematical works, we will visit three. The first is a short treatise, called *Measurement of the Circle*. In the tradition of studying the squaring of the circle, Archimedes proves that the area of the circle is equal to one-half the radius times the circumference. If we define π as the ratio of the circumference of circle to its diameter, $\pi = C/2r$, this gives

$$A = \frac{1}{2}rC = \frac{1}{2}r(2\pi r) = \pi r^2.$$

It was previously known that the area was proportional to r^2, i.e., $A = kr^2$. Archimedes showed that constant of proportionality is π.

He then goes on to numerically approximate π. He bounds the area of the circle above and below by inscribing polygons in the circle, and inscribing the circle in other polygons, as in the method of exhaustion. By considering polygons with an increasing number of sides (ultimately 96 sides), he arrives at the bounds $3\frac{10}{71} < \pi < 3\frac{1}{7}$. For many years after this, the estimate of $3\frac{1}{7}$ was commonly used for π; it is called the Archimedean value of π.

Archimedes' masterpiece was *On the Sphere and Cylinder*. In it he studied a sphere and its circumscribed cylinder (Figure 1.22). He proved that the surface area of the sphere is $2/3$ that of the cylinder (including the two caps), and the volume of the sphere is $2/3$ that of the cylinder.

Let us see how we can use Archimedes' results to find formulas for the surface area and volume of a sphere. First, we figure the surface area of the cylinder. Here is the approach:

$$A = A(\text{top}) + A(\text{bottom}) + A(\text{sides}).$$

If the radius of the sphere is r, each of the caps (top and bottom) is a circle of radius r, so has area πr^2. To figure the area of the sides of the cylinder, mentally unroll it. You will get a rectangle whose height is the height of the cylinder, $2r$, and width is the circumference of the circle, $2\pi r$. So the area of the side is the product of the width and height, or $(2\pi r)(2r)$. Putting this together,

$$A = \pi r^2 + \pi r^2 + (2\pi r)(2r) = 6\pi r^2.$$

Thus the surface area of the sphere is $2/3(6\pi r^2) = 4\pi r^2$.

The volume of the cylinder is the area of the base times the height, or $\pi r^2(2r) = 2\pi r^3$. Hence, Archimedes' theorem tells us that the volume of the sphere is $\frac{4}{3}\pi r^3$. As with the circle, before this theorem, the Greeks knew that the volume of the sphere was proportional to r^3 but did not know the constant of proportionality.

Figure 1.22   A sphere and its circumscribed cylinder.

Archimedes was so proud of these results that he requested the diagram of the sphere and cylinder be carved on his gravestone.

Archimedes proved the above results in the Greek fashion, in a way that did not reveal how he discovered them. In particular, he perfected the method of exhaustion. This method is useful, however, only after you know what the answer is. His method of discovery was mysterious until about 100 years ago. In 1899 a treatise containing a copy of his work called *The Method* was discovered in Constantinople, in the library of a Greek monastery. It had been written on a parchment in the 10th century, but then overwritten by a religious work in the 13th century. The practice of reusing parchments was not uncommon, since parchment was quite valuable. Such a reused parchment even has a name, palimpsest.

In *The Method* Archimedes gives an ingenious technique for computing areas and volumes, one that combines his work in geometry and physics. He mentally slices the shape he is studying into thin cross sections, then balances these against cross sections of a known shape, using the law of the lever. Some modern mathematicians have seen hints of the integral calculus (developed in the 17th century) in this process. It was not, however, rigorous, so after he used this method to discover results, he proved them in the usual deductive way.

Marcellus, the general who commanded the army that overran Syracuse, was upset to find that Archimedes had been killed by one of his soldiers. He erected a small column to mark Archimedes' grave, and had the figure of the sphere inscribed in the cylinder carved on it, as Archimedes had wanted. Cicero, the Roman philosopher and statesman, upon being appointed governor of Sicily in 75 BCE, 137 years later, found the grave neglected and overgrown. He cleaned it up and restored the marker. But it was again neglected, to be found only in 1965 during excavation for the construction of a hotel.

Conic Sections

Recall that conic sections were originally studied by Menaechmus to solve the Delian problem, doubling the cube. From the algebraic point of view, they are a natural topic of study, after straight lines. Specifically, a straight line is the set of points solving an equation of the form

$$ax + by + c = 0.$$

If we allow second powers of the variables, we get conic sections, which are solutions of equations of the form

$$ax^2 + bxy + cy^2 + dx + ey + f = 0.$$

The Greeks did not have this algebra, so they studied conic sections geometrically.

Apollonius (c. 240–174 BCE)

Apollonius was reportedly born in the Greek city of Perga, in modern-day Turkey, and lived as a young man in Pergamum. In the late 3rd century, Pergamum, also in Turkey, became a major intellectual center, modeled after Alexandria. It housed the second largest library in the Greek world.

After Pergamum, Apollonius moved to Alexandria, where he spent most of his career. He published a number of works on astronomy, geometry, and arithmetic, most of which have been lost.

Apollonius' masterwork was the *Conics*. It was the first known mathematical text to systematically and exhaustively treat a single topic—conic sections. It contains eight books and 487 propositions. The first part covers what was known to previous mathematicians, and the second part consists of original contributions.

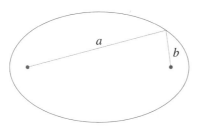

Figure 1.23   An ellipse and its foci.

Apollonius gave us the terms for the three conics: ellipse, hyperbola, and parabola. Among the many results in the *Conics*, the ellipse is shown to be the set of points the sum of whose distances from two fixed points is constant. The two points are called the *foci* (singular *focus*) of the ellipse. An ellipse and its foci are shown in Figure 1.23; the sum $a + b$ is the same for any point on the ellipse. As the two foci get closer together, the ellipse becomes less elongated, until, when the foci are the same point, the ellipse is a circle.

You can use the idea of the foci to construct an ellipse physically. Take a pad of paper, and stick two thumb tacks in where the foci are to be, as in Figure 1.24. Then take a length of string, and tie the two ends to the tacks. If you stretch the string taut, the two parts of the string will give the distances to the two tacks. The sum of these distances is the length of the string. So, if you take a pencil and trace out the curve you get by moving the pencil, always keeping the string taut, the resulting figure will be an ellipse.

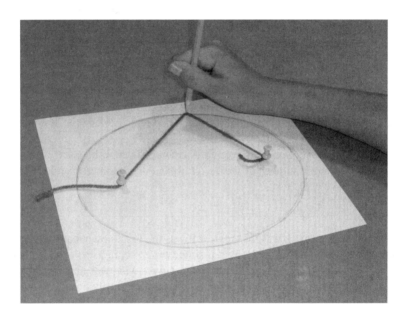

Figure 1.24   Drawing an ellipse.

Apollonius gave a similar result for the hyperbola. In that case, the *difference* of the distances to the foci is a constant. (See Figure 1.25.)

Apollonius did not have a similar result for a parabola, but his contemporary Diocles showed that the parabola is the set of points equidistant from a point (its focus) and a line (its *directrix*). See Figure 1.26.

At the heart of every large, modern telescope is a mirror in the shape of a paraboloid, a solid figure obtained by rotating a parabola. This uses an important reflection property, illustrated in Figure 1.27. In the figure, we have a cross-section of a paraboloid, i.e., a parabola, with the focus marked by a dot. All light that comes from a direction perpendicular to the directrix, after bouncing off the mirror, passes through

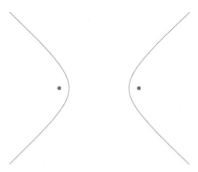

Figure 1.25   A hyperbola and its foci.

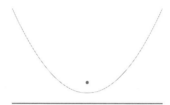

Figure 1.26   A parabola with focus and directrix.

the focus. This property is used to focus the light to form an image of distant objects. The word "focus" is from the Latin, meaning fireplace or hearth.

The *Conics* was very influential for many years. In the 17th century, Kepler drew on it when he discovered that planetary orbits are ellipses, with the Sun located at one focus.

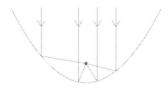

Figure 1.27   Reflection property of a parabola.

EXERCISES

1.39   In the text, we restated Common Notion 1 in algebraic terms. Do the same for Common Notion 2.

1.40   In the text, we restated Common Notion 1 in algebraic terms. Do the same for Common Notion 3.

1.41   Given that ABC is a right angle, find θ.

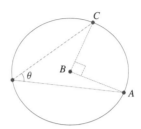

1.42   Given that the two triangles below are similar, find x and y.

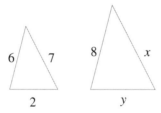

1.43   The two right triangles below are similar. Find x and y. (Hint: the Pythagorean Theorem may help.)

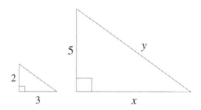

1.44   Euclid's Proposition II-4 is a geometric version of the binomial identity

$$(a + b)^2 = a^2 + 2ab + b^2.$$

In this problem, we consider the analogous trinomial identity

$$(a + b + c)^2 = a^2 + b^2 + c^2 + 2ab + 2ac + 2bc.$$

   a) Draw a version of Figure 1.18 to represent the trinomial identity.
   b) By considering areas, show that your diagram illustrates the trinomial identity.

1.45   The first proposition of the *Elements* is the construction of an equilateral tri-angle using straightedge and compass. Show how to carry out such a construction.

1.46   Given three points in the plane, not all on a line, show how to construct, using straightedge and compass, the circle that passes through them.

1.47   The law of the lever, proved by Archimedes, applies to playground teeter totters. Suppose that a 50 pound child sits 9 feet from the fulcrum (center) of the teeter totter. How far from the fulcrum would a 75 pound child sit in order to balance the other child?

1.48   The Earth weighs about 1.3×10^{25} pounds (13 followed by 24 zeroes). Sup-pose that Archimedes had a fulcrum placed 8000 miles from the Earth (about one diameter), and he can manage 200 pounds of weight on his end of the lever. How long should he be from the fulcrum to "move the world?"

1.49   Suppose that a sphere is inscribed in a cylinder, and that we measure the surface area of the cylinder to be 27.1434 cm^2 and the volume to be 10.8573 cm^3. Using Archimedes' theorem, find the surface area and volume of the sphere.

1.50   Find the surface area and volume of a sphere of radius 2 feet.

1.51   An ellipse has foci at $(2,0)$ and $(-2,0)$, and contains the point $(3,0)$, as in the figure below.
    a) What is the sum of the distances from $(3,0)$ to the two foci?
    b) If the point $(0,y)$ is on the ellipse, what is y?

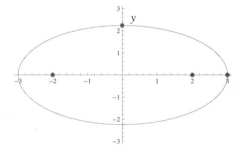

1.52   Below is a parabola with the x-axis as its directrix. Its equation is $y = \frac{1}{8}x^2 + 2$. Find the coordinates of the focus. (Hint: the point on the parabola with $x = 0$ is equidistant from the focus and the directrix.)

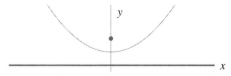

1.53   Conic sections can be created by folding papers. Try typing "folding conic sections" in your favorite Internet search engine to find out how to do this.

1.5 The Slow Decline

> The Greeks held the geometer in the highest honor, and, to them, no one came
> before mathematicians. But we Romans have established as the limit of this art,
> its usefulness in measuring and reckoning. The Romans have always shown more
> wisdom than the Greeks in all their inventions, or else improved what they took
> over from them, such things at least as they thought worthy of serious attention.
>
> CICERO (106–43 BCE)

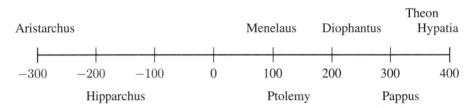

Late ancient Greek mathematics.

Progress is not inevitable. Mesopotamian mathematics reached a peak in the Old
Babylonian period (c. 1900–1550 BCE), then stagnated for more than a thousand
years. Egyptian mathematics also stagnated after the early part of the second millennium BCE.

Ancient Greek mathematics achieved its greatest heights in the 3rd century BCE.
The brilliance of this period was not matched again until the 17th century. Research
did not entirely cease, however, at least for six hundred years or so. There were
notable developments in trigonometry and number theory, and some very influential
texts appeared in this period.

The Roman Empire

Roman citizens overthrew their Etruscan rulers and established a republic in 509
BCE. In the ensuing centuries, they gradually extended their power, to the rest of
Italy, then the western Mediterranean, then the eastern Mediterranean. By the middle
of the first century BCE, they ruled the entire Mediterranean, including Asia Minor
and Egypt. By that time, they had also lost their republic, in which much of the power
was entrusted to an elected senate, to be replaced ultimately by an imperial system.
Starting with Julius Caesar (100–44 BCE), the Romans proceeded to conquer what
is now France, southern Germany, and Britain.

The height of the Roman empire was in the years of the Pax Romana ("Roman
peace") from 27 BCE to 180 CE. In these years, Rome was ruled by a succession of
emperors. The "peace" didn't mean no wars. There were still wars of conquest, often
quite brutal. The historian Tacitus quotes a barbarian chieftain from a conquered
Germanic tribe: "They make a wilderness and call it peace."

The Romans were great builders, constructing a vast system of roads, some of
which are still in use, and an impressive network of aqueducts to carry water to a city

that eventually housed about a million people. Culturally, they were no match for Greece. They did learn much from the Greeks, but only in some areas. In particular, as indicated in the Cicero quote above, they did not value higher mathematics. Certainly, their architects and engineers mastered practical mathematics, but the Romans produced no notable theoretical mathematics.

Greek culture did not disappear in the Roman era. The Latin language dominated only the western part of the Roman empire, while Greek remained the lingua franca[1] of the eastern Mediterranean. Furthermore, some Greek centers of learning, including Alexandria, continued their intellectual tradition, if at a somewhat reduced level of achievement.

Astronomy and Trigonometry

Astronomy has long been a stimulus for mathematics. In this period, it gave rise to a major area of math—trigonometry. The term comes from the Greek, *trigon*, meaning triangle, and *metron*, measure. The subject has developed beyond that in modern times, but early trigonometry was all about measuring triangles, that is, the sides and angles in a triangle.

Trigonometry is based on similarity of triangles. Consider a right triangle (Figure 1.28), where the angle ACB is a right angle. The ratio of the lengths of two corresponding sides will be the same in any triangle with the same angles. For example, suppose that in the triangle $A'B'C'$ the angle $A'C'B'$ is a right angle, and the angle α' is the same as the angle α. Since the sum of the angles in a triangle is always the same, this means that the angles ABC and $A'B'C'$ are also equal; in other words, the two triangles are similar. Hence the ratios of corresponding sides will be equal.

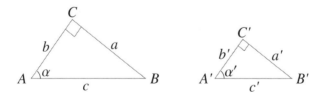

Figure 1.28   Two similar right triangles.

Consider the ratio a/c. Since the triangles are similar, $a/c = a'/c'$. In general, in any right triangle which also shares the angle α, the ratio of its corresponding sides will be the same number a/c. In trigonometry, we give that ratio a name: the sine of

---

[1] From the Italian, literally "Frankish language," lingua franca refers to a common language used to communicate by people with different native tongues. For example, English is now the lingua franca of the scientific community.

α, or $\sin \alpha$. Similarly, we name other ratios. The most common names are

$$\sin \alpha = \frac{a}{c}, \quad \cos \alpha = \frac{b}{c}, \quad \tan \alpha = \frac{a}{b}.$$

The symbol cos is short for cosine, and tan is short for tangent. A convenient way to remember these is to rename the sides: a is called opp, because it is opposite the angle, b is adj, for adjacent, and c is hyp, for hypotenuse.

$$\sin \alpha = \frac{\text{opp}}{\text{hyp}}, \quad \cos \alpha = \frac{\text{adj}}{\text{hyp}}, \quad \tan \alpha = \frac{\text{opp}}{\text{adj}}$$

Why is this useful? It allows you to learn about an unknown triangle by studying a known similar triangle. Aristarchus used this idea, albeit not with the modern terms, to determine the ratio of the distances from the Earth to the Sun and Moon. First he measured the angle between the Sun and Moon when the Moon is half-full, α in Figure 1.29. (The diagram is not to scale.) He found it to be $87°$. He also noted that, because the Moon is half-full, the Earth-Moon-Sun angle was a right angle. Then he used the argument above to deduce that the ratio of the distances to the Moon and the Sun, m/s, is $\cos 87°$ (m is adj, s is hyp). Finally, using various properties of triangles, he estimated that $\frac{1}{20} < \cos 87° < \frac{1}{18}$, so that $\frac{1}{20} < \frac{m}{s} < \frac{1}{18}$. Therefore, he concluded, the Sun is between 18 and 20 times as far away as the Moon.

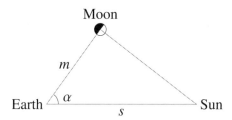

Figure 1.29   Sun and Moon distances.

Aristarchus' argument is perfect, but unfortunately his answer was off by a factor of about 20. The difficulty lay with the measurement; the angle is not $87°$ but about $89.8°$. The small difference in angle makes a big difference in the cosine, because the angle is near $90°$.

Aristarchus (c. 310–230 BCE)

Aristarchus was born on the Greek island of Samos. He probably studied and worked in Alexandria, in the early 3rd century BCE. His only surviving work is *On the Sizes and Distances of the Sun and Moon*, from which the previous argument is taken. He also estimated the actual distances, not just their ratio, and from these the sizes of the Sun and Moon.

Aristarchus is most famous for being the first astronomer known to have posited that the Earth revolves about the Sun, and not vice versa. Other scholars of his day had a scientific problem with this, namely, if the Earth moves, why is it that the stars don't appear to change during the year? His answer, that the stars are very far away, was not popular, although it has proven to be correct. Like Galileo 1900 years later, he was accused of impiety.

The most difficult part of Aristarchus' argument, besides the measurement, was the approximation of $\cos 87°$. Of course, he could have constructed a small triangle with the proper angles, and physically measured its sides. But, being a mathematician, he preferred more mathematical methods which held at the least the promise of greater accuracy. (He was actually quite accurate: $\cos 87° \approx 1/19.1073$.)

If astronomers were to use such arguments regularly, it would be useful to have tables giving values, say, of the cosine of every angle between 1 and 89 degrees. Then the astronomer need merely look up the answer, saving a lot of time. In fact, this is what happened. In order to study the heavens, they found such tables to be a great aid. In order to make these tables, however, they had to develop the mathematics of trigonometry.

The first mathematician known to have computed trigonometric tables was Hipparchus.

Hipparchus (c. 190–120 BCE)

Hipparchus was one of the greatest astronomers in history. He was born in Nicaea, Bythnia (now Iznik, Turkey), where he made his first astronomical observations. Later on he moved to the Greek island of Rhodes in the Aegean Sea. Little else is known of his life.

Hipparchus wrote at least a dozen works, of which only one minor commentary survives. Most of what we know of his work comes from references in the texts of others, most notably Ptolemy.

Hipparchus' work in astronomy built on earlier work by the Babylonians and Greeks, notably Eudoxus and Apollonius. Hipparchus made careful observations, and compiled a catalog of 850 stars. He estimated the distances to the Sun and Moon.

Much of mathematical astronomy in this period was dedicated to predicting the motions of the Sun, Moon, and planets. The word "planet," which originally included the Sun and Moon, comes from the Greek *planasthai*, to wander, because planets wander against the unchanging background of the stars. It was this wandering that the astronomers wanted to model mathematically.

The details of the mathematical models used by Hipparchus are too involved to go into here, but basic to the mathematics was the understanding of triangles, both in the plane and on spheres. To assist in the computations, Hipparchus reportedly constructed a table of *chords* subtended by arcs of a circle of standard radius r. For example, the chord $\operatorname{crd} \alpha$ is the length l in Figure 1.30. Hipparchus computed approximations of $\operatorname{crd} \alpha$, for α a multiple of $7\frac{1}{2}^\circ$, up to $360°$, a total of $48 \ (= 360/7.5)$ numbers. (He used a standard radius of 3438.)

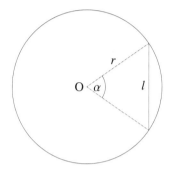

Figure 1.30   Chord subtended by arc.

What, you might ask, has this to do with trigonometry? The answer can be found in Figure 1.31, which includes the triangle from the previous diagram, cut in half. Notice that $\dfrac{l/2}{r} = \sin(\alpha/2)$. Since $\operatorname{crd}\alpha = l$, we have

$$\operatorname{crd}\alpha = 2r\sin(\alpha/2).$$

So if we know the chord l, we know the sine, and vice versa.

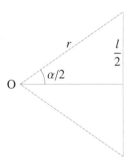

Figure 1.31   Relation of chord to sine.

The greatest astronomical discovery of Hipparchus was the precession of the equinoxes. As the Earth revolves around the Sun, it also rotates on its axis, like a spinning top. The axis of this spin always points at the north star, Polaris, or so it seems. In fact, the Earth wobbles, so the axis points in different directions, completing a circle in about 26,000 years. So, in about 12,000 years, the north star will be Vega, not Polaris. (Actually, the axis will point about 5° from Vega, but close enough.) Since this change is so slow, it is no surprise that it took many centuries of observations to notice it. The way Hipparchus spotted it was by noticing that the location of the Sun against the background stars at the Spring and Fall equinoxes had changed, hence the name precession of the equinoxes.

The next major developments in trigonometry, over two hundred years after Hipparchus, were due to Menelaus.

Menelaus (c. 70–130 CE)

Virtually nothing is known of the life of Menelaus. He worked in Alexandria and in Rome. Only one of his works, *Sphaerica*, survives, in an Arabic translation.

Menelaus also made a table of chords. Although it has been lost, it was probably more extensive than that of Hipparchus, for it was in six books.

In the *Sphaerica*, Menelaus treats trigonometry as a science separate from astronomy. Most importantly, he deals with spherical trigonometry, triangles on a sphere, proving a number of results that were important to mathematical astronomy. Astronomers were interested in spherical trigonometry because they studied the celestial sphere. The mathematics of spherical triangles is different from that of flat triangles. For example, the sum of the angles of a spherical triangle is not always 180 degrees. In fact, it is not hard to construct spherical triangles where each of the three angles is 90 degrees, so that the sum is 270 degrees. (See the exercises.)

The apex of ancient trigonometric studies, as well as ancient astronomy, was in the work of Claudius Ptolemy.

Claudius Ptolemy (c. 100–178)

Little is known of the life of Ptolemy, the greatest of the ancient astronomers. He may have been born in Egypt; he certainly worked at the Museum in Alexandria. Some of the astronomical observations he recorded can be dated to the period 124–142 in Alexandria.

Ptolemy wrote many works on science, mathematics, and astrology (yes), including two that were the standards in their fields for many centuries: the *Geography*, and his astronomical masterpiece, the *Almagest*.

Ptolemy's *Almagest* is the astronomical equivalent of Euclid's *Elements*. It was the culmination of Greek astronomy, not to be superseded until the work of Copernicus, Kepler, Galileo, and Newton. The book was originally called *Mathematiki Syntaxis—Mathematical Collection*. Later, it became known as *Megisti Syntaxis*—the greatest collection, and then in Arabic *al-magisti*, which morphed into the *Almagest*, by which it is commonly known today.

In the *Almagest*, Ptolemy gives a table of chords for all angles from one-half of a degree to 180 degrees, in intervals of one-half of a degree. The numbers, following the Greek astronomical tradition, are in the Babylonian sexagesimal notation.

In order to approximate these numbers, Ptolemy had to develop a variety of mathematical techniques. For example, using symmetry, it was easy to find the chords of $45°$ and $30°$. Then Ptolemy demonstrated how, given chords for two angles, he could find the chord for the difference of those two angles. This allowed him to compute the chord of $15°$. The formula Ptolemy discovered is equivalent to the following modern version.

$$\sin(\alpha - \beta) = \sin\alpha\cos\beta - \cos\alpha\sin\beta$$

The computations were not easy; it has been suggested that he employed (human) calculators to finish his table.

Given his table, Ptolemy could solve any triangles needed. He also applied these results to spherical trigonometry, using theorems of Menelaus and others, as well as his own. Finally, he presented a detailed mathematical model for each of the planets. The goal was to be able to predict their movements, e.g., to predict the time that Mars would rise on any given date, at any place on the Earth. In fact, he computed a number of such predictions, then compared them against actual observations to confirm the theory.

Ptolemy also produced another influential book, the *Geography*. In this work, Ptolemy listed 8000 locations, and drew a large map of the known world and twenty-six regional maps. There was a mathematical issue here: how does one represent the spherical Earth on a flat piece of paper? (Scholars were aware long before Columbus that the Earth was round. In fact, Eratosthenes had quite a good estimate of its radius in the 3rd century BCE.) Ptolemy came up with two solutions to this, two *projections* from a sphere to a plane. These were not improved upon for more than a thousand years.

Even though Ptolemy was aware of Eratosthenes' work, he used an inferior (too small) estimate of the Earth's size. Columbus used this size in the 15th century, one reason he thought he had reached Asia when he landed in the Americas.

The Silver Age of Hellenistic Mathematics

After the 3rd century BCE, the pace of mathematical research slowed. While there were advances, such as in trigonometry, much of the scholarly work was directed at preserving, not enhancing, mathematical knowledge.

There was a spurt of activity, however, from the middle of the 3rd century through the 4th century. In particular, works by Diophantus and Pappus stand out.

Diophantus (c. 210–290 CE)

About all that is known of Diophantus is that he worked in Alexandria in the 3rd century, and that he wrote the *Arithmetica*, in thirteen books.

A famous puzzle about him, written a couple hundred years after his death, asks for the number of years he lived.

> "Here lies Diophantus," the wonder behold. Through art algebraic, the stone tells how old: "God gave him his boyhood one-sixth of his life, One twelfth more as youth while whiskers grew rife; And then yet one-seventh ere marriage begun; In five years there came a bouncing new son. Alas, the dear child of master and sage after attaining half the measure of his father's life chill fate took him. After consoling his fate by the science of numbers for four years, he ended his life."

ANTHOLOGIA PALATINA

The *Arithmetica* is in thirteen books. We have versions of six books in Greek, from a 13th century copy. In the 1970s four of the other books were discovered in Arabic translations. These books are not arranged in the step-by-step logical fashion of Euclid's *Elements*. Rather, they consist of a series of problems, 290 in the surviving books, more like the style of ancient Babylon or Egypt.

The *Arithmetica* is different from Euclid in another way; it has little geometry. The problems are about algebra and elementary number theory. This was apparently original with Diophantus. For this he has been called the "father of algebra."

Here is one of the more elementary problems.

> To divide a given number into two having a given difference.

Diophantus explained how to solve this for the case where the given number is 100 and the difference is 40. Before giving his solution, let us translate this into our modern notation. We are looking for numbers, say x and y, such that $x + y = 100$ and $y - x = 40$. From the second equation, we can solve for y, getting $y = x + 40$. Substituting this into the first equation, we get $2x + 40 = 100$. This yields $x = 30$ and $y = x + 40 = 70$.

Diophantus solved this in a similar way, starting with: if x is the smaller number, then $2x + 40 = 100$. His style was to state a general problem, then demonstrate a procedure for solving a particular case of the problem. He did not give a general solution, although his method could often be used to solve the general problem.

The problem above is called *determinate*: there are two equations in two unknowns, so there is only one solution. Most of Diophantus' problems are *indeterminate*: there are more unknowns than equations, so there might be many solutions. For example:

> To divide a given number into two squares.

The example he worked with had 16 as the given number. So the problem is to find two squares which sum to 16, that is, to solve $x^2 + y^2 = 16$. In these problems, Diophantus contented himself with finding a single solution, although again his method might serve to find others. He only accepted positive rational solutions. His methods were often ad hoc, and clever. In this case, the two numbers he found were $\frac{12}{5}$ and $\frac{16}{5}$. (Check that they work!)

One of the most important innovations in the *Arithmetica* is the use of notation. Before this, algebraic problems were described entirely in words. This was called rhetorical algebra. Diophantus invented a system of notation that he used to make the solution of a problem easier. For example, he used ς for the variable name, what we would call x, and Δ^v for x^2. Thus, he would write $\Delta^v\beta$ for $2x^2$, since β, the second letter of the alphabet, was standard then for 2. Another example:

$$\Delta^v\beta\varsigma\gamma$$

means the same as our $2x^2 + 3x$, since γ is 3.

His system was not complete, so he needed to use some words still, in a style historians call syncopated algebra. Fully symbolic algebra didn't arrive for another 1300 years.

The *Arithmetica* remained influential for many years, inspiring other mathematics well into the 17th century.

The other influential work from this period is the *Synagoge*, or *Collection*, of Pappus.

Pappus (c. 290–350)

Pappus was the most important geometer since Apollonius, some five hundred years earlier. Of the life of Pappus, we know only that he worked in Alexandria in the first half of the 4th century, and he had a son named Hermodorus. We know when he worked, because a solar eclipse that he observed can be dated to October 18, 320.

In addition to mathematics, Pappus wrote on astronomy, geography, and hydro-statics, but little survives of these works.

Pappus' *Collection* is exactly that, a collection of separate books, written by Pappus, but perhaps put together by a student. The books are on different geometric topics, and of varying quality. Many of the books contain surveys of the work of his predecessors. In fact, the *Collection* is our best source for the lost works of many Greek geometers.

Book 5 concerns *isoperimetric* figures, those with different shapes but equal perimeters or surface areas. It contains a proof that, of all regular solids with the same surface area, the sphere has the largest volume. Concerning a similar problem in the plane, he has praise for honey bees, whose honeycombs are composed of hexagons.

> Bees were endowed with a certain geometrical forethought...There being, then, three figures which of themselves can fill up the space round a point, viz. the triangle, the square and the hexagon, the bees have wisely selected for their structure that which contains the most angles, suspecting indeed that it could hold more honey than either of the other two.

Bees, then, know just this fact which is useful to them, that the hexagon is greater than the square and the triangle and will hold more honey for the same expenditure of material in constructing each. But we, claiming a greater share in wisdom than the bees, will investigate a somewhat wider problem, namely that, of all equilateral and equiangular plane figures having an equal perimeter, that which has the greater number of angles is always the greater, and the greatest of them all is the circle having its perimeter equal to them.

Book 7, "On the Domain of Analysis," is the most influential. It discusses the "analytic" method Greek mathematicians used to solve problems or discover proofs, as opposed to the formal "synthetic" method of proof Euclid made famous. The latter is logically more rigorous but hides the actual discovery process. Book 7 also contains a number of theorems that were important in the development of projective geometry in the 17th century.

The *Collection* has been called the requiem for ancient Greek geometry. The next time geometry of this quality was developed was in the 17th century.

The Decline of Rome and the Rise of Christianity

Greek mathematics was dead by 500 CE. One is tempted to ask why it died, just as Edward Gibbon, in his *The History of the Decline and Fall of the Roman Empire*, famously asked why the Roman empire died. Perhaps the better question is why each, the empire and the mathematics, survived for so long. What was distinctive about the Roman empire was not that it ended, but that it was so successful for so long. Similarly, Greek mathematics is characterized by its long run of brilliance. So let us ask what sustained this brilliance.

An intellectual tradition usually requires a continuity of practitioners. It is not easy to pick up and read a copy of Euclid. Students then, as now, relied on an oral tradition to introduce them to the subtleties of mathematics. Hence the tradition might not survive the absence of mathematicians for a generation or two. The rise of larger institutions in which mathematicians could work was of considerable importance. Greek examples include the Pythagorean society, Plato's Academy, and the Alexandrian Museum. Unlike now, however, the community of theoretical mathematicians in this period was never large, so it was susceptible to interruption.

The support of mathematics has, until recent times, always been rather tenuous. The cutting edge of theoretical mathematics is usually not very practical. Even though history has demonstrated many times that what begins as purely theoretical eventually develops practical applications, this connection is not always evident to those in power. Ancient mathematicians were heavily reliant on the patronage of the elite. They enjoyed this patronage in Greece before Alexander, and under the Ptolemies afterwards. As we have already seen, though, the Romans did not value mathematics highly, so their conquest of Egypt in the first century BCE was a blow to mathematics. Also notable in this conquest, in a skirmish between Julius Caesar's army and local troops in 48 BCE, the library at the Alexandrian Museum was mostly destroyed by a fire. The Roman Marc Antony donated the library of Pergamum to Cleopatra as a replacement. It was stored at the nearby temple of Serapis.

Greek mathematics arose as part of a larger philosophical movement, in an attempt to find rational explanations for natural phenomena. The religious and philosophical climate underwent major changes in Roman times, which weakened this motivation. The most important change was the rise of Christianity, which had different philosophical priorities. Its ascendance was also accompanied by religious conflicts. It took some time for Christianity to become a major force, but by the 4th century it had risen to the status of official religion. This was a further blow to the Greek mathematical tradition.

As important as these cultural concerns were social conditions, which took a turn for the worse after 180 CE, when the emperor Marcus Aurelius died. What followed was a period of civil unrest, economic decline, and plague, especially in the western part of the empire. In the period 235–284 occurred what historians call the Crisis of the Third Century. During this time, dozens of men were declared emperor by some part of the Roman army, usually to die shortly thereafter. The constant civil unrest weakened the borders, so raids from the north and east became more frequent. A plague starting in 251 decimated the population in many places.

The empire stabilized again in the early 4th century, but in a quite different form. In particular, the emperor Constantine made two momentous changes: he converted to Christianity, and he moved the capital of the empire to the city of Byzantium, later renamed Constantinople (now called Istanbul). The western empire did not recover, however, continuing to suffer invasions from the north. Rome itself was sacked in 410 by the Visigoths. The traditional end of the western empire is usually dated to 476, when the German Odovacer deposed the western emperor Romulus (who was not recognized in Constantinople).

The eastern, Greek-speaking, part of the empire fared much better, with Constantinople remaining a major cultural and political center for many centuries.

Egypt's Alexandria was not immune to the troubles of this time. Its economy suffered as trade declined. Support for mathematics also declined. In 391 a Christian mob attacked and destroyed the pagan temple of Serapis and much of its library, the one established in Cleopatra's reign.

Theon (c. 335–405) and Hypatia (c. 355–415)

Theon was a mathematician and Neoplatonist who worked in Alexandria. He published work on astronomy and mathematics, as well as astrology. His most distinguished student was his daughter Hypatia, who became a leading scholar and teacher in her own right.

Hypatia is the most famous woman mathematician of antiquity. When she was sixty years old, on her way to a lecture, she was taken from her carriage by a mob, stripped, dragged to a neighboring church, and brutally murdered. She had become involved in a nasty political struggle and was accused of sorcery.

Theon wrote a number of commentaries on earlier works. The commentary was a popular form of scholarly writing for many centuries. The commentaries might include original research, or only exposition. Theon's commentary on Ptolemy's astronomical works *Almagest* and *Handy Tables* are mainly explanatory. Although a leading scholar of his day, it does not appear that he was a very original thinker.

Theon is most famous for his edition of Euclid's *Elements*. This edition added little to the original, but replaced previous editions and became the standard for many years. In Western Europe, this was the source of all subsequent editions of Euclid until late in the 19th century.

Hypatia helped her father with his work and produced commentaries of her own. Our knowledge of her work is spotty, but it appears that she surpassed her father. She was certainly a famous teacher, offering instruction in philosophy and religious literature as well as mathematics and astronomy. Her edition of Archimedes' *Measurement of the Circle* was the source of most subsequent editions.

Work at the Alexandrian Museum continued after the time of Hypatia, but did not produce any distinguished mathematics. In general, in the Greek-speaking eastern part of the empire, scholarship declined. In 529 the emperor Justinian ordered all pagan schools closed. The Academy in Athens was taken over by the state, and folded shortly thereafter. The memory of Greek mathematics did not die, however. We shall see later that Islamic mathematicians were able to learn from scholars trained in this tradition.

The story was different in the west, which suffered catastrophic decline. The population diminished. Rome in its heyday had about one million inhabitants; by the 8th century, no western city had more than about 50,000 people. Trade also declined precipitously. This was the heart of the Dark Ages.

Mathematics was still sometimes taught, usually in schools associated with Benedictine monasteries. There monks, often from Ireland, copied and preserved Greek and Latin manuscripts. The audience for these manuscripts was not large; literacy was not widespread even among the nobility.

The monks probably did not understand much of the mathematics. The most popular mathematics text was *De Institutione Arithmetice* of Boethius (480–524), which was based on the *Introductio Arithmetica* of the first century Alexandrian scholar Nicomachus. Although Nicomachus' work was an elementary text at the time it was written, with no proofs, it was the most advanced arithmetic known in western Europe for many centuries.

Another influential handbook, by the Italian monk Flavius Magnus Aurelius Cassiodorus (c. 480–575), justified the study of arithmetic by quoting Jesus: "the very hairs of your head are all numbered." In general, mathematical works from the time after Boethius often substituted the citing of authority for proofs. All but practical mathematics was gone from the West.

EXERCISES

The following table can be used in the exercises.

α	$\sin \alpha$	$\cos \alpha$
$45°$	.707	.707
$60°$	.866	.5

1.54   Consider the right triangle below.

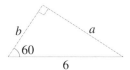

a) Find lengths a and b.

b) What is $\tan 60°$?

1.55   Consider the right triangle below.

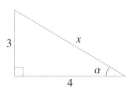

a) Find x.

b) Find $\tan \alpha$.

c) Find $\cos \alpha$.

1.56   Suppose that a right triangle has a hypotenuse of length 6, and the sine of one of its angles (other than the right angle) is .5. Find the lengths of the two legs.

1.57   If $r = 100$ and $\operatorname{crd} \alpha = 75$, what is $\sin(\alpha/2)$?

1.58   Suppose that a circle has $r = 100$. Find $\operatorname{crd} 90°$.

1.59   Find $\sin 15°$. (Hint: $15 = 60 - 45$.)

1.60   Find $\sin 75°$.

1.61   Find $\cos 15°$ and $\cos 75°$.

1.62   Suppose that you have three points on the Earth, with A and B on the equator, a quarter of the way around the Earth, and C at the North Pole. What are the angles in the triangle ABC? (Hint: the angles don't add up to $180°$.)

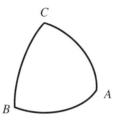

1.63   Geometry on the sphere is different from that in the plane. Try this exercise.

Hang your right arm down by your side. Curl your fingers into a fist, with your thumb pointing forward.

Lift your arm straight out away from your body, keeping your hand as it was, and not twisting your wrist.

Now move your arm so that it points forward. (Your thumb will point to the left.)

Finally, lower your arm straight down again.

Is your thumb now pointing the same direction it was at the start? What does this have to do with a sphere? (Hint: a sphere is the set of points at the same distance, say arm's length, from a center, say a shoulder.)

1.64   According to the puzzle in the text, how long did Diophantus live? (Hint: the age seems to be divisible by 7 and 12.)

1.65   How might Diophantus have written $3x^2 + 2x$?

1.66   Use Diophantus' method to find the general solution of the first Diophantus problem, that is, to find x and y such that $x + y = s$ and $y - x = d$, for any sum s and difference d.

1.67   Another problem Diophantus considered is this: to find a square number between $5/4$ and 2. Find such a number. (Recall that Diophantus only allowed positive rational numbers.)

CHAPTER 2

THE GROWTH OF MATHEMATICS TO 1600

[The universe] cannot be read until we have learnt the language and become familiar with the characters in which it is written. It is written in mathematical language, and the letters are triangles, circles and other geometrical figures, without which means it is humanly impossible to comprehend a single word.

GALILEO GALILEI (1564–1642)

So far, we have looked at the ancient roots of mathematics, primarily around the Mediterranean Sea. In this chapter we will look further east, to the two great Asian cultures of China and India, before returning to Western Asia and Europe. The mathematics developed in these cultures was in each case influenced by others, although less so in the case of China. We will trace some of the connections among the different traditions, but must acknowledge up front that much remains unknown about this fascinating topic.

Mathematics for the Liberal Arts.
By Donald Bindner, Martin Erickson, Joe Hemmeter Copyright © 2012 John Wiley & Sons, Inc.

2.1 China

> The Master said: Learning without thinking is useless. Thinking without learning is dangerous.
>
> CONFUCIUS (551–479 BCE)

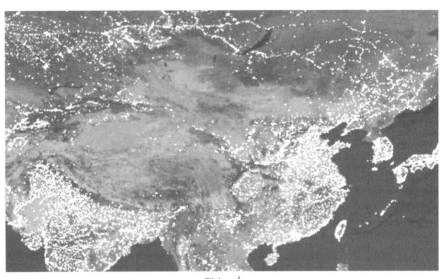

China.[1]

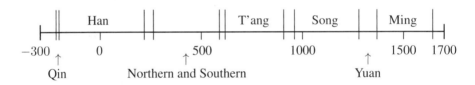

Major Chinese dynasties, from the Qin to the Ming.

History to 1600: An Overview

China is dominated by two great river systems. The Yellow River and the Yangtze River both originate in the Tibetan Plateau, and flow eastward to the Pacific Ocean. The Yellow River is in northern China, the Yangtze further south.

The earliest known Chinese civilizations arose in the Yellow River valley. By 4500 BCE villagers there were growing millet and raising pigs. The earliest known

---

[1]From "Earth at Night." C. Mayhew and R. Simmon (NASA/GSFC), NOAA/NGDC, DMSP Digital Archive.

writing dates from the Shang dynasty (c. 1550–1046 BCE), on "oracle bones"—animal bones and turtle shells sometimes used for divination—and on bronze. The scribes also wrote on bamboo strips, but these didn't last very long. The Shang dynasty only encompassed a small area in northern China.

The successor to the Shang was the Zhou dynasty (c. 1046–256 BCE), a semi-feudal society with a somewhat larger geographical extent. Work on an early version of the Great Wall of China, intended to protect the northern border against invasion by nomads, commenced in the 8th century BCE. Culture flourished in the 6th century BCE, with the founding of several academies of scholars. The most famous philosopher of this time was Confucius (*Kung Fu-zi*, "Master Kung"), who valued education, loyalty, moderation, and his version of the Golden Rule: "What you do not wish for yourself, do not do to others." Also dating from this period is the Taoism of Lao-Tzu, a more mystical philosophy and an important influence on early Chinese science.

After the Zhou dynasty was the short-lived but important Qin dynasty (221–206 BCE). This dynasty was initiated by one of the most remarkable rulers of any time and place, Qin Shi Huangdi ("First August and Divine Emperor of Qin"). Qin united all of China for the first time. He reorganized the government, destroying the power of the feudal lords, and he established a central bureaucracy. He engaged in massive infrastructure projects, building roads and irrigation systems, constructing a canal linking the Yangtze and Pearl River systems in the south, and completing the Great Wall in Inner Mongolia. The current Great Wall was built later on the foundations of this one. The construction of the wall cost the lives of thousands of workers and helped lead to the end of the Qin dynasty a few years after its founder's death.

Qin Shi Huangdi standardized the Chinese system of weights and measures, and the Chinese written language too. After this time, regardless of the local spoken language, educated Chinese used the same written characters. The emperor distrusted scholars and banned private ownership of books. For his tomb, he employed about 700,000 workers over a period of more than 30 years to build a massive underground palace in wood, clay, and bronze. In 1974 local farmers made a remarkable discovery at his tomb. He had "protected" his underground palace with an army of 8000 life-size soldiers and horses modeled in clay. The army was complete with generals and chariots, and each soldier was unique. Qin Shi Huangdi died at age 49 in 210 BCE. His dynasty was overthrown four years later.

The Han dynasty (206 BCE–220 CE) succeeded the Qin. The most notable period in this dynasty was the rule of Emperor Wu, which lasted over 50 years, from 140–87 BCE. During this time, Wu extended his empire from Vietnam in the south to northern Korea, and westward into central Asia. The Silk Road became a major trading route, reaching all the way to Rome. Chinese culture flourished. In the year 2 CE, a census put the population of China at almost 60 million people. Sometime in the first century paper was invented. Education, based on Confucianism, became the route to advancement in the bureaucracy, via a new civil service examination system. This system was maintained, with brief disruptions, into the 20th century.

The period after the Han dynasty (220–598 CE) was a time of political disunity, characterized by a succession of short-lived dynasties and invasions from the north. In fact, much of Chinese history was characterized by invasions from the north; the Great Wall was built and rebuilt in an effort to protect against northern invaders. Although the wall was not always successful, the invaders did not replace the native Chinese culture, instead typically adopting it themselves.

There were several notable developments in this period, which comprised the Three Kingdoms Period (220–265) and the Northern and Southern Dynasties (265–589). One was the invention of paired stirrups in western China, which made the cavalry a more effective fighting force. This was also the time that the first porcelains were made. Although Buddhism was first imported from India in the Han dynasty, it became more important in these centuries. Also in this time, there was a great migration of people from northern China to the south, many as refugees. Thus the southern, rice-growing region of China became more important.

The empire was restored in the short-lived Sui dynasty (589–618). Like the Qin dynasty, the Sui is noted for its huge projects and its unpopularity. In addition to work on the Great Wall, the Sui reportedly conscripted five million peasants to work on the Grand Canal linking the Yangtze valley to northern China. This canal, over 1000 miles long and still the longest in the world, was used mainly to transport grain from the south to the north.

Block printing was invented around 600. In this method, a wooden block was carved in such a way that the desired characters stood out. The block was then coated with ink and pressed upon paper or other substance. With this method the same page could be printed repeatedly.

Next up was the T'ang dynasty (618–907), one of the high points in Chinese civilization. Poetry and the fine arts flourished. The capital at Chang'an was the largest city in the world, with almost two million people. The Imperial Academy was founded in 754 to prepare scholars for public service. The curriculum was based on classical Confucian literature. Gunpowder was invented in the 8th century.

After the T'ang ensued a half-century of disunity, ending with the establishment of the Song dynasty (960–1279), another brilliant period. The Song rule of northern China was somewhat tenuous; a northern invasion forced them to move their capital south to Hangzhou. Nonetheless, the arts, trade, and urban culture flourished. Agricultural production doubled. Tea and cotton cultivation expanded. Neo-Confucianism expanded its influence at the expense of Buddhism.

A couple of important advances in shipping date from the Song dynasty. One is the sternpost rudder, which provided increased maneuverability for sailing ships. Another is the magnetic compass, which may actually have been invented earlier but was first widely used in shipping in this time.

Movable type printing also dates from the Song dynasty. In block printing an entire page is carved at one time. The new method involved carving individual characters, which then could be assembled into a page for printing. The characters could be reused.

Starting around 1200 the Mongols under Genghis Khan built one of the greatest empires ever, covering most of northern and western Asia, stretching into Eastern Europe and the Middle East. His empire was larger in area and population than the Roman empire. Kublai Khan, Genghis's grandson, moved the capital of his empire to what is now Beijing, in 1264. From there, he completed the conquest of southern China, and formed the Yuan dynasty (1279–1368).

Chinese contacts with the West increased during Mongol rule. Moslems, Tibetan Buddhists, Nestorian Christians, and Roman Catholics were all invited to China. The most famous European visitor was the Venetian Marco Polo, who visited Kublai Khan in Beijing 1275–1292 and wrote about this amazing place upon his return to Europe, which was at the time backward compared to China. Chinese mathematics attained its greatest achievements at this time.

The foreign rule of the Yuan was never popular, and it was overthrown by an ex-Buddhist monk named Zhu Yuanzhang, who established the Ming dynasty (1368–1644). The Ming rulers rebuilt the Great Wall and built a new southern capital at Nanjing.

In the early 14th century, there was a remarkable series of naval expeditions launched from China, seven in all, led by the eunuch Zheng He (c. 1371–1433). These preceded by several decades the more famous European voyages of exploration. Zheng He traveled throughout the Indian Ocean, as far as the Persian Gulf, the Red Sea, and eastern Africa. These were not entirely unknown lands to the Chinese; they had traded with them before. So Zheng He's were not entirely voyages of discovery. Instead, he was interested in advertising the might of China, and collecting treasure and tribute. His fleet contained as many as 300 ships with 28,000 crewmen. Some of the ships were huge, up to 400 feet in length. By comparison, Columbus' flagship was 85 feet long. In addition to treasure, Zheng He collected exotic animals, such as giraffes, lions, and zebras.

In the end, these voyages were terminated by politics. The conservative Confucian faction won out over the eunuch faction at court. If that had not happened, one can only imagine the different course of history if the first Europeans voyagers had encountered Chinese imperial power in the spice islands.

Later in the Ming dynasty, Chinese farmers began growing new crops from America, such as maize (corn), potatoes, and peanuts. There was increasing contact with western culture. A notable figure at this time was the Jesuit missionary Matteo Ricci, who traveled to Macao in southern China in 1582 and to Beijing in 1601. He became a Confucian scholar himself, and he introduced China to many western ideas and inventions. He was valued at court for his knowledge of astronomy. After this time, Chinese mathematics was no longer isolated from Western mathematics.

Early Chinese Mathematics

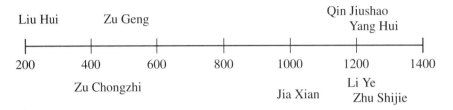

Chinese mathematicians.

The Chinese number system was decimal, based on 10. The earliest writings, on oracle bones, had separate symbols for 10, 100, and so on, but by the Han period a place-value system like ours was used, with separate symbols only for 1, ... , 9. Zero was represented by a space, later by a dot, then finally by a circle in the 12th century.

Calculation was usually carried out on a counting board, using small rods to represent the numerals. There were two sets of counting rod numerals, shown in Figures 2.1 and 2.2.

Figure 2.1   Vertical counting rods.

Figure 2.2   Horizontal counting rods.

These rods were arranged in horizontal rows on the counting board, one row to a number. These numbers could be manipulated to carry out whatever computation was required. To avoid confusion, the horizontal and vertical numbers were alternated within a row, and the rightmost digit also alternated between horizontal and vertical. Some examples are shown in Figure 2.3. Negative numbers could be handled easily: a different color rod was used, black for negative, red for positive.

The earliest mathematical texts we have are from the Han period, at which time Chinese mathematics was already relatively sophisticated. The first work, the *Suan shu shu* (*Book of Numbers and Computation*) was only discovered in 1984, in a

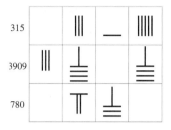

Figure 2.3   Numbers on a counting board.

tomb. Two others of importance are the *Zhoubi suanjing* (*Arithmetical Classic of the Gnomon and the Circular Paths of Heaven*), an astronomical text, and the *Jiuzhang suanshu* (*Nine Chapters on the Mathematical Art*). The latter, often referred to as simply the *Nine Chapters*, remained important in China for many centuries.

These early books, and many later ones, were primarily collections of problems, along with their solutions and commentary. The problems concerned practical matters such as surveying, engineering, and taxation. They were probably used in the education of bureaucrats.

Liu Hui (lyoo hwā) (c. 220–280)

Liu Hui worked in the Three Kingdoms period immediately following the end of the Han dynasty. He lived in the northern Kingdom of Wei. Nothing else is known of his life.

Liu edited the standard version of the *Nine Chapters*, adding his own commentaries on the material and also a tenth chapter, the *Sea Island Mathematical Manual*. His reputation as one of the greatest mathematicians of ancient China rests on this work.

Geometry

The Chinese had formulas for the area and volume of a number of geometrical figures. The *Nine Chapters* gives several methods of computing the area of a circle, all of which use the value of 3 for π. In his commentary, however, Liu Hui noted that this was incorrect. He used a succession of polygons inscribed in a circle to obtain a better approximation: 3.141024. (Compare to what we now know: 3.141592653....) This method is essentially the same as that used by Archimedes (see Section 1.4 and Figure 1.15 in Section 1.3), although Liu Hui did not provide the rigorous proof that Archimedes did.

The *Nine Chapters* makes clear that Chinese mathematicians were quite familiar with the Pythagorean Theorem and knew how to apply similar triangles to surveying problems. It also contains a rule for determining the diameter of a sphere from its volume. Here again, Liu notes that this rule was incorrect but he was not able to determine the correct formula.

Zu Chongzhi (429–500) and Zu Geng (c. 480–525)

Zu Chongzhi came from a family of distinguished astronomers and mathematicians. He himself was a prominent astronomer, engineer, and mathematician. He spent most of his life as a court official, working mainly in Jiankang (modern Nanjing). He also wrote ten novels.

Zu Geng was the son and collaborator of Zu Chongzhi. He followed in the family tradition as astronomer, mathematician, and court official (as did Zu Geng's son).

Zu Chongzhi improved on Liu Hui's estimate for π, obtaining the inequalities $3.14159127 < \pi < 3.1415926$. This estimate was not bested for more than a thousand years.

Zu Chongzhi is best known for revising the calendar. His revision was based on a new estimate of the length of the year, an estimate that was accurate to 50 seconds. He had trouble implementing the new calendar, however. A court rival accused him of "distorting the truth about heaven and violating the teaching of the classics." Zu Chongzhi answered that his calendar came

> not from spirits or from ghosts, but from careful observations and accurate mathematical calculations. . . . People must be willing to hear and look at proofs in order to understand truth and facts.

Zu Geng continued his father's campaign to revise the calendar, and finally succeeded in having it adopted by the emperor, in 510. His main claim to fame, however, was finding the correct formula for the volume of a sphere, thereby completing the task that Liu Hui was unable to accomplish.

Chinese astronomers prepared trigonometric tables in the 8th century. Apparently, these tables were based upon the work of Indian mathematicians. There was considerable contact between China and India at this time, particularly in the form of Buddhist monks. As we shall see later in this section, Indian astronomers were more advanced, partly based on earlier Greek influence.

Although Chinese geometers gave proofs for some of their results, they never built up the careful logical structure that characterized Greek mathematics.

Algebra

One of the problems in the *Nine Chapters* is this.

> There are three classes of grain, of which three bundles of the first class, two of the second, and one of the third make 39 measures. Two of the first, three of the second, and one of the third make 34 measures. And one of the first, two of the

second, and three of the third make 26 measures. How many measures of grain are contained in one bundle of each class?

If we denote the measures of grain in the three bundles by x, y, and z, we get the following three equations.

$$3x + 2y + z = 39$$
$$2x + 3y + z = 34$$
$$x + 2y + 3z = 26$$

(Check!) The problem is to solve for x, y, and z. The text gives instructions for the solution, using a counting board. First, write the coefficients of the three equations in columns, from right to left. Using modern notation, this yields the following arrangement.

1	2	3
2	3	2
3	1	1
26	34	39

Multiply the middle column by 3,

1	6	3
2	9	2
3	3	1
26	102	39

and subtract twice the right-hand column from the middle column.

1	0	3
2	5	2
3	1	1
26	24	39

Note that we now have a 0 at the top of the second column. We next obtain a 0 at the top of the first column: multiply the first column by 3,

3	0	3
6	5	2
9	1	1
78	24	39

then subtract off the third column.

0	0	3
4	5	2
8	1	1
39	24	39

Multiply the first column by 5, then subtract off 4 times the second column.

$$
\begin{array}{ccc}
0 & 0 & 3 \\
0 & 5 & 2 \\
36 & 1 & 1 \\
99 & 24 & 39
\end{array}
$$

(Again, check.) This gives us another 0. Now, we write the equations that correspond to these columns, reversing how we originally got the columns.

$$3x + 2y + z = 39$$
$$5y + z = 24$$
$$36z = 99$$

Notice that the third equation is easy: $z = 99/36 = 2\frac{3}{4}$. Substituting this into the second equation gives $5y + 2\frac{3}{4} = 24$, which we can solve for $y = 4\frac{1}{4}$. Finally, substituting our values for y and z into the first equation and solving gives $x = 9\frac{1}{4}$. (As always, check.)

Do you see the pattern? By appropriately adding or subtracting multiples of the columns, we produced enough 0s so that the resulting set of equations was easy to solve. Chinese mathematicians noticed that this way of combining columns did not change the solutions. In other words, the new set of equations has the same solution as the old one. This method can be generalized, and was in the *Nine Chapters*, to solve more complicated problems with more unknowns and equations. This algorithm is essentially the same as what is now known as Gaussian elimination, independently invented by Carl Friedrich Gauss in 1800.

As early as the *Nine Chapters*, Chinese mathematicians also had methods for solving some quadratic and cubic equations. Recall that a general quadratic equation is of the form $ax^2 + bx + c = 0$ for some numbers a, b, and c. Similarly, a cubic equation is of the form $ax^3 + bx^2 + cx + d = 0$. The Chinese had methods for approximating x using counting rods. Their method for solving quadratic equations was not the same as our quadratic equation, however. It used the binomial expansion $(x + y)^2 = x^2 + 2xy + y^2$.

In the 11th century, Jia Xian generalized this method to approximate roots of higher order equations, i.e., when the highest power of x is more than 2 or 3.

Jia Xian (jyä shē'an) (c. 1010–1070)

Almost nothing is known of Jia Xian's life, other than that he was a government official and studied with the mathematician and astronomer Chu Yan. He reportedly wrote at least two books, but they are now lost. His work is known to us primarily through later texts.

Jia Xian's method of extracting roots was based on the expansion of the binomial $(x + y)^n$. Here are the expansions for $n = 0, 1, 2, 3$.

$$(x + y)^0 = 1$$

$$(x + y)^1 = x + y$$

$$(x + y)^2 = x^2 + 2xy + y^2$$

$$(x + y)^3 = x^3 + 3x^2y + 3xy^2 + y^3$$

Consider the coefficients of the terms on the right of these equations. If we write only the coefficients, we get this triangle.

$$1$$

$$1\ 1$$

$$1\ 2\ 1$$

$$1\ 3\ 3\ 1$$

These are the first few rows of what is now known as *Pascal's triangle*. (Blaise Pascal was a 17th century mathematician.) Jia Xian discovered this triangle, and the following easy method to generate it. Except for the 1s at the ends of each row, every number is the sum of the two numbers above it. Thus the next row is 1 4 6 4 1, and the corresponding binomial expansion is

$$(x + y)^4 = x^4 + 4x^3y + 6x^2y^2 + 4xy^3 + y^4.$$

Jia's method of approximating roots of polynomials was essentially the same as that developed in Europe by William Horner and Paolo Ruffini some 750 years later.

Li Ye (lē yŏo) (1192–1279)

Li Ye was born in Zhending, in Hebei Province in northern China. He passed the civil service examination and worked in the government until it fell to the Mongols. He then moved to the foot of Mt. Fenglong, where he lived in seclusion, and often in poverty. Later in his life, he was invited by Kublai Khan to serve in the Mongol government, which he did only briefly, after which he returned to Mt. Fenglong.

Li Ye published several works, of which the most famous is *Ceyuan haijing* (*Sea Mirror of Circle Measurements*), which he wrote in 1248. The *Sea Mirror of Circle Measurements* contains 170 problems based on one geometric diagram of a circular city wall circumscribed by a right-angled triangle. Although the problems are of geometric origin, each of them led to a quadratic equation which Li would then solve, both algebraically and geometrically.

Yang Hui (yäng hwā) (c. 1238–1298)

Yang Hui was from southern China. Almost nothing else is known about him, although it is likely that he was a minor civil servant.

Yang Hui wrote two major works which still survive: *Xiangjie jiuzhang suanfa* (*A Detailed Analysis of the Mathematical Methods in the 'Nine Chapters'*) from 1261, and *Yang Hui suanfa* (*Yang Hui's Methods of Computation*) from 1274–75. The former work is our main source of the algebra of Jia Xian on solving equations, mentioned earlier.

Yang Hui's Methods of Computation is a collection of seven volumes. It contains work on a variety of mathematical topics, including the solution of quadratic equations. His writing is notable for its careful exposition of methods. It also provides examples of magic squares (see the exercises).

Zhu Shijie (jōo shē jē) (c. 1260–1320)

Zhu Shijie was born near modern-day Beijing. After extensive travels as an itinerant mathematics teacher, he settled down in modern-day Yangzhou, where he attracted students "like clouds from the four quarters."

Zhu Shijie wrote two major works: *Suanxue qimeng* (*Introduction to Mathematical Science*) and *Siyuan yujian* (*Precious Mirror of Four Elements*). The first was an elementary book.

The *Precious Mirror* is considered the summit of Chinese algebra. It is most notable for its treatment of equations with more than one variable. Zhu adapted earlier methods of solving polynomials to equations with up to four unknowns. He also dealt with series (sums), for example, giving the rule for adding the first n counting numbers,

$$1 + 2 + 3 + \cdots + n = \frac{n(n+1)}{2},$$

and the first n squares,

$$1^2 + 2^2 + 3^2 + \cdots + n^2 = \frac{n(n+1)(2n+1)}{6}.$$

Congruences

Some time between 300 and 500 CE, a book called *Sunzi suanjing* (*Mathematical Classic of Master Sun*) was written. In it is the following problem.

> Suppose we have an unknown number of objects. When counted in threes, 2 are
> left over, when counted in fives, 3 are left over, and when counted in sevens, 2 are
> left over. How many objects are there?

A method of solving this problem is given.

> Multiply the number of units left over when counting in threes by 70, add to the
> product of the number of units left over when counting in fives by 21, and then add
> the product of the number of units left over when counting in sevens by 15. If the
> answer is 106 or more then subtract multiples of 105.

Let's use the method. First we multiply 2 by 70, to get 140. To this we add 3 times
21, getting $140 + 63 = 203$, then add $2 \cdot 15$, giving us 233. This is more than 105,
so subtract 105 to get 128. This is still too large, so subtract 105 again to get our
answer: 23. It is easy to check that 23 works.

This problem uses what we now call *congruences*. Let a and b be integers, and m
a positive integer. We say that a is *congruent to b modulo m* if a and b have the same
remainder when divided by m. We write this as $a \equiv b \pmod{m}$.

Using this notation, we can write the problem this way. Find a number x that
satisfies the following three congruences.

$$x \equiv 2 \pmod 3, \qquad x \equiv 3 \pmod 5, \qquad x \equiv 2 \pmod 7$$

Congruence problems arise naturally when working with calendars. Say that your
calendar has 365 days in the year, 30 days to a month. If we assume that day 1 is the
beginning of a year and a month, then when will the 15th of the month fall on the
100th day of the year? In this case, we want a number x such that $x \equiv 15 \pmod{30}$
and $x \equiv 100 \pmod{365}$.

Qin Jiushao (chin jyoo shou) (c. 1202–1261)

Qin Jiushao is one of the more colorful characters in the history of mathematics. He
was born in southern China late in the Song dynasty, and lived in the time when
the Mongols were in the process of conquering China. His father was a government
official in the imperial capital, and Qin Jiushao was able to study at the Imperial
Astronomical Bureau.

As well as being one of the best Chinese mathematicians ever, Qin was an accomplished poet and expert in music and architecture. He was in the military for a while,
and was known for his skill in fencing, archery, and riding. He served in a number
of government positions, which he used to make himself rich. He was known for his
corruption, apparently including poisoning his opponents. He was also known for
his love affairs. According a Chou Mi, a contemporary biographer, in his mansion
Qin had a "series of rooms for lodging beautiful female musicians and singers."

Qin is best known for his *Shushu jiuzhang* (*Mathematical Treatise in Nine Sections*). In it is the first extant explanation of Jia Xian's method of solving polynomials. It is most famous, however, for Qin's work on congruences.

Qin addressed this problem: given integers $m_1, m_2, \ldots, m_n$ and remainders r_1, $r_2, \ldots, r_n$, find a number x such that $x \equiv r_i \pmod{m_i}$ for each $i = 1, 2, \ldots, n$. An example of this problem is the one at the beginning of this section, with $r_1 = 2$, $m_1 = 3, r_2 = 3, m_2 = 5, r_3 = 2, m_3 = 7$.

Such systems of congruences do not always have a solution. For example, consider the system $x \equiv 1 \pmod{2}$ and $x \equiv 2 \pmod{4}$. If x is a solution of the first congruence, it must be odd. But no odd number can solve the second congruence. Qin gave conditions for when a solution exists, in what became known later as the *Chinese Remainder Theorem*. He also invented a method that could solve any problem for which there was a solution.

More about congruences can be found in Chapter 5.

The 13th century, with the work of Li Ye, Qin Jiushao, Yang Hui, and Zhu Shijie, was the acme of medieval Chinese mathematics. The only notable event after this time was the invention of the modern form of the Chinese abacus, which appeared sometime before 1400. In an abacus, numbers were represented by beads on strings or rods. The beads in an upper section represent five, in a lower section one. Each column holds one digit. Beads count if they are moved to the divider (see Figure 2.4). The abacus allowed for much faster computations than the counting board, and was widely used in Asia until very recently.

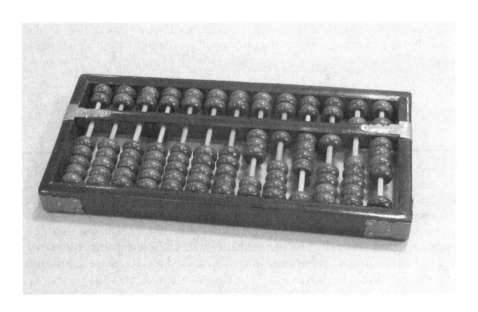

Figure 2.4   An abacus showing $314,159$.

During the Ming dynasty, there were no major mathematical developments. Not only did the field not advance, but some of the earlier mathematics was lost. Then, in

1607, Matteo Ricci and Xu Gaunqi translated the first six books of Euclid's *Elements* into Chinese, ushering in a new age of Chinese mathematics.

EXERCISES

2.1   What would the following numbers look like on a counting board? Arrange them as in Figure 2.3.

    a) 23

    b) 517

    c) 890

    d) 6004

2.2   Use the method from the *Nine Chapters* to solve this system of equations for x, y, and z.

$$-x + \frac{1}{2}y - z = 2$$
$$2x - 2y + 4z = -10$$
$$3x + 2y - z = 9$$

2.3   There are three classes of grain, of which three bundles of the first class, one of the second, and two of the third make 29 measures. Four of the first, one of the second, and one of the third make 28 measures. And two of the first, three of the second, and five of the third make 46 measures. How many measures of grain are contained in one bundle of each class?

2.4   Write out the first six rows of Pascal's triangle. Use your results to expand $(x + y)^5$ and $(x + y)^6$.

2.5   Use the formulas of Zhu Shijie to compute.

    a) $1 + 2 + 3 + \cdots + 100$

    b) $1^2 + 2^2 + 3^2 + \cdots + 10^2$

2.6   Decide whether each of the following congruences is true or false.

    a) $24 \equiv 3 \pmod 7$

    b) $6 \equiv 10 \pmod{17}$

    c) $242 \equiv 2 \pmod{10}$

    d) $240001 \equiv 3 \pmod 2$

2.7   Find a value of x that satisfies the following congruences.

$$x \equiv 2 \pmod 3, \qquad x \equiv 2 \pmod 5, \qquad x \equiv 3 \pmod 7$$

2.8   Find a value of x that satisfies the following congruences.

$$x \equiv 15 \pmod{30}, \qquad x \equiv 100 \pmod{365}$$

2.9   Suppose we have an unknown number of objects. When counted in threes, 2 are left over, when counted in fives, 2 are left over, and when counted in sevens, 6 are left over. How many objects are there?
  a) Write this problem using the $a \equiv b \pmod{m}$ notation.
  b) Solve the problem.

2.10   A band director wants his band to march in rows of 7, but finds that, when he lines them up, he has only 5 people in the last row. So, after some careful figuring, he thinks maybe rows of 15 are OK. But when he lines them up in rows of 15, the last row has only one person in it.
  a) Write this problem using the $a \equiv b \pmod{m}$ notation.
  b) Assuming that there are no more than 100 people in the band, how many people are there?
  c) What if there might be more than 100 people?

As early as about 2800 BCE, a type of pattern called a magic square appears in Chinese literature. A *magic square* of order n is a square array of the numbers 1, 2, ..., n^2 with the property that all rows, all columns, and the two diagonals have the same sum. Below is a magic square of order 3, called the Lo Shu Square. All rows, columns, and diagonals add up to 15.

4	9	2
3	5	7
8	1	6

2.11   Find another magic square of order 3.

2.12   Show that there is no magic square of order 2. (There are magic squares for every order greater than 2.)

2.13   Show that every magic square of order 3 has row sum 15. (Hint: the sum of all rows is $1 + 2 + 3 + \cdots + 9$.)

2.14   Find a magic square of order 4.

2.15   Find a formula for the row sum of a magic square of order n. (Hint: the sum of all rows is $1 + 2 + 3 + \cdots + n^2$.)

2.2 India

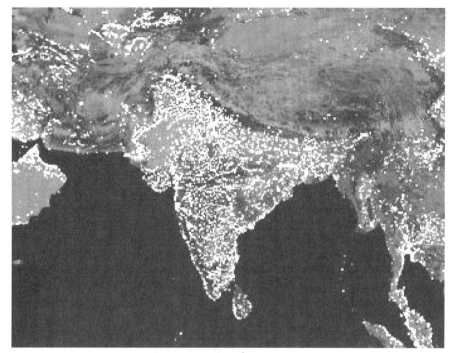

India.[1]

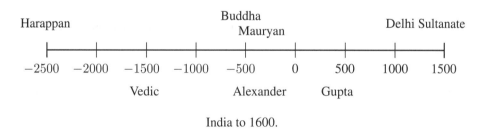

India to 1600.

History to 1600: An Overview

The Indian tectonic plate began crashing into Asia about 55 million years ago, a collision that continues, deforming both India and Asia. Major results of the collision are the Himalayan Mountains, the tallest in the world, and the Tibetan Plateau, with an area about half that of the continental United States and an *average* height above that of the highest peak in the lower 48 states. Most of the major rivers in Asia

[1]From "Earth at Night." C. Mayhew and R. Simmon (NASA/GSFC), NOAA/NGDC, DMSP Digital Archive.

originate in either the Himalayans or the Tibetan Plateau, including China's Yellow and Yangtze rivers.

The rivers that dominate Pakistan and northern India are the Indus and the Ganges. (Though separate countries now, India and Pakistan have not always been so. For the purposes of this section, we will refer to the whole area as India.) The Indus rises in Tibet and flows through China and Pakistan into the Arabian Sea, west of India. The Ganges rises in India and flows through northern India into Bangladesh, to empty into the Bay of Bengal to the east of India. Much of northern India lies in the Ganges plain.

By 3000 BCE agriculture had come to the Indus River valley, influenced by the Sumerians. The staples were wheat and barley. By 2500 BCE there was a flourishing civilization in modern-day Pakistan and western India, called the Harappan civilization. This urban civilization is known primarily through excavations of two of its cities: Mohenjo-daro and Harappa. Even though the Harappan civilization covered an area larger than the ancient civilizations of Mesopotamia, Egypt, or China, less is known of it. Virtually nothing was known until the excavations of Mohenjo-daro and Harappa began in 1920. Its script has still not been deciphered.

About 1500 BCE the Indus River valley was invaded by Aryans from central Asia. Their military power was based on chariots. The subsequent civilization is usually called Vedic, after the vedas, which were early Hindu religious works. This was the time that Hinduism, and the caste system with its rigid social strata, were developed. The language of the Aryans was Sanskrit, from which most current Indian languages derive. Sanskrit is in a family of languages called Indo-European, which contains almost all modern European languages. Indeed, it was the similarity of Sanskrit to classical Greek and Latin that inspired the British Chief Justice of India, Sir William Jones, to invent historical comparative linguistics in 1786. From roughly 800–300 BCE, the Aryans extended their domain throughout the Ganges River basin in northern India. In the 6th century, Siddhartha Gautama, Buddha, was born in modern-day Nepal.

In 326 BCE Alexander the Great invaded India. He conquered most of the Indus River valley by the following year, then headed back west. After his death, one of his generals established the Seleucid dynasty, which ruled much of western Asia. Then Chandragupta Maurya, starting in eastern India, established the Mauryan empire (317–184 BCE), the first to unify virtually all of India. Eventually, he ruled both the Ganges and the Indus River valleys. Chandragupta's grandson Ashoka extended the empire from Afghanistan to Bangladesh, from Nepal almost to the southern tip of India. Ashoka is one of the most famous figures in Indian history. Early in his reign, he was known to be quite ruthless. However, in his conquest of Kalinga (modern Orissa) in eastern India, he witnessed such destruction and suffering that he experienced a change of heart. He renounced conquest thereafter, converted to Buddhism, and spent much of the rest of his life promoting it.

After the Mauryan empire, India was divided into a collection of smaller states. The next great empire was the Gupta dynasty (320–535 CE), founded by another Chandragupta. This was a brilliant period in Indian culture, including literature, music, architecture, and art. The Shakespeare of Sanskrit drama, Kalidasa, flourished around the year 500. Indian culture was spread by merchants and missionaries throughout southeast Asia, as far as Indonesia. This time is also known as the golden age of Indian science, particularly in astronomy and mathematics.

The Gupta dynasty was overthrown in 535 by the White Huns, called the Rajputs. There followed another period of smaller states. In 711 the lower Indus valley was conquered by the Muslims. Later, Muslim rulers conquered virtually all of India; notable is the Delhi Sultanate period (1206–1526), when India was ruled by a succession of Muslim dynasties.

Early Indian Mathematics

Indian numerals were decimal, grouped by tens. Eventually, Indian mathematicians developed the number system that we use today, but early on their system was not positional; they did not only have nine or ten symbols whose interpretation was based on their position. In addition to the symbols for 1, ..., 9, they had separate symbols for 10, 20, and so on.

Much of our knowledge of early Indian mathematics comes from *Śulbasūtras* written between 800 and 200 BCE. These were a type of poem called a *sutra*. The *Śulbasūtras* contained compact instructions for building altars for sacrifices. They had formulas (written in words, not symbols) for a variety of areas and volumes, some accurate, some not. There was no notion of proof. These poems did reveal a knowledge of basic geometry, including the Pythagorean Theorem; one had an approximation of the square root of 2 accurate to five decimal places.

India was sufficiently close to Mesopotamia and the Mediterranean to at least occasionally exchange mathematical ideas. There is evidence of an influx of Mesopotamian astronomy around 600 BCE. When Alexander invaded in 326 BCE, he brought with him not only soldiers, but also scholars. There were other transmissions of knowledge, especially astronomy. Native Indian astronomy and mathematics were inferior to the imports until the time of the Gupta empire.

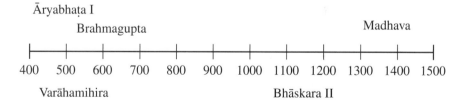

Indian mathematicians.

The Golden Age

During the Gupta empire, northern India was one of the richest and most brilliant civilizations in the world. The capital of the empire was at Pāṭaliputra, a city on the Ganges in eastern India, near modern-day Patna. The Gupta kings supported learning, founding the great Nalanda University, near Pāṭaliputra, which was one of the world's major centers of learning until it was sacked and burned in 1193.

Indian mathematics of this time was developed mainly in service to astronomy. Indian astronomy was based on Greek models. The Indian tradition was concentrated in schools called *paksas*, of which the most famous were the *Brahmapaksa* in Ujjain in central India, and the *Āryapaksa* in Kusamapura in northern India.

Āryabhaṭa I (476–550)

The first Indian mathematician of whom we have any knowledge is Āryabhaṭa I. He lived in Kusamapura, near Pāṭaliputra, where he founded the *Āryapaksa* school of astronomy.

Although he wrote at least two texts, only one, the *Āryabhaṭīya*, survives. The *Āryabhaṭīya* was written in verse couplets, and was completed when the author was twenty-three years old. It is one of the most influential astronomical works in history, not only in India but subsequently in the Muslim world.

The first Indian satellite, launched in 1975 (on a Russian rocket), was named Aryabhata in his honor.

$$\sim$$

Although the *Āryabhaṭīya* was primarily an astronomical text, a sizable portion of it, 33 out of 118 verses, deals with mathematics. It has trigonometric tables to be used in computation, and a good approximation to π, namely, 3.1416.

Āryabhaṭa also gave formulas for the sum of squares and cubes, in modern notation:

$$1^2 + 2^2 + \cdots + n^2 = \frac{n(n+1)(2n+1)}{6}$$

$$1^3 + 2^3 + \cdots + n^3 = (1 + 2 + \cdots + n)^2.$$

The *Āryabhaṭīya* contained a novel method for solving certain types of indeterminate equations. Using the notation introduced in the section on China, we can write the problem as follows: given numbers a, b, r, s, find a number N such that $N \equiv a$ (mod r) and $N \equiv b$ (mod s). This is a special case of the Chinese Remainder Theorem. As in China, such problems arose in astronomical calculations.

There are no proofs in the *Āryabhaṭīya*.

Varāhamihira (505–587)

Varāhamihira was a leading astronomer of the *Brahmapaksa* school, working in Ujjain. His major work is the *Pancha-siddhantika* (*Five Treatises*). This work summarizes the astronomical knowledge of his time, including much western (Babylonian, Egyptian, Greek) astronomy, as well as native Indian astronomy.

Varāhamihira was also interested in astrology, about which he wrote extensively.

The *Pancha-siddhantika* improved on the trigonometric tables given in the *Āryabhaṭīya*. In addition, Varāhamihira discovered a number of trigonometric identities, given below in modern notation.

$$\sin \alpha = \cos(90^\circ - \alpha)$$

$$\sin^2 \alpha + \cos^2 \alpha = 1$$

$$\sin^2 \alpha = \frac{1 - \cos 2\alpha}{2}$$

The word *identity* means that each equation holds for every value of α. For example, if we know that $\cos 60^\circ = .5$, then the first identity tells us that $\sin 30^\circ = .5$.

In another work, the *Brihat Samhita*, Varāhamihira dealt with combinatorics. Among the many topics in the book, he gave formulas for making various perfumes. One question he addressed was how many perfumes could be mixed using a choice of four ingredients from a total of 16: "If a quantity of 16 substances is varied in four different ways, the result will be 1820." He did not tell us how he arrived at that number, but it seems likely that he had a procedure equivalent to the following formula. The number of ways to choose r items from a total of n (these are called *combinations*) is

$$C_r^n = \frac{n(n-1)(n-2)\cdots(n-r+1)}{r(r-1)(r-2)\cdots 1}.$$

In his example,

$$C_4^{16} = \frac{16 \cdot 15 \cdot 14 \cdot 13}{4 \cdot 3 \cdot 2 \cdot 1} = 1820.$$

Brahmagupta (598–c. 665)

Brahmagupta lived most of his life in Bhillamala, modern-day Bhinmal, in Rajasthan in northwestern India. He belonged to the *Brahmapaksa* school, and headed the astronomical observatory in Ujjain. He produced four texts on astronomy and mathematics, of which the most famous is the *Brāhmasphuṭasiddhānta* (*Correct Astronomical System of Brahma*). This text, written in 628, was translated into Arabic in Baghdad in 771, where it had a major influence on Islamic astronomy and mathematics.

Brahmagupta, unlike many mathematicians until quite recently, was comfortable dealing with negative numbers. He gave correct rules for doing arithmetic on negative numbers; for example, a negative times a negative equals a positive. He was one of the first to deal with 0 as a number, although he thought that one could get away with dividing by 0, which modern mathematics disallows because it leads to logical inconsistencies.

Brahmagupta also studied *Diophantine equations*, indeterminate equations with integer solutions. An example is the equation $2x + 3y = 17$, where x and y must be integers. There are many solutions; two of them are $x = 1$, $y = 5$ and $x = 4$, $y = 3$. Diophantus had studied these as well (hence the name), but only gave one solution per equation. Brahmagupta completely solved the linear Diophantine equation $ax + by = c$, where a, b, and c are given integers. More on linear Diophantine equations can be found in Sections 5.9 and 5.10.

Another equation that Brahmagupta studied was of the form $y^2 = ax^2 + b$, where a and b are given integers. The case where $b = 1$ is now called the *Pell equation*.

A simple example of a Pell equation is $y^2 = 2x^2 + 1$. It is easy to find one solution, $x = 2$, $y = 3$, which we will write $(2, 3)$. Brahmagupta gave a novel method for generating new solutions from old. First, he wrote two solutions, one above the other. See Figure 2.5, where we use the solution $(2, 3)$ twice.

Figure 2.5   Generating a new solution to $y^2 = 2x^2 + 1$.

Then he generated a new solution by doing crosswise multiplication to get a new x, namely, $x = 2 \cdot 3 + 2 \cdot 3 = 12$, and vertical multiplication (doubling the first product) to get a new y, namely, $y = 2(2 \cdot 2) + 3 \cdot 3 = 17$. Thus we have a new solution, $(12, 17)$. Let's check: $2 \cdot 12^2 + 1 = 289 = 17^2$. In general, if we start with two (possibly identical) solutions of the equation, this crosswise and vertical multiplication will produce a new one. As another example, let us start with $(2, 3)$ and $(12, 17)$ (Figure 2.6). The generated values are $x = 2 \cdot 17 + 12 \cdot 3 = 70$ and $y = 2(2 \cdot 12) + 3 \cdot 17 = 99$. Our new solution is $(70, 99)$. Check!

Figure 2.6   Generating a third solution to $y^2 = 2x^2 + 1$.

European mathematicians did not achieve Brahmagupta's level of understanding of the Pell equation until the 18th century, more than one thousand years later.

In geometry, Brahmagupta found a formula for the area of a cyclic quadrilateral, which is a quadrilateral that can be inscribed in a circle. Using the notation of Figure 2.7,

$$A = \sqrt{(s-a)(s-b)(s-c)(s-d)}$$

where s is one-half of the perimeter: $s = \frac{1}{2}(a+b+c+d)$.

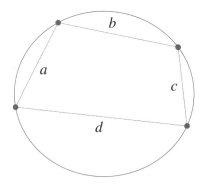

Figure 2.7 A cyclic quadrilateral.

If we let $d = 0$, then we have a triangle, which can always be inscribed in a circle, and the formula reduces to the famous Heron's formula for the area of a triangle: $A = \sqrt{s(s-a)(s-b)(s-c)}$. (Heron was a Greek scholar who worked in Alexandria in the first century CE. He is most famous as an engineer and inventor.)

Notation and Computation

Perhaps India's most important contribution to mathematics is our modern decimal place-value system, with 0. This system developed gradually. By around 600 CE a place-value system was employed, using only nine symbols (corresponding to our digits $1, \ldots, 9$). The use of 0 may have come a bit later, but was probably in place by the early 8th century. This system spread to the Arab world, and later to Europe. It is now usually called the Hindu-Arabic system.

The symbols we use for the digits are not the same as the original Hindu ones, although there is some overlap. The symbols themselves evolved as the system passed from India to the Arab world to Europe, being given their modern form by Albrecht Dürer in the early 16th century.

The Indians did not use decimals for fractions; this was introduced later in the Muslim world.

The great value of this numerical system was how easy it made computation. The Indians themselves developed a method of multiplying numbers, later called *gelosia*, or lattice, multiplication. (*Gelosia* was a type of iron grill placed over windows in medieval Italy.) Here is an example, computing 425×629. First, we draw a 3×3 grid (because each of the numbers has three digits), then place the two numbers above and to the right of the grid (see Figure 2.8).

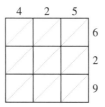

Figure 2.8   Gelosia multiplication: 425×629.

Next, multiply each of the digits of the first number by each one of the second number, placing the result in the associated grid square. For example, since $4 \times 6 = 24$, we place a 2 and a 4 in the top left square, the 2 above the diagonal, the 4 below. This gives us the left diagram in Figure 2.9. If there is only one digit in the product, place it below the diagonal. Doing this for all digits gives us the right diagram in Figure 2.9.

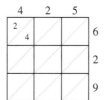
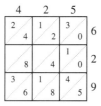

Figure 2.9   Gelosia multiplication: placing products inside the squares.

Now we look at the diagonals. Beginning at the lower right, we add the numbers in each diagonal, writing the sum below the grid. If the sum is more than 9, we carry the tens digit to the next diagonal. The first three steps in this process are shown in Figure 2.10.

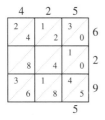
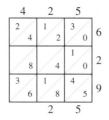
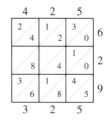

Figure 2.10   Gelosia multiplication: summing the diagonals.

We continue summing the diagonals, writing the rest of the digits up the left side of the grid (Figure 2.11). We can then read the answer, starting down the left side and across the bottom: $425 \times 629 = 267{,}325$.

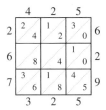

Figure 2.11 Gelosia multiplication: $425 \times 629 = 267{,}325$.

This method of multiplication spread from India to the Arab world, and then to Italy, introduced by Fibonacci in 1200. It was widely used in Europe until the 17th century. Printing may have helped lead to its demise, since the diagrams were not easy to typeset.

Medieval Indian Mathematics

Bhāskara II (1114–1185)

Bhaskara II was perhaps the greatest medieval Indian mathematician. (Bhaskara I was a 7th century astronomer/mathematician.) The son of a noted astrologer, Bhaskara II headed the astronomical observatory in Ujjain. He was thus a successor to Varāhamihira and Brahmagupta, and the last great scholar of the Brahmapaksa.

Bhaskara wrote at least a half-dozen works; the most influential are the *Siddhanta Siromani* (*Head Jewel of an Astronomical System*), *Lilavati* (*The Beautiful*), and *Bijaganita* (*Seed Counting* or *Root Extraction*). The latter two are sometimes considered part of the first text.

Bhaskara produced improved trigonometric tables and, more than most of his predecessors, explained the theory behind the computations. He states the equivalent to the following trigonometric identities.

$$\sin(a + b) = \sin a \cos b + \cos a \sin b$$

$$\sin(a - b) = \sin a \cos b - \cos a \sin b$$

It is possible that Bhaskara's work was influenced by Islamic astronomy, which was very sophisticated at this time.

Bhaskara is also notable for his use of algebraic symbols. For example, he used the initial syllables for various colors where we would use the symbols x, y, or z.

The highlight of Bhaskara's mathematical work was his study of indeterminate equations of the form $y^2 = Dx^2 \pm b$, where D and b are given positive integers. Extending Brahmagupta's earlier work, he wrote down a general method for solving the Pell equation $y^2 = Dx^2 + 1$, although he did not give a proof that it always worked. (The first such proof didn't appear until 1929.)

After the time of Bhaskara II, northern India entered a turbulent period with the invasion of Islamic armies, and scholarship suffered. On the southwest coast of India, however, a cultural center flourished at Kerala, sheltered by mountains to its east. Its leader was Madhava.

Madhava (1340–1425)

Madhava of Sangamagramma was born in a town near Cochin, Kerala. He worked on both astronomy and mathematics. Some of his astronomical works have survived, but we know of his mathematics only through commentaries of his successors.

The most important mathematics developed in Kerala originated in attempts to improve the accuracy of trigonometric tables, which were used in astronomy and navigation. The issue addressed was how to obtain accurate approximations to trigonometric values. Below, we present in modern notation what Madhava discovered.

First, we make a convenient change of unit. Instead of using degrees, we measure angles in *radians*. Instead of 360 degrees to a circle, we have 2π radians per circle. The conversion is easy:

$$\theta^\circ = \frac{2\pi}{360} \cdot \theta = \frac{\pi}{180} \cdot \theta \text{ radians.}$$

So, for example, 90° is $\frac{\pi}{180} \cdot 90 = \frac{\pi}{2}$ radians. One advantage of this unit is in figuring the length of an arc of a circle, for example l in Figure 2.12. Recall that the circumference of a circle is $2\pi r$, where r is the radius. Using 360 degrees to a circle, the length of the arc of θ degrees is then $l = 2\pi r \frac{\theta}{360}$. In a similar way, if we use radians, we have 2π radians to a circle, so an arc of x radians has length $l = 2\pi r \frac{x}{2\pi} = rx$, a simple formula.

Madhava discovered that polynomials can give good approximations of some trigonometric functions. For example, if x is small, then $\sin x \approx x$. (The symbol $\approx$ means "approximately equal to.") A better approximation can be obtained by using a third degree polynomial: $\sin x \approx x - x^3/6$. This can be continued, getting better and better approximations by using higher degree polynomials. Of course, we can't use just any polynomials. Madhava found the pattern required for the polynomials. Here are the first four.

$$x$$

$$x - \frac{x^3}{6}$$

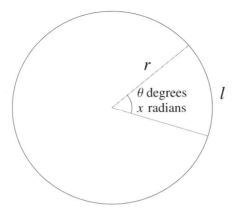

Figure 2.12   The length of a circular arc.

$$x - \frac{x^3}{6} + \frac{x^5}{120}$$

$$x - \frac{x^3}{6} + \frac{x^5}{120} - \frac{x^7}{5040}$$

Part of this pattern should be clear; each term has odd degree, and we alternate adding and subtracting terms. But where do the denominators come from? It turns out that they are factorials. A *factorial* is defined, for any positive integer, as the product of the integers up to and including that one. For example, $5! = 5 \cdot 4 \cdot 3 \cdot 2 \cdot 1 = 120$. Using this notation, we can write the four polynomials above in this way.

$$x$$

$$x - \frac{x^3}{3!}$$

$$x - \frac{x^3}{3!} + \frac{x^5}{5!}$$

$$x - \frac{x^3}{3!} + \frac{x^5}{5!} - \frac{x^7}{7!}$$

For small values of x, these give good approximations to $\sin x$. If you continue this process, these numbers, for any real number x, get as close as you like to $\sin x$. We write this as follows.

$$\sin x = x - \frac{x^3}{3!} + \frac{x^5}{5!} - \frac{x^7}{7!} + \cdots$$

Who would have thought that trigonometric values would be so closely related to polynomials?

There is a similar pattern for the cosine. In this case,

$$\cos x = 1 - \frac{x^2}{2!} + \frac{x^4}{4!} - \frac{x^6}{6!} + \cdots .$$

The expressions on the right are called *power series* expansions of the functions.

The tangent function has a series expansion too. (Recall that the tangent is the sine divided by the cosine.) First, let us introduce something called an arctangent. For a number x, the *arctangent* of x, abbreviated $\arctan x$, is defined by

$$\arctan x = y \ \text{ if } \ x = \tan y.$$

In other words, the arctangent of x is the angle whose tangent is x. As an example, let us compute $y = \arctan 1$, so that $\tan y = 1$. Recall the formula for the tangent in the triangle: opposite over adjacent. (See Figure 2.13.) If $\tan y = 1$, then opposite equals adjacent, so by symmetry the angle is $45°$, which in radians is $45 \cdot (\pi/180) = \pi/4$. Hence $\arctan 1 = \pi/4$.

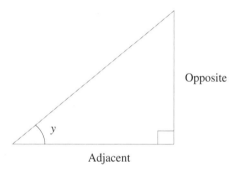

Opposite

y

Adjacent

Figure 2.13 $\tan y = \text{opp}/\text{adj}$

Madhava obtained a power series expansion for the arctangent. In modern terms, we can write it thus:

$$\arctan x = x - \frac{x^3}{3} + \frac{x^5}{5} - \frac{x^7}{7} + \cdots .$$

Let us apply this to $\pi/4 = \arctan 1$:

$$\frac{\pi}{4} = 1 - \frac{1}{3} + \frac{1}{5} - \frac{1}{7} + \cdots .$$

This amazing formula for π was rediscovered by Gottfried Leibniz in the 17th century, when it became deservedly famous. But Madhava had it first.

Historians of mathematics have only studied Kerala fairly recently, so our knowledge is not complete. In particular, little is known of the possible transfer of knowledge to and from Kerala. Kerala played a major part in the spice trade between the

East Indies and the West. In his historic voyage of exploration, the first place in India where the Portuguese explorer Vasco Da Gama landed was at Kappad in Kerala, on May 27, 1498. It is possible that the mathematical discoveries of Kerala made their way to Europe. Certainly, trigonometric power series played an important part in the development of calculus in the 17th century. It is intriguing to think that India may have contributed to this great advance in mathematics.

EXERCISES

2.16   Find $1^2 + 2^2 + \cdots + 100^2$ and $1^3 + 2^3 + \cdots + 100^3$.

2.17   How many perfumes could be mixed using a choice of three ingredients from a total of twelve?

2.18   A poker hand is a set of five cards chosen from a deck of 52 cards. How many poker hands are there?

2.19   Two solutions to the Diophantine equation $2x + 3y = 17$ are given in the text. Find two more solutions. (Hint: either use trial and error or look for a pattern.)

2.20   Show that $2x + 4y = 101$ has no integer solutions. (Hint: is $2x + 4y$ odd or even?)

2.21   Two solutions to the Pell equation $y^2 = 2x^2 + 1$ are given in the text. Use Brahmagupta's method to find another solution.

2.22   Find four solutions of the Pell equation $y^2 = 6x^2 + 1$.

2.23   Find the area of the cyclic quadrilateral shown below.

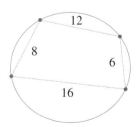

2.24   Find the area of the triangle whose sides have lengths 1, 2, and 3.

2.25   Use Gelosia multiplication to compute.
   a) 726×512
   b) 2247×52

2.26   Use Varāhamihira's trigonometric identities and the table to compute.

α	$\cos \alpha$
$30°$	.866
$45°$	.707

a) $\sin 60°$
b) $\sin 45°$
c) $\sin 15°$
d) $\sin 75°$ (Hint: $75 = 45 + 30$.)

2.27   Use the definitions of the sine of an angle as opposite over hypotenuse and the cosine as adjacent over hypotenuse to prove Varāhamihira's identity $\sin^2 \alpha + \cos^2 \alpha = 1$. (Hint: use the Pythagorean Theorem.)

2.28   Find the radian measure of each angle.
a) $60°$
b) $45°$
c) $15°$
d) $720°$

2.29   Find the degree measure of each angle.
a) $\pi/6$ radians
b) 2π radians
c) 5 radians
d) 360 radians

2.30   What is $\arctan 0$? (Hint: $\sin 0 = 0$.)

2.31   Use the first three terms of the power series expansion to approximate.
a) $\sin .5$
b) $\cos .5$
c) $\arctan .5$

2.3 Islam

The Big Dipper in the sky has eight stars: Alkaid, Mizar and Alcor (a famous double star), Alioth, Megrez, Phecda, Merak, and Dubhe. These names have something in common: they are Arabic. Many of our star names are Arabic. This is because Europe learned its science from the Muslims. In the centuries around the year 1000 CE, the Muslims were the greatest scientists in the world. They had translated and studied the ancient Greek texts, and adopted the superior Indian decimal system. They had learned astronomy from Greek, Mesopotamian, and Indian sources. They eventually taught all of these to Europe, but not until they had added considerably to this inherited tradition.

History to 1600: An Overview

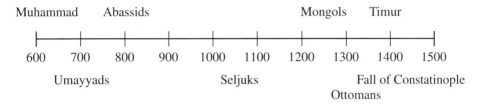

Muslim history.

Muhammad (570–632) lived in the Arabian desert in what is now Saudi Arabia, where he founded a new religion called Islam, Arabic for "submission to God." Within twenty years of the prophet Muhammad's death, the armies of Islam had conquered Syria, Mesopotamia, Persia, Egypt, and Libya.

The political successors of Muhammad were called caliphs (from the Arabic *khalifah*, "successor"). In the middle of the 7th century, there was a struggle over who would be the next caliph. Eventually, the Umayyad clan emerged as the victors, establishing the Umayyad caliphate, or dynasty, which lasted 661–750. This succession battle produced the major schism in Islam, with the minority Shi'a Muslims supporting a different caliph from the majority Sunni Muslims.

The Umayyad caliphate moved their capital from Saudi Arabia to Damascus, in Syria, where they used the old Persian and Roman bureaucracies. In the late 7th and early 8th centuries, after the internal politics were settled, Islam expanded once more, through northern Africa into Spain in the west, and as far as the Indus valley in the east.

The Umayyad dynasty was overthrown by the Abbasid family, except in the west, where the Umayyads retained their power for several more centuries, centered in Córdoba, Spain. The Abbasids, descendants of an uncle of Muhammad, established the Abbasid dynasty (750–1258). In 762 they moved their capital to the newly-founded city of Baghdad in modern-day Iraq, not far from where previous capital cities had served long-gone Mesopotamian states.

The following centuries were the golden age of Muslim culture. For a while, Baghdad was one of the largest, richest, and most civilized places in the world. Its scholars drew upon the preceding cultures of Mesopotamia, Persia, Greece, and India. Paper making was learned from the Chinese. Arabic became the lingua franca of science. Mathematics flourished.

The Abbasids gradually lost power. Other dynasties arose, some of them centered further east, particularly in Persia, modern-day Iran. In the 11th century, the Seljuk Turks established an empire that ruled lands from Egypt to India. In the 13th century, Mongol invaders conquered much of this area. Throughout this period Islamic mathematics continued, although at a gradually decreasing level.

The last great Islamic empire was that of the Ottomans (1326–1920), who eventually ruled much of Northern Africa, the Middle East, and Southeastern Europe, finally expiring in World War I. In 1453, after a long struggle, the Ottomans succeeded in conquering Constantinople, making it their capital. The empire's most famous figure was Suleiman the Magnificent, who was sultan in 1520–1566.

During the period 1000–1500, Islam continued expanding south and east. Muslims converted much of Sub-Saharan Africa. Their merchants and missionaries also spread the faith to southeast Asia, as far as the Philippines. The most populous Muslim country today is Indonesia.

In the following pages, we trace the main accomplishments of the Islamic mathematics of this time. This history is hardly complete; there are still many manuscripts from a number of cultural centers that historians have not studied, and some that haven't even been read.

Transmission

Scholars move more than most people. Major centers of learning usually are populated with many immigrants. Very few of the top philosophers of 4th century Athens were actually born in Athens. A sizable percentage of American winners of Nobel science prizes emigrated from other countries.

When Alexandria became inhospitable to learning, some of its scholars, particularly Nestorian Christians, moved to Syria. Late in the 5th century, they had to move again, this time to Persia, where they were joined after 529 by others from newly-intolerant Greece. The major center was at Jundishapur in modern-day southwestern Iran, which flourished in the 6th and 7th centuries, known especially for its medicine.

Around 800 the intellectual center of western Asia moved to Baghdad. An important person in this development was the caliph Harūn al-Rashīd,[1] who ruled 786–809. His caliphate, and that of his son al-Ma'mūm (r. 813–33), was a golden age for learning. The court of al-Rashīd was immortalized in *The Thousand and One Nights*. The Almanon crater on the Moon is named for al-Ma'mūm, in recognition of his support for astronomy. These rulers were eager to learn from other cultures and welcomed scholars who were not necessarily Muslims. Tolerance was not carried to extremes, however; later in his life, al-Ma'mūm established an inquisition to enforce, through torture and execution, his enlightened philosophy.

Starting in the late 8th century, many scientific classics were translated into Arabic, the first being Indian astronomical texts. Harūn al-Rashīd established a library in Baghdad, collecting manuscripts from throughout the region. Delegations were sent by al-Ma'mūm to Syria, Constantinople, and Syracuse in search of Greek writings. Euclid was translated in the year 800, followed by works of Archimedes, Apollonius, Ptolemy, Diophantus, and others, including major medical and philosophical texts. By the end of the 9th century, most of the major works had been translated. In fact, many of the classic Greek texts are available to us today only in Arabic translations.

---

[1] Some Arabic names contain "*al-*." The word *al* means "the." In names, it often refers to the person's origin, e.g., Muhammad al-Khwārizmī is from Khwārizm, or some characteristic, e.g., Harūn al-Rashīd is literally "Harun the Just." Other common elements of Arabic names include *ibn*, son of, and *abu*, father of.

In 830 al-Ma'mūm established a research institution called Bayt al-Hikmah (The House of Wisdom). With its associated library and astronomical observatory, it became a major center of Islamic scholarship. Many of the translations mentioned above were done here, but also original research in the sciences, including mathematics.

In the 10th century, Baghdad began declining, and much of the subsequent mathematics moved farther east. Major centers were established at various times in Nishapur and Isfahan in Iran, in Samarkand in Uzbekistan, and other places. Farther west, Cairo hosted important scholars, and work was done in Damascus, Syria, in Córdoba, Seville, and Toledo in Islamic Spain, and in the Maghreb of northern Africa.

Numeration

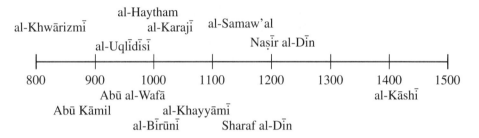

Muslim mathematicians.

The Hindu system of numbers developed gradually into the Hindu-Arabic system—the system we use—over a number of centuries. The Hindu system, a decimal place-value system with a zero but no decimal fractions, was known in western Asia as early as 662, when it was described in a Syrian text by the Nestorian bishop Severus Sebokht. (By decimal fraction we mean, for example, .2 instead of $1/5$.) The Arabs at this time had a number of systems. Merchants used a form of finger reckoning. There was also a system that represented numbers using letters. Astronomers were familiar with the Babylonian sexagesimal place-value system.

The earliest known Arabic text on the Hindu system is the *Kitāb al-jam'wal-tafrīq bi ḥisāb al-Hind* (*Book on Addition and Subtraction after the Method of the Indians*), by Muhammad al-Khwārizmī. The original Arabic text has been lost, but is available in Latin versions made in the 12th century. In this book, al-Khwārizmī explained how to write numbers using the nine digits plus zero, and also how to add, subtract, multiply, divide, and extract square roots. This is an elementary, practical book, emphasizing the value of the new system in writing large numbers and in doing calculations. The calculations are for the dust board, a flat surface on which sand was spread.

Muḥammad ibn Mūsā al-Khwārizmī (c. 780–850)

It is not known where al-Khwārizmī was born. Given his name, he or his ances-
tors may have came from Khwārizm, a region in what is now Uzbekistan and Turkmenistan. It is known that he worked in Baghdad and was one of the first scholars at
the House of Wisdom.

In addition to his work on mathematics, al-Khwārizmī published on geography,
and reportedly was called upon to cast a horoscope for the caliph al-Wathīq in 847.
He prudently predicted another 50 years for the caliph, who died 10 days later.

∼

Hindu arithmetic was gradually adopted in Arabic scientific circles. The earliest
arithmetic text which still survives in Arabic is the *Kitāb al-fuṣūl fi-l-ḥisāb al-hindī*
(*The Book of Chapters on Hindu Arithmetic*), written about 952 by Abu l-Ḥasan al-
Uqlīdīsī. This has the first recorded use of decimal fractions. Al-Uqlīdīsī showed
how to divide by 10 (move the number one place to the right) and by 2 (move to the
right and multiply by 5). The Hindus did not use decimal fractions, so this was an
invention of the Arabs, who had the Babylonian sexagesimal fractions as a model.
Al-Uqlīdīsī also made a pitch to replace the dust board by paper.

Abu l-Ḥasan al-Uqlīdīsī (c. 920–980)

Little is known of al-Uqlīdīsī. We do know that he wrote at least two works, and
composed *The Book of Chapters on Hindu Arithmetic* in Damascus. He claimed to
have traveled widely, and may have worked in Baghdad.

The al-Uqlīdīsī part of his name came from "Euclid." He apparently earned his
living by making and selling copies of Euclid's *Elements*.

∼

There is some question about how thoroughly al-Uqlīdīsī understood decimal
fractions, since he gave only limited examples. The first text that displays an unambiguously thorough understanding is the *al-Qiwāmī fī al-Ḥisāb al-Ḥindī* (*Treatise on Arithmetic*) of 1172, by Ibn Yaḥyā al-Maghribī al-Samaw'al (biography in
the next section). In particular, he understood that, in the case where the decimal fraction could not be exact, e.g., dividing 210 by 13, we could get decimal
fractions that approximate the answer as closely as desired. We now would write
$210/13 = 16.153846\ldots$.

Finally, we mention the *Muftah al-Ḥisāb* (*Key to Arithmetic*), written in the 15th
century by the Persian Ghiyāth al-Dīn Jamshīd al-Kāshī (biography in the section on
trigonometry). This work introduced a better notation, a vertical line to separate the
integer part and the decimal part of a number, e.g., $210/13 = 16|153846\ldots$.

Our debt to the Islamic world for this number system is enshrined in our language.
The word for zero in Arabic is *sifr*, meaning "empty," which gave us our words

"zero" and "cipher," an old-fashioned word for zero. The name al-Khwārizmī is also the origin for our word "algorithm," a well-defined method of solving a problem. An example of an algorithm is long division.

Algebra

Around 825 al-Khwārizmī wrote a book on algebra, *Al-kitāb al-muḫtaṣar fī hisāb al-jabr wa-l-muqābala* (*The Condensed Book on the Calculation by Restoring and Comparing*). The word *al-jabr* refers to the algebraic process of moving a negative quantity on one side of an equation to the other, where it becomes positive, e.g., replacing the equation $2x + 7 = 8 - x$ by $3x + 7 = 8$. The word *al-muqābala* refers to the process of subtracting a number from both sides of the equation, e.g., replacing $3x + 7 = 8$ by $3x = 1$.

In this work al-Khwārizmī explained how to solve linear and quadratic equations. He classified these equations into six types. He did not allow negative numbers, so for example he considered the equations $2x^2 + 3x = 4$ and $5x^2 + 6 = 7x$ to be of different types. Nowadays, we would consider them both instances of $ax^2 + bx + c = 0$, where any of a, b, or c can be negative. Although al-Khwārizmī did not allow negative solutions, he did, unlike many previous mathematicians, recognize that some quadratic equations had two solutions.

All problems were stated in words; there was nothing like our modern notation. Here is one problem he solved. "What must be the square which, when increased by ten of its own roots, amounts to 39?" We would write this as: Solve $x^2 + 10x = 39$ for x. Here is al-Khwārizmī's solution.

> The solution is this: You halve the number of roots, which in this case gives five. This you multiply by itself, the product is twenty-five. Add this to thirty-nine, the sum is sixty-four. Now take the root of this, which is eight, and subtract from it half the number of roots, which is five, the remainder is three. This is the root of the square you sought for, and the square itself is nine.

Translated, this is $\sqrt{\left(\frac{10}{2}\right)^2 + 39} - \frac{10}{2}$, which works out to $x = 3$.

Al-Khwārizmī viewed his book as more practical than theoretical, but did include some proofs. Under the influence of the Greeks, at this time all proofs were geometric. To explain his solution above, he used a diagram like Figure 2.14. Note that the area of the larger square is $x^2 + 5x + 5x + 25 = (x^2 + 10x) + 25 = 39 + 25 = 64$. Since the area is also $(x + 5)^2$, we have $(x + 5)^2 = 64$, or $x + 5 = 8$, hence $x = 3$.

This text was very influential in Europe after it was translated in the 12th century. The word *al-jabr* was not translated, and eventually morphed into the word *algebra*.

One of the important developments in Islamic mathematics was the gradual "arithmetization" of algebra, treating algebraic problems as being about numbers, not tied to geometry. Recall the Greek notion of magnitude. In the equation $x^2 + 10x = 39$, for example, x^2 would be considered a different type of magnitude than x, since x^2 represents an area and x a line segment. But by the time of the Egyptian mathematician Abū Kāmil, that distinction was being eroded. Both expressions were often treated the same; they were just numbers.

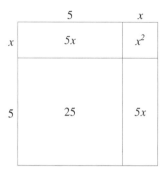

Figure 2.14   Solving $x^2 + 10x = 39$.

Abū Kāmil ibn Aslam (c. 850–930)

We know little of the life of Abū Kāmil. He lived in Egypt, and was influenced by the works of Euclid, Heron, and al-Khwārizmī, although he was too young to have met the latter.

Abū Kāmil wrote a number of mathematical books, including several which did not survive.

In *Kitāb fī al-jabr wa'l-muqābala* (variously translated as *Algebra* or *Book on Algebra*), Abū Kāmil treated more complicated problems than had al-Khwārizmī. In particular, he was at ease with expressions involving irrational roots, *surds*. In one of his problems, a solution was $10 + \sqrt{2} - \sqrt{42 + \sqrt{800}}$ (which he expressed in words).

Abū Kāmil also used the technique of substitution to reduce a problem to an easier one. In one example, he treats the equation

$$\left(\frac{x}{10-x}\right)^2 - \left(\frac{10-x}{x}\right)^2 = 2.$$

First he substitutes for $\frac{10-x}{x}$. Let us call it y; then $1/y^2 - y^2 = 2$. Multiplying by y^2, we get $1 - y^4 = 2y^2$. Now we can substitute $z = y^2$ to get $1 - z^2 = 2z$. Since this is a familiar quadratic, we can solve for z, then y, then x. Details are left to the exercises.

The next major developer of Islamic algebra was Abū Bakr al-Karajī. In *al-Fakhrī* (*The Marvelous*), al-Karajī introduced some of the algebra of polynomials. In particular, he wrote in a general way about higher powers, what we would write as x, x^2, x^3, ..., and their reciprocals, $\frac{1}{x}$, $\frac{1}{x^2}$, $\frac{1}{x^3}$, ... He developed rules for adding, subtracting, and multiplying polynomials, although the only division he recognized was by monomials (polynomials with only one term, e.g., $2x^3$).

Abū Bakr al-Karajī (c. 980–1030)

It is uncertain where al-Karajī was born. The most popular view among historians is that he was Persian, but some place his birth near Baghdad. What is certain is that he worked in Baghdad in the early 11th century. He wrote a number of works on mathematics and engineering,

---

Al-Karajī also studied irrational numbers. He produced a number of results such as the following.

$$\sqrt[3]{A} + \sqrt[3]{B} = \sqrt[3]{3\sqrt[3]{A^2B} + 3\sqrt[3]{AB^2} + A + B}$$

Ibn Yaḥyā al-Maghribī al-Samaw'al extended al-Karajī's algebra of polynomials. He took the important step of allowing negative coefficients. He also provided charts for doing operations on polynomials, apparently based on the dust board. These charts, reminiscent of Chinese counting boards, made complicated computations easier. In them, he represented each term of an extended polynomial, which could include negative powers of x, by a column. In a somewhat simplified example, the chart below would represent the expression $2x^3 - 4x^2 + 1 - 8x^{-1} + x^{-2}$.

3	2	1	0	−1	−2
2	−4	0	1	−8	1

Al-Samaw'al explained how to add, subtract, multiply, *and* divide such expressions. In particular, he discovered long division. You may recall that al-Samaw'al was also the first to completely master decimal fractions. This is intimately related to his work in algebra, for a decimal number may be thought of as a type of extended polynomial with the unknown being 10. For example, if we let $x = 10$ in the expression $7x^2 + 8x + 2 + 3x^{-1} + 8x^{-2}$, we get the number 782.38.

Ibn Yaḥyā al-Maghribī al-Samaw'al (1125–1174)

Al-Samaw'al was born in Baghdad, the son of a Jewish scholar from Morocco. At the time he grew up, Baghdad was no longer a cultural hotbed, and he was required to learn mathematics on his own.

Al-Samaw'al reportedly wrote 85 works, few of which survive. His major mathematical work, *Al-Bāhir fi'l-ḥisāb* (*The Shining Book of Calculation*) was written when he was nineteen. Later, he traveled extensively and was a famous physician. The only one of his medical texts to survive is *The Companion's Promenade in the Garden of Love*, basically a sex manual.

When al-Samaw'al was about forty years old, he converted to Islam, writing a famous book that has since been used as a Muslim polemic against Judaism.

With the works of al-Karajī and al-Samaw'al and their predecessors, algebra was freed from geometry. For al-Karajī, the goal of algebra is the "determination of unknowns starting from known premises." Al-Samaw'al clearly understood the intimate connections between algebraic and arithmetic manipulations.

Another theme in Islamic algebra was the study of cubic equations, ones of the form $ax^3 + bx^2 + cx + d = 0$, where a, b, c, and d are given numbers. The problem is to find all values of x for which the equation holds. A leading figure in this study was 'Umar ibn Ibrāhīm al-Khayyāmī.

'Umar ibn Ibrāhīm al-Khayyāmī (Omar Kayyam) (1048–1131)

The Persian scholar 'Umar ibn Ibrāhīm al-Khayyāmī was born in Nishapur, a major cultural center of that time, located in modern-day northeastern Iran. His family's background may be indicated by his name; in Arabic, *al-kayyāmī* means "the tent-maker."

Al-Khayyāmī spent his working life in the empire of the Seljuk Turks, relying on their sometimes unreliable patronage. From Nishapur, he moved to Samarkand in Uzbekistan, and after that to Isfahan, in central Iran, where he headed the astronomical observatory for eighteen years. Political difficulties ended his support in Isfahan, and he eventually made his way to Merv in current Turkmenistan. At that time, Merv was one of the largest cities in the world. Al-Khayyāmī eventually returned to Nishapur, where he died in 1131.

In addition to mathematics and astronomy, Al-Khayyāmī wrote on music and philosophy. In the West, he is best known as Omar Kayyam, the author of the famous collection of poems called the *Rubaiyat*.

Building on the work of Greek and Islamic predecessors, al-Khayyāmī classified all cubic equations and solved them using geometric methods. Al-Khayyāmī did not allow negative coefficients, so, for example, he treated $x^3 + x = 0$ as a different type of equation from $x^3 = x$. He also considered cubics different if they had different nonzero terms: $x^3 - x + 1 = 0$ was in a different class than $x^3 + 1 = 0$. (All of this was stated in words, since he did not have our notation.) In all, he identified fourteen different classes of cubics.

Once he had his classification, al-Khayyāmī proceeded to solve each class in a geometric fashion, as the intersection of two conic sections. For example, one of his classes was $x^3 + cx = d$. He constructed a parabola, $x^2 = \sqrt{c}y$, and a circle,

$x(d/c - x) = y^2$, then showed that the intersection of the parabola and circle solved the cubic, in modern terms, that the x coordinate of the intersection satisfied the equation.

Al-Khayyāmī's approach was not unlike the way that the Greeks had solved some specific cubics many years earlier. His method was, however, more general. In a sense, he solved all cubics. But his solution was quite unlike the solution of quadratic equations by the quadratic formula. It was geometric; he did not obtain an equation for x in terms of the coefficients, to arrive at a specific number. He did make it clear that he would like such a solution, but was unable to obtain one.

Al-Khayyāmī was aware that cubics could have anywhere from zero to three solutions, and expressed this fact by the number of intersections of his conic sections. A different approach was taken by Sharaf al-Dīn al-Ṭūsī, who was interested in the relationship between the number of solutions and the coefficients.

Sharaf al-Dīn al-Ṭūsī (c. 1135–1213)

Sharaf al-Dīn al-Ṭūsī was born in the region of Tus in northeastern Iran. He studied astronomy and astrology as well as mathematics. He was famous as a teacher, teaching in Damascus and Aleppo in Syria, Mosul and Baghdad in Iraq.

Recall the quadratic formula

$$x = \frac{-b \pm \sqrt{b^2 - 4ac}}{2a}$$

used to solve the quadratic equation, $ax^2 + bx + c = 0$. How many solutions are there? The key is to look at the *discriminant*, $b^2 - 4ac$. If this is negative, we have no solutions, since we would have to take the square root of a negative number. If $b^2 - 4ac = 0$, then we have one solution, $-b/2a$. If the discriminant is positive, we have two solutions. Al-Dīn studied cubic equations in a similar way. He made his own classification of cubics, and, in some cases, identified what we now call the discriminant of a cubic as crucial to the number of solutions. He did not, however, have any cubic equivalent of the quadratic formula.

Sharaf al-Dīn also gave a method for approximating the roots of a cubic equation. It was similar to that of Jia Xian, discussed in Section 2.1.

Geometry

Islamic mathematicians were more interested in practical geometry than the Greeks were. They produced works on geometry intended for surveyors and artisans. Some of that work was sophisticated, for example, presenting constructions that artisans could use to produce interesting geometric figures.

There was also work on theoretical geometry, based on Greek geometry. One area of interest was Euclid's parallel postulate. Recall (from Section 1.4) that this postulate was the wordy one which Euclid used to prove results about parallel lines.

In his *Maqāla fī sharḥ muṣādarāt kitāb Uqlīdis* (*Commentary on the Premises of Euclid's Elements*), al-Haytham attempted to remove this postulate by proving it from the others. This effort was not completely successful; he introduced a new definition of parallel lines that, unknown to him, implicitly contained the postulate. Along the way, however, he proved some important geometric theorems.

Abū ʿAlī al-Ḥasan ibn al-Ḥasan ibn al-Haytham (Alhazen) (c. 965–1040)

Abū ʿAlī al-Ḥasan ibn al-Ḥasan ibn al-Haytham, known in Europe as Alhazen, was one of the preeminent scientists of the Muslim world. He was born in Basra, in southern Iraq, but worked much of his life in Cairo, Egypt, at the *Dār al-Ilm* (House of Knowledge), Cairo's rival to Baghdad's House of Wisdom.

Al-Haytham's main claim to fame is his *Optics*, in seven books. This was later translated into Latin, and was very influential for several centuries. Mathematically, he is best known for "Alhazen's problem": given a reflecting surface and two points not on the surface, find the point(s) on the surface where light emanating from the first point will, after reflection, pass through the second point. He studied this problem for several types of surfaces. It is also called Alhazen's billiards problem: given a billiard table of a given shape (e.g., a circle), and positions of a cue ball and another ball, where should the cue ball be aimed so as to hit the other ball after one bounce?

Omar Kayyam later studied the parallel postulate, in the *Sharḥ mā ashkala min muṣādarāt kitāb Uqlīdis* (*Commentary on the Problematic Postulates of the Book of Euclid*). He managed to replace it with a new postulate and show that the resulting system was equivalent to Euclid's.

In 1250 Naṣīr al-Dīn wrote *Al-risāla al-shīyaʿan al-shakk fi-l-khuṭūt al-mutawā-ziya* (*Discussion Which Removes Doubt about Parallel Lines*), commenting on the preceding work and attempting to prove the parallel postulate. Again, this was not successful, but it did advance understanding of the problem.

Naṣīr al-Dīn al-Ṭūsī (1201–1274)

Naṣīr al-Dīn al-Ṭūsī, not to be confused with Sharaf al-Dīn al-Ṭūsī, was also from Tus in Iran. He studied there and at Nishapur. This was a time of political upheaval, culminating in the Mongol conquest of the area, so Naṣīr al-Dīn traveled extensively. He eventually settled in Maragha in northwestern Iran, were he founded an important astronomical observatory.

Naṣīr al-Dīn was prolific, reportedly authoring more than 150 texts on many scientific and religious subjects. He was one of the most influential astronomers in the time between Ptolemy and Copernicus.

Another area of Euclid that Islamic mathematicians studied was incommensurables. In Book X of the *Elements*, Euclid dealt with incommensurable magnitudes, those which have no common measure. In modern terms, their ratio is an irrational number. Magnitudes were not the same as numbers, however. One could not compare areas and lengths, for example.

In practice, Islamic mathematicians dealt with numbers, not magnitudes, but they were aware of the logical difficulties this caused, particularly when dealing with irrationals. Some attempts were made to justify the Islamic approach in ways of which Euclid would have approved. One such was in *Risāla fi'-maqādir al-mushtaraka wa'l-mutabāyana* (*Treatise on Commensurable and Incommensurable Magnitudes*), published around 1000 by Abū 'Abdallāh al-Ḥasan ibn al-Baghdādī.

In this work, ibn al-Baghdādī developed a logical foundation in which he could treat all magnitudes as lengths of lines. His system worked for rational numbers and square roots, as well as square roots of square roots, and so on. Along the way, he proved an intriguing result about the density of irrationals, namely, that between any pair of rational numbers there exist an infinite number of irrational numbers.

Mathematical Induction

As mentioned in Section 2.2, Āryabhaṭa gave the following formula for the sum of cubes:

$$1^3 + 2^3 + \cdots + n^3 = (1 + 2 + \cdots + n)^2.$$

He did not, however, write down a proof. The first to do this was al-Karajī, at least for $n = 10$. It is clear that his argument would work for any n. Here it is.

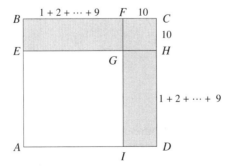

Figure 2.15   Proving that $1^3 + 2^3 + \cdots + 10^3 = (1 + 2 + \cdots + 10)^2$.

Consider the square $ABCD$ in Figure 2.15. We will compute its area S in two ways. First, since one side has length $1 + 2 + \cdots + 10$, we have

$$S = (1 + 2 + \cdots + 10)^2.$$

Now consider the shaded area. It is composed of three pieces. The area of the rectangle $BFGE$ is $10(1 + 2 + \cdots + 9)$, as is the area of $GHDI$. Finally, the little square $FCHG$ has area 10^2. Putting this together, we have the shaded area equaling

$$2 \cdot 10(1 + 2 + \cdots + 9) + 10^2.$$

Since S is just the shaded area plus the unshaded area, which is a square of side $1 + 2 + \cdots + 9$, we can write

$$S = (1 + 2 + \cdots + 9)^2 + 2 \cdot 10(1 + 2 + \cdots + 9) + 10^2.$$

Next, we use the formula $1 + 2 + 3 + \cdots + n = n(n + 1)/2$ to write

$$1 + 2 + \cdots + 9 = \frac{9 \cdot 10}{2}.$$

Therefore

$$
\begin{aligned}
S &= (1 + 2 + \cdots + 9)^2 + 2 \cdot 10(1 + 2 + \cdots + 9) + 10^2 \\
&= (1 + 2 + \cdots + 9)^2 + 2 \cdot 10\frac{9 \cdot 10}{2} + 10^2 \\
&= (1 + 2 + \cdots + 9)^2 + 9 \cdot 10^2 + 10^2 \\
&= (1 + 2 + \cdots + 9)^2 + (9 + 1)10^2 \\
&= (1 + 2 + \cdots + 9)^2 + 10^3.
\end{aligned}
$$

Combining this with our first equation for S yields

$$(1 + 2 + \cdots + 10)^2 = (1 + 2 + \cdots + 9)^2 + 10^3.$$

Note that we have now reduced the problem to studying $(1 + 2 + \cdots + 9)^2$. But the same argument shows that

$$(1 + 2 + \cdots + 9)^2 = (1 + 2 + \cdots + 8)^2 + 9^3,$$

so

$$(1 + 2 + \cdots + 10)^2 = (1 + 2 + \cdots + 8)^2 + 9^3 + 10^3.$$

Repeating, we get

$$(1 + 2 + \cdots + 10)^2 = (1 + 2 + \cdots + 7)^2 + 8^3 + 9^3 + 10^3.$$

Continuing in this way, we finally arrive at

$$
\begin{aligned}
(1 + 2 + \cdots + 10)^2 &= (1)^2 + 2^3 + 3^3 + \cdots + 10^3 \\
&= 1 + 2^3 + 3^3 + \cdots + 10^3 \\
&= 1^3 + 2^3 + 3^3 + \cdots + 10^3,
\end{aligned}
$$

which is what we sought to prove.

A notable feature of this argument is how al-Karajī reduced the case with $n = 10$ to the one with $n = 9$, then $n = 8$, etc., until the case $n = 1$, which is trivial. This is the essence of the type of argument we now call mathematical induction, one of the most powerful tools of modern mathematics.

A similar type of argument was used by al-Haytham to prove a formula for sums of fourth powers, which he needed in his study of Alhazen's problem. Let us recall a couple of formulas we have seen before.

$$1 + 2 + 3 + \cdots + n = \frac{n(n+1)}{2}$$

$$1^2 + 2^2 + 3^2 + \cdots + n^2 = \frac{n(n+1)(2n+1)}{6}$$

These formulas were known well before al-Haytham's time, in word form, without our notation. Known as well was the formula for the sum of cubes:

$$1^3 + 2^3 + 3^3 + \cdots + n^3 = \frac{n^2(n^2 + 2n + 1)}{4}.$$

Al-Haytham extended this to the sum of fourth powers:

$$1^4 + 2^4 + 3^4 + \cdots + n^4 = \left(\frac{n}{5} + \frac{1}{5}\right)n\left(n + \frac{1}{2}\right)\left[(n+1)n - \frac{1}{3}\right].$$

His argument could easily be extended, with a lot of computation, to find a formula for the sum of fifth powers, then sixth, and so on, but he stopped at fourth powers.

Although al-Karajī, al-Haytham, and others used arguments that were much like mathematical induction, an explicit statement of mathematical induction, and its widespread use, was centuries away.

Trigonometry

Islamic scholars learned of trigonometry from the Greeks and the Indians. As with these earlier cultures, the important motivation came from astronomy. Muslims had extra reasons to study astronomy. They were (and are) required to pray five times a day, facing Mecca. The astronomers needed to determine the direction of Mecca, and the time of day. Some of the early "Islamic" astronomers were also associated with the Sabean religion, which was centered around celestial objects. In addition, as always, there was astrology.

There are six standard trigonometric functions in modern mathematics. We introduced the sine, cosine, and tangent in Section 1.5. The other three are the cotangent, secant, and cosecant.

$$\cot \alpha = \frac{1}{\tan \alpha}, \quad \sec \alpha = \frac{1}{\cos \alpha}, \quad \csc \alpha = \frac{1}{\sin \alpha}$$

The full list of these six functions first appeared in the 9th century. Before then, only the sine and cosine were in use, although the tangent had made a brief appearance in

China in the previous century. (Actually, the Muslims' trigonometric functions were constant multiples of ours, a minor difference.)

An important early text was an astronomy book, the *Zīj al-Majisti*, by Abū al-Wafā. This work contained, in addition to the six trigonometric functions, some fundamental results in spherical trigonometry. Recall that spherical trigonometry is concerned with triangles on a sphere. This was important in astronomy both because the Earth is (nearly) a sphere, and because the sky can be treated as a sphere. In fact, the simpler plane trigonometry, as taught in high school courses now, mostly developed after the more complicated spherical trigonometry, because astronomy was the central application.

Abū al-Wafā al-Būzjāni (940–998)

Abū al-Wafā al-Būzjāni was born in Khorasan, in the northeastern part of modern-day Iran. He worked in Baghdad, under the Buyid dynasty, which for a time was a great supporter of science. Abū al-Wafā helped design and build a major observatory there in 988.

In addition to his theoretical work, Abū al-Wafā wrote a business text called *Book on What Is Necessary from the Science of Arithmetic for Scribes and Businessmen*. Even though he was conversant with Hindu numerals, Abū al-Wafā wrote numbers in this book in words, and computations were done using fingering-reckoning. This was what businessmen of his day used. Also notable is his use of negative numbers, very rare in Islamic mathematics. They were useful in business to represent debts.

A younger contemporary of Abū al-Wafā, Abū l-Rāyḥan Muḥammad ibn Aḥmad al-Bīrūnī, also mastered the various trigonometric functions, and established a number of relations between them. He applied his knowledge to astronomy, among other things, giving a method to calculate the direction to Mecca from any location, and inventing a new method to estimate the radius of the Earth, obtaining a result about 20% too small.

Abū l-Rāyḥan Muḥammad ibn Aḥmad al-Bīrūnī (973–1055)

Abū l-Rāyḥan Muḥammad ibn Aḥmad al-Bīrūnī was born in Khwarizm, near Biruni in modern-day Uzbekistan. He lived in a time of political turmoil, and had to spend considerable time courting royal sponsors or fleeing from them. He was employed as an astrologer, although he privately rejected it, and was an accomplished diplomat. He wrote, "I was compelled to participate in worldly affairs, which excited the envy of fools, but made the wise pity me."

Al-Bīrūnī wrote over 140 works, most importantly on astronomy, mathematics, and geography. His book *India*, in which he addressed all aspects of Indian culture,

is still important. In astronomy, he notably proposed that the Earth travels around
the Sun.

Islamic mathematicians also made tables of trigonometric functions. Abū al-Wafā
invented a new method to approximate sine values, which allowed him to create a
sine table accurate to four sexagesimal places. (Astronomers were still using the
Babylonian sexagesimal system.)

Eventually, trigonometry came to be considered a mathematical subject in its own
right, not solely as part of astronomy. This culminated in the *Kitāb al-Shakl al-qattā*
(*Treatise on the Complete Quadrilateral*), by Naṣīr al-Dīn al-Ṭūsī, a comprehensive
work on plane and spherical trigonometry. It has the first statement and proof of
the law of sines for plane triangles. Referring to Figure 2.16, the law of sines is the
following.

$$\frac{a}{\sin A} = \frac{b}{\sin B} = \frac{c}{\sin C}$$

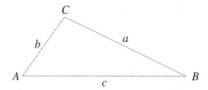

Figure 2.16   Notation for the law of sines.

The word *sine* came to us by a roundabout path. In Sanskrit, *jyā-ardha* means
"chord-half." Āryabhaṭa abbreviated this to *jyā*, or its synonym *jīvā*. When his works
were translated into Arabic, *jīvā* became *jiba*, which does not mean anything, but
sounds about the same. In Arabic, however, vowels are not written, so this word
was written *jb*. Later Arabic writers confused this with the word *jaib*, which means
"breast." When Arabic trigonometry was translated into Latin, the Latin word for
"breast," *sinus*, was used, leading to our *sine*.

The law of sines is a useful tool for solving a common type of problem in geometric applications. This problem occurs when we have measurements of some
sides and angles in a triangle, and wish to determine the unknown sides and angles.
This is called *solving the triangle*. In addition to the law of sines, other tools for
solving triangles include the Pythagorean Theorem, and the fact the sum of angles in
a triangle is $180°$.

As an example, we will solve the triangle in Figure 2.17, i.e., find sides b and c,
and angle B. We first note that, since the sum of angles is $180°$, angle B must be
$25°$. Then we use the law of sines to write

$$\frac{5}{\sin 55°} = \frac{b}{\sin 25°} = \frac{c}{\sin 100°}.$$

Solving these equations, we get

$$b = 5 \left(\frac{\sin 25^\circ}{\sin 55^\circ} \right) \approx 2.58$$

and

$$c = 5 \left(\frac{\sin 100^\circ}{\sin 55^\circ} \right) \approx 6.01.$$

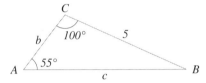

Figure 2.17   Solving a triangle.

Finally, we mention the work of al-Kāshī in the 15th century. He came up with a better approximation method, one that enabled Ulūgh Beg to approximate the sine of one degree, accurate to 16 decimal places. Ulūgh Beg, al-Kāshī's patron and ruler of a sizable central Asian kingdom, was also an astronomer. He used al-Kāshī's method to produce tables for sines and tangent for every minute of arc (one-sixtieth of a degree), to five sexagesimal places. This must have been an enormous effort, producing $10,800 (= 60 \cdot 90 \cdot 2)$ entries. Perhaps he was more interested in this work than in governing. He wasn't very successful at the latter; his son eventually had him killed.

Ghiyāth al-Dīn Jamshīd al-Kāshī (c. 1380–1429)

Ghiyāth al-Dīn Jamshīd al-Kāshī was born in Kashan in central Iran. He was one of the greatest of the Islamic astronomers and mathematicians. His early years were quite difficult; this was the time when Timur (also known as Tamerlane or Tamburlaine) was conquering much of western Asia.

Timur's grandson, Ulūgh Beg, built a major center of learning in his capital city of Samarkand, and invited al-Kāshī to join him there. Al-Kāshī did his most important work in Samarkand.

In the years 800–1400, Islamic mathematicians were the best in the world. During much of this time, however, as at many other times and in many other places, support for higher mathematics was uncertain. Scholars depended on the patronage of the rulers, and many rulers did not consider mathematics important, beyond the obvious practical applications. By the 11th century, there was diminished support for theoretical mathematics. Some in Islam considered it a "foreign" science. It always was thought of as inferior to religious studies. Nonetheless, advances continued, if at a slower pace. It was not until the 16th century that European mathematicians clearly outshone their Islamic counterparts.

EXERCISES

2.32   Solve.

  a) $2x + 7 = 8 - x$
  b) $2x^2 + 2 = 5x$
  c) $x + x^3 = 3x$

2.33   Al-Khwārizmī solved $x^2 + 10x = 39$ to get $x = 3$.
  a) There is another solution. Find it.
  b) Why do you think al-Khwārizmī did not give this solution?

2.34   Complete Abū Kāmil's solution of

$$\left(\frac{x}{10 - x}\right)^2 - \left(\frac{10 - x}{x}\right)^2 = 2.$$

2.35   Evaluate $2x^2 + 3x + 4 + 5x^{-1} + 6^{-2}$ at $x = 10$.

2.36   We saw how Al-Khwārizmī solved $x^2 + 10x = 39$. Use a diagram like Fig 2.14 to find a positive solution to $x^2 + 6x = 91$. Include an argument like Al-Khwārizmī's to show how your diagram leads to your solution.

2.37   Classify cubics like al-Kayyāmī did. Do you get fourteen classes?

2.38   Find out, by looking at the discriminant, how many solutions each of the following equations has.

  a) $x^2 - 3x + 7 = 0$
  b) $2x^2 - 5x + 2 = 0$
  c) $3x^2 - 11 = 0$
  d) $4x^2 - 17x + 5 = 0$

2.39   Consider the cubic equation $x^3 + 4x = 12$. In the text, we mention that al-Kayyāmī solved this cubic as the intersection of a parabola and a circle.
  a) What are the equations of the parabola and the circle in this case?
  b) Show that the intersection does actually solve the cubic. (Hint: use the equation of the parabola to solve for y in terms of x, then plug this value for y into the equation of the circle, and simplify.)

2.40   Use the formula $1^3 + 2^3 + \cdots + n^3 = (1 + 2 + \cdots + n)^2$ to calculate $1^3 + 2^3 + \cdots + 10^3$. (Hint: recall that $1 + 2 + \cdots + n = n(n + 1)/2$.)

2.41   Use al-Haytham's formula to calculate $1^4 + 2^4 + \cdots + 10^4$.

2.42 Solve the triangle.

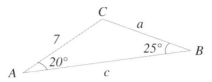

2.43 Solve the triangle.

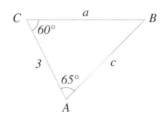

2.4 European Mathematics Awakens

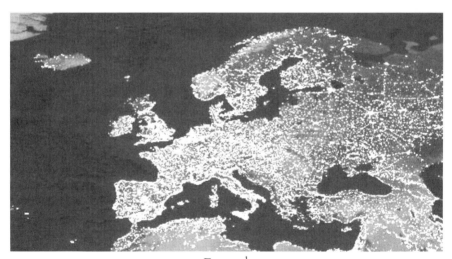

Europe.[1]

History to 1600: An Overview

In the year 1000, Western Europe was a primarily agricultural society. A common organizational pattern in Southern Europe, gradually spreading north over the next

---

[1]From "Earth at Night." C. Mayhew and R. Simmon (NASA/GSFC), NOAA/NGDC, DMSP Digital Archive.

couple of centuries, was the manor. The manor was a large estate, ruled over by a lord, and worked by serfs, also called villeins. Serfs were free men, but not free to move; they were bound to the land. Although they were not slaves, their status was not very high; the word *villein* gave us our word "villain." This was not a time with strong kings or great empires. There were no large metropolitan cities. Trade was limited. The ruling class was a military elite, the knights.

The great unifying force in Europe was the Catholic Church. It had gradually expanded its range over the preceding centuries, completing the conversion of Europe (or at least the princes of Europe) in the 11th century. The popes were major rulers in their own right, and the Church exercised much influence throughout Europe.

In 1095 Pope Urban II called for a holy war to take the Holy Sepulcher, Jesus' burial place, in Jerusalem. This led to the First Crusade, which conquered Jerusalem in 1099. The Arabs under Saladin retook the city in 1187. More crusades followed, the last in 1271–72. In the end, the crusades were unsuccessful, as the Muslims regained control of the whole area.

Although the crusades were a failure militarily, they increased contacts between Western Europe and Muslims. This resulted in an expansion of European cultural awareness, and more trade. Also in the 13th century, under Mongol domination, trade increased along the Silk Road, the set of routes across central Asia through which Chinese silk and other goods were transported to the west. The Silk Road was used to head east, as well; Marco Polo and other Europeans traveled to China to learn about that great society. In addition to the Silk Road, there were important sea routes from southern Asia and the Spice Islands to the Persian Gulf and Red Sea along which spices, especially, were traded. These trade goods then traveled overland, were loaded onto ships in the Mediterranean, thence to Italian ports such as Venice and Genoa.

Within Europe itself, the population was gradually increasing, as was the internal trade. Craftsmen in a particular trade—for example, carpenters—formed guilds, associations to promote their interests. In the commune movement, spreading from north-central Italy in the 11th century, people in a town banded together for mutual protection (often against local nobility or church officials). Fairs held in some towns further stimulated trade.

The new towns and merchants asserted themselves politically. Parliaments arose throughout Europe, bodies in which elected representatives of the towns, as well as nobles and clergymen, advised the king. Modern parliamentary systems, in which the center of power *is* the parliament, were centuries away, but the early parliaments allowed the new merchants to represent their interests to the rulers.

Along with silk and spices, the trade routes were conduits of technology. From China, in particular, came three technologies without which the subsequent European development would be unthinkable: paper, gunpowder, and printing.

The first important European invention from this time was the mechanical clock, invented in the late 13th century. This was probably based on a Chinese invention, the escapement, in which a continuous motion is interrupted, broken into pieces of equal time. (The ticking of a clock arises with the escapement.) The Chinese escapement was invented in the 8th century, when the turning of a water wheel was segmented into discrete steps. In the early European clocks, the escapement acted on a falling weight.

Subsequent European inventions based on Chinese technology proved to be crucial. In particular, we note the cannon from the early 14th century, and Gutenberg's printing press from the 15th century. It is difficult to overestimate the importance of these in history.

Beginning with University of Bologna 1088, universities began springing up over Western Europe. Notable examples included the University of Paris (c. 1150), the University of Oxford (1167), and the University of Cambridge (1209). These schools fostered the study of Greek and Muslim learning. The scholastic movement, whose most famous representative is Thomas Aquinas, grew out of an effort to meld this imported knowledge, especially Aristotle, with Christian philosophy.

Another result of the renewed interest in classical—Roman and Greek—societies was the artistic flowering of the Renaissance, beginning in 14th century Italy, with such artists are Botticelli, da Vinci, and Donatello.

In 1347 Europe was struck by one of the deadliest plagues in history: the Black Death. The disease was the bubonic plague, endemic (but not so deadly) to rodents in Asia. In 1330 there was an outbreak among humans in China. The disease spread along trade routes, entering Europe originally on Italian trading ships. In the following five years, roughly one-third of the population of Europe, about 25 million people, died. This had profound effects throughout the continent. In particular, there was suddenly a shortage of labor, which improved the lot of the peasants.

Gradually, larger states with more centralized power began to emerge, notably France, England, and Spain. Europe began a long period of expansion. In the 15th century, Portuguese explorers sailed further and further down the west coast of Africa, culminating in the voyage of Vasco da Gama, arriving in India in 1498 after sailing around the southern tip of Africa. In the following century, the Portuguese cemented their grip on the spice trade.

Meanwhile, in Spain, the monarchs had expelled the Muslims, finishing in 1492, and in the same year they financed Columbus' first trip to America. The subsequent exploitation of the American empire, particularly the silver mines, made Spain the richest country in Europe.

The effect in America was less benign. In 1496 a census was taken in Hispaniola, the island now home to Haiti and the Dominican Republic. At that time, there were 1.1 million people. Another census, taken a mere 18 years later, showed a population of 22,000. The native Americans had no immunity to Eurasian diseases. One result of this disaster was another disaster, the importation of African slave labor.

In the 15th and 16th centuries, in response to massive corruption, there were a number of efforts to reform the Catholic Church. These culminated in the Protestant Reformation, launched by Martin Luther in 1517. Also notable in the movement were the founding of the Church of England in 1532, and the establishment of a theocratic state in Geneva by John Calvin in 1541. Protestantism was adopted by much of Northern Europe.

The 16th century saw several important scientific advances. The Swiss Paracelsus (1493–1541) insisted on observation in medicine, and created medical chemistry.

His real name was Philippus Aureolus Theophrastus Bombastus von Hohenheim; his tirades against Aristotle and others gave us our word "bombast." In 1543 the Flemish physician Andreas Vesalius (Andries van Wesel) (1514–1564) revolutionized the study of human anatomy with his text *De humani corporis fabrica* (*On the Structure of the Human Body*). The same year, the year of its author's death, saw the publication of Copernicus' blockbuster, *De revolutionibus orbium coelestium* (*On the Revolutions of the Celestial Spheres*), which presented his Sun-centered cosmology.

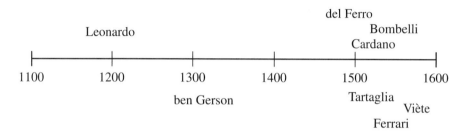

Medieval European mathematicians.

Transmission

Mathematics never completely disappeared from Western Europe. There was always the practical mathematics of everyday life. In addition, the calendar was important to the Church, especially for determining the date of Easter. (The rule now: Easter is the first Sunday after the first full moon after the vernal equinox.)

Scholars knew that there was an ancient tradition of higher mathematics, but did not have access to much of its literature. The most influential mathematical works were introductory ones, especially those of Boethius (see Section 1.5).

In the 10th century, the scholar Gerbert d'Aurillac (945–1003) reintroduced the teaching of mathematics to the cathedral school at Rheims. In his work is the earliest reference in the west to the Hindu-Arabic numerals, although without a zero and without the algorithms that demonstrated the usefulness of the system. He had studied in Spain, where he presumably learned from Muslin scholars. Gerbert became Pope Sylvester II in 999.

The 12th century saw the translation in Latin of many important mathematical classics. The first Latin translation of Euclid's *Elements*, from an Arabic version, was completed by Adelard of Bath (1075–1164) around 1130. Many other works followed, including both Greek and Arabic texts; those of al-Khwārizmī were particularly influential. The most prolific of the translators were Gerard of Cremona (1114–1187) and his team, who translated more than 80 works, including the first translation of Ptolemy's *Almagest* and a new translation of the *Elements*, from the Arabic version of Thābit ibn Qurra.

The most important center of translation was the city of Toledo in central Spain. It had recently been retaken by Christians from the Muslims and had major Arabic libraries. Toledo was home to three cultures: Muslim, Jewish, and Christian. Many of the translations were done by pairs of translators, from Arabic to Spanish by Jews, then from Spanish to Latin by Christians.

As the Europeans became more familiar with the Greek and Muslim mathematics, they began to publish more works, although it was to be a couple of centuries before they produced important new mathematics. Many of the earliest works were by Jews, who were more familiar with the Muslim culture.

One of these texts (written in Hebrew), the *Treatise on Mensuration*, by Abraham bar Ḥiyya (c. 1070–1136), includes an interesting derivation of the formula $A = (1/2)C(d/2)$ for the area of the circle, where C is the circumference and d is the diameter. (See Figure 2.18.) If we think of the circle as being composed of a lot of concentric circles, we can cut this circle along a radius, and unfold its constituent circles into straight lines that form a triangle. The base of the triangle is the length of the longest circle, namely $b = C$. The height of the triangle is the radius of the circle, $h = d/2$. Applying the usual formula for the area of a triangle, $A = (1/2)bh$, gives the result.

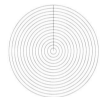

Figure 2.18   The area of a circle.

Leonardo of Pisa (Fibonacci) (c. 1170–1240)

Leonardo was the son of a Pisan merchant. His father lived for a time in the north African city of Bugia (now Bejaia, Algeria), where Leonardo spent much of his youth, studying with Arabic teachers. It was there that he first learned of the Islamic and ancient Greek mathematics. Afterwards, he traveled extensively around the Mediterranean, often meeting with Islamic scholars. He returned to Pisa in northwestern Italy around the year 1200, where he wrote most of his texts.

Leonardo wrote five texts which are still extant. His work was famous in its own time; Pisa acknowledged his contributions with a stipend in 1240.

Leonardo is today known by the name Fibonacci ("son of Bonaccio"), a name he never used. The name was given to him by a 19th century editor of his works, Baldassarre Boncompagni.

Leonardo's *Liber Abacci* (*Book of Calculation*), published in 1202, was the most influential European work on Islamic mathematics.

The *Liber Abaci* began with an introduction to the Hindu-Arabic numerals. Leonardo explained the place value system, and gave algorithms for the usual arithmetic operations, on whole numbers and fractions. These algorithms are not much different from our modern ones. He also showed how to use the new system to solve practical problems such as currency conversions and calculations of profit.

Much of the text involves solving a wide variety of problems. Many of these would be familiar to any reader of a modern algebra text. Some problems lead to quadratic equations, which he shows how to solve. Others involve congruences (Chinese remainder problems), indeterminate equations, and systems of linear equations. Some, though not most, of the problems were taken verbatim from Islamic texts.

Here is one of Leonardo's problems, taken from ibn al-Haytham. Find a number that, when divided by 2 has remainder 1, when divided by 3 has remainder 2, when divided by 4 has remainder 3, when divided by 5 has remainder 4, when divided by 6 has remainder 5, and when divided by 7 has remainder 0.

Here is how Leonardo solved it. Consider first division by 6. If n is any number divisible by 6, then $n - 1$ will have remainder 5. Similarly, if n is divisible by 5, $n - 1$ has remainder 4. This sort of argument works for each the first five conditions (excluding dividing by 7). In each case, the remainder is one less than the divisor, so if n is a number divisible by all of 2, 3, 4, 5, 6, then $n - 1$ will give us the correct remainders. Now, 60 is the smallest common multiple of 2, 3, 4, 5, 6, so $59 = 60 - 1$ will have the right remainders for 2, 3, 4, 5, 6. Unfortunately, dividing 59 by 7 leaves remainder 3, *not* 0. However, if we use the next common multiple of 2, 3, 4, 5, 6, namely 120, we hit pay dirt, since $119 = 120 - 1$ also gives us the right remainder upon division by 7.

The most famous problem from the *Liber Abacci* concerns rabbits. "A certain man had one pair of rabbits together in a certain enclosed place, and one wishes to know how many are created from the pair in one year when it is the nature of them in a single month to bear another pair, and in the second month those born to bear also."

Let us solve Leonardo's problem using modern notation. Denote by F_n the number of pairs of rabbits at the start of the nth month. We start out with one pair: $F_1 = 1$. This pair does not produce in the first month, so $F_2 = 1$, but does produce in the second month: $F_3 = 2$. In the third month, the first pair produces another, but the second pair is not yet old enough: $F_4 = 2 + 1 = 3$. In the fourth month, the youngest pair is too young, but the older two produce: $F_5 = 3 + 2 = 5$.

What happens in the nth month? We have all the rabbits that were around in the month before: F_{n-1} (the rabbits never die). Plus, each of the pairs alive *two* months before will reproduce, giving us F_{n-2} new pairs. Therefore, $F_n = F_{n-1} + F_{n-2}$, i.e., the number of pairs in any month will be the sum of the two preceding months. If we apply this rule, we get the following sequence:

$$1, \ 1, \ 2, \ 3, \ 5, \ 8, \ 13, \ 21, \ 34, \ 55, \ 89, \ 144, \ 377, \ \ldots .$$

After twelve months, we have $F_{13} = 377$ pairs of rabbits. This sequence is now known as the *Fibonacci sequence*, one of the most famous in all of mathematics. It has many interesting properties, most discovered long after Leonardo.

There was little new mathematics in the *Liber Abacci*, and Leonardo did not include all of the latest Islamic developments. Nonetheless, he clearly mastered the earlier Islamic mathematics and demonstrated its power.

Numeration

Before the arrival of the new Hindu-Arabic numerals, Western Europe used Roman numerals. This was a decimal system, but not positional. Different versions occur. Below are the symbols in one version, still occasionally used, as in Super Bowl XLVI.

I	V	X	L	C	D	M
1	5	10	50	100	500	1000

Other numbers were represented by repeating symbols, e.g., III for 3 or CCXX for 220. The exception was when the number approached the next higher symbol, when the smaller symbol would be put before the larger, e.g., IV for 4 or CD for 400. Here are some more examples.

VIII	XV	XLIV	DC	MM	MCMXCII
8	15	44	600	2000	1992

Roman numerals were not very convenient for computations. People often did their arithmetic with their fingers or on a counting board, then recorded the results in Roman numerals.

One of the great advantages of the Hindu-Arabic numerals was the ease with which calculations could be made. It was also important that cheap paper was available for the calculations. (Europe had learned to make paper in the 10th century.)

Fibonacci's *Liber Abacci* was influential in converting Europe to Hindu-Arabic numerals, but the process was not swift, and not without controversy. A number of cities, including Florence in 1299, outlawed the use of the numerals. One of the perceived drawbacks of the new system was the ease with which numerals could be changed, for example from a 1 to a 4. This is why we still write the amounts on checks in words as well as in numerals. It took several centuries for the new numerals to come into general use.

The main audience for the new mathematics were merchants, especially in Italy. The nature of their business was changing. In the Middle Ages, merchants traveled to distant places, bought their goods, then returned home to sell them. In Italy in the 13th and 14th centuries, a new business model arose. The new merchants stayed at home and hired others to do the traveling. This required different mathematics, to handle, for example, letters of credit, bills of lading, and interest calculations. Double-entry bookkeeping was invented. Not to be confused with keeping two sets of books, double-entry bookkeeping was a system whereby each accounting entry was matched by a second, one entry adding to an account, the other subtracting an equal amount from another account. The system makes it easier to balance the books.

A new class of mathematicians, the Italian *abacists* (*maestri d'abbaco*), arose to serve this new need. The abacists ran a number of schools for the children of merchants. One account from the 1320s reports that at least one thousand children were studying the new math in six schools in Florence, a city of about 90,000.

The abacists wrote many textbooks, from which we can tell what they taught. The texts consisted mainly of problems, with solutions but no theory. The problems illustrated the use of the Hindu-Arabic numerals and Islamic algebra, along with some elementary geometry. Here is an example of a problem about simple interest.

> The lira earns 3 denarii a month in interest. How much will 60 lire earn in 8 months?

One thing the abacists did not adopt from Islam was the use of decimal fractions (e.g., 3.5 instead of $3\frac{1}{2}$). We saw in Section 2.3 that Al-Samaw'al understood decimal fractions, but the abacists continued to use regular fractions. The European conversion to the easier decimal fractions was largely due to Simon Stevin (1548–1620).

Combinatorics and Induction

The study of how sets of items could be arranged or combined, part of the field we now call combinatorics, has been pursued in many cultures. We have already mentioned the work of Varāhamihira in Section 2.2, counting the number of perfumes. In the 11th century, the Spanish-Jewish philosopher Abraham ben Meir ibn Ezra (1090–1167) computed the possible number of conjunctions of the seven "planets" (Sun, Moon, Jupiter, Saturn, Mars, Venus, and Mercury), in a work on astrology.

In 1321 Levi ben Gerson published the *Maasei Hoshev* (*The Art of the Calculator*), in Hebrew, in which he proved a variety of combinatorial identities, and gave applications of them. The theorems were not necessarily new, but some of his proofs were. In particular, he mastered the application of what we now call mathematical induction.

Levi ben Gerson (1288–1344)

Levi ben Gerson was born in Bagnols-sur-Cèze near the city of Orange in what is now France. Orange was not in France when Levi lived there, which was fortunate, since the king of France in 1306 expelled all Jews and confiscated their property.

Levi himself had good relations with the Christians, in fact dedicating one of his works to Pope Clement VI. He was also aware of much of Islamic and Greek science.

He was a philosopher, astronomer, and Talmudic scholar, as well as a mathematician. He invented the Jacob Staff, a device used to measure the angular separation between celestial objects, which was an important navigation instrument for sailors for several centuries.

Levi ben Gerson is also known by several other names, including Gersonides and Levi ben Gershon.

We saw in Section 2.3 how Abū Bakr al-Karajī used induction to prove the formula for the sum of cubes: $1^3 + 2^3 + \cdots + 10^3 = (1 + 2 + \cdots + 10)^2$. Here is how Levi used induction to count the number of permutations of n objects.

A *permutation* of n objects is a linear arrangement. For example, the permutations of the letters a, b, and c are abc, acb, bac, bca, cab, and cba. Let us denote by P_n the number of permutations of n elements, e.g., $P_3 = 6$. Levi proved the formula $P_n = n!$. (Recall that $n! = 1 \cdot 2 \cdot 3 \cdot \ldots \cdot n$.)

Note first that the formula holds trivially for $n = 1$, since there is only one permutation of 1 object. Now suppose that we know that $P_n = n!$ for some fixed n. Consider P_{n+1}, the number of permutations of $n + 1$ objects.

Suppose we have a set of $n + 1$ objects. Pick one, say a. One way to arrange the objects is to start with a, then finish with any permutation of the remaining n objects. So the number of permutations of the larger set which start with a is just P_n. But there is nothing special about a: the number of permutations of the larger set which start with any fixed element is just P_n. Since there are $n + 1$ choices for the first element, we have $P_{n+1} = (n + 1)P_n$.

Recall that we started out assuming that $P_n = n!$. Then

$$\begin{aligned} P_{n+1} &= (n + 1)P_n \\ &= (n + 1)n! \\ &= (n + 1)(1 \cdot 2 \cdot 3 \cdots n) \\ &= 1 \cdot 2 \cdot 3 \cdots n \cdot (n + 1) \\ &= (n + 1)! \end{aligned}$$

so the formula works for $n + 1$.

So what? Well, suppose that we wanted to know that the formula worked for $n = 10$. We started out noting that it works for $n = 1$. But the last argument shows that, since it works for $n = 1$, it works for $n + 1 = 2$, so $P_2 = 2!$. But then since $P_2 = 2!$, the argument shows that $P_3 = 3!$ We can repeat this until we get $P_{10} = 10!$. Nothing is special about 10, so the formula is true for any positive integer. Levi called this process "rising step by step without end."

Geometry

The geometry of Western Europe at this time continued to be dominated by the ancient Greek classics. Euclid's *Elements* was widely taught in the universities. After the conquest of Constantinople by the Ottomans in 1453, a number of scholars fled to the West, bringing along with them a deeper knowledge of Greek geometry, and in some cases Greek manuscripts. In the 16th century, Euclid's work became more widely available, with its first translations into Italian, German, French, and English.

One area of geometry which the Europeans developed, starting in 15th century Italy, was the mathematics of perspective. This new work was done primarily by artists, in some cases artist-mathematicians. The basic problem was to develop techniques that gave the illusion of three-dimensional depth to a two-dimensional drawing. For example, parallel lines receding from the viewer have the appearance, in a painting, of converging to a point in the distance (Figure 2.19).

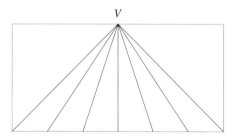

Figure 2.19   Parallel lines in perspective.

The Italian artist and goldsmith Filippo Brunnelleschi (1377–1446) created a number of paintings illustrating perspective. The first text on the subject was produced by Leon Battista Alberti (1404–1472) in 1435. In it, he showed how to render a checkerboard in perspective. Figure 2.20 is an example of a 6×4 checkerboard in perspective.

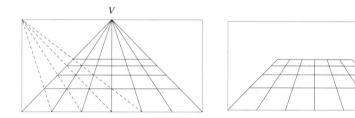

Figure 2.20   Checkerboard in perspective.

Piero della Francesca (1420–1492) extended Alberti's work, showing how to use perspective in drawing a variety of three-dimensional figures. Finally, we mention the work of Albrecht Dürer, who instructed artists in Northern Europe in the new techniques. Along the way, Dürer produced the first geometry text in German.

This new work on perspective produced some stunning pictures but was not particularly theoretical. A sound theoretical basis for the mathematics of perspective came later.

Algebra

It was in algebra that the new Europe produced its first major advances over the mathematics it inherited.

One crucial development was the widespread introduction of a better notation. Compare these problems.

> A thing and its square, decreased by three times the thing, gives eight. What is the thing?

> Solve $x + x^2 - 3x = 8$.

It is difficult to imagine modern algebra without its notation. This notation developed in Europe, beginning in the 15th century (although not completed for a couple of centuries).

The early developments came from Italy, like most of the important changes, economic and mathematical, taking place in Europe in this time. By the early 15th century, the *abacists* had started to abbreviate words in algebra. For example, the word *cosa* (thing) was their word for the unknown, what we now would write as x. In the 15th century, this was sometimes shortened to c. Similarly, *censo* (square) was written ce. Later in that century, Luca Pacioli (1445–1517) used the symbols $\bar{p}$ and $\bar{m}$ for plus and minus, from the Italian *più* and *meno*.

By the late 15th century, the new math began to spread northward first to France, then Germany, then England. Christoff Rudolff (1499–1545) wrote the first important algebraic text in German, the *Coss*. (In Germany, algebraists were called Cossists, from the Italian *cosa*.) In it, he introduced our symbols $+$ and $-$ for addition and subtraction. The Englishman Robert Recorde (1510–1558) created our symbol $=$ for equality.

The second major algebraic advance of this time was the solution of the cubic. Recall that a cubic equation is one of the form $ax^3 + bx^2 + cx + d = 0$, where $a \neq 0$. Cubic equations had been studied as long ago as Babylonian times, and later by the Greeks, Chinese, and Muslims. As mentioned in Section 2.3, cubics had been classified, and solved geometrically, by Al-Khayyāmī in the 11th century. But no one was able to find an algebraic solution similar to the quadratic formula for degree two equations. This problem was finally solved in the 16th century.

The first important step was taken by Scipione del Ferro. He was able to solve equations of the form $x^3 + cx = d$, sometime in the first two decades of the 16th century. He did not, however, publish his results. It was common at this time to keep discoveries secret. They could be useful in winning public challenges. In such a challenge, each of two contenders would present the other with a list of problems to solve. The contenders would later meet publicly to present their solutions to the problems. These challenges were sometimes used to fill university professorships. There might also be considerable sums of money riding on the outcome.

Scipione del Ferro (1465–1526)

Scipione del Ferro was born in Bologna, in northern Italy. His father Floriano was in the paper making business. Scipione probably studied at the University of Bologna, a major center of learning in his time. He was appointed a lecturer in arithmetic and geometry at the university in 1496. He was also a businessman. None of his writings have survived.

Del Ferro did reveal his solution to his student, Antonio Maria Del Fiore. In 1535, after Niccolò Tartaglia announced that he had solved the cubic of the form $x^3 + bx^2 = d$ (keeping his solution secret), Fiore challenged Tartaglia to a public contest. He apparently hoped to win using del Ferro's work, since all thirty of the problems he submitted were of the form $x^3 + cx = d$ that del Ferro had solved. In response to the challenge, Tartaglia managed to figure out the solution to this form as well, and handily won the contest. He prudently declined the prize of thirty banquets prepared by the loser.

Niccolò Tartaglia (1499–1557)

Niccolò Fontana was born in Brescia, in northern Italy. His father Michele earned his living making deliveries on horseback between Brescia and neighboring towns. Michele Fontana was killed when Niccolò was six years old, and the family descended into poverty. In 1512 a French army invaded Brescia and massacred an estimated 45,000 citizens. During the massacre, Niccolò received several saber cuts, including one to his jaw and palate that left him with a speech impediment. This was the source of his nickname *Tartaglia* (*stammerer*), which he later used in his publications.

Tartaglia studied in Padua, although his formal training was minimal. He taught mathematics in Verona and, after 1534, in Venice, where he gave lessons in local churches. Even though he had a good reputation for his mathematical ability, he was poor throughout his life.

Tartaglia was an engineer as well as a mathematician. He published many works, including the first Italian translations of Archimedes and Euclid.

Working in Milan, Gerolamo Cardano heard about this contest. He wrote Tartaglia, asking permission to include Tartaglia's work in a text that he was writing. Tartaglia refused, since he intended to publish this work himself. But Cardano was persistent, and in 1939 Tartaglia gave him a cryptic version of the solution of three types of cubic equation, without proof, phrased as a poem. Cardano pledged not to publish Tartaglia's work.

Gerolamo Cardano (1501–1576)

Gerolamo Cardano was born in Pavia in northern Italy, the illegitimate son of Fazio Cardano, a lawyer and friend of Leonardo da Vinci. Gerolamo was trained as a doctor, practiced medicine for several years in a small town near Padua, then moved to Milan in 1534. His early career was difficult; he was repeatedly denied admission to the Milan College of Physicians due to his illegitimate birth.

Eventually, Cardano overcame his origins, and became one of the most celebrated physicians in all of Europe. In 1543 he became professor of medicine at the University of Pavia, where he was a popular lecturer. In the 1550s he traveled extensively throughout Europe. In one incident, he reportedly consulted with the archbishop of Scotland about a worsening case of asthma. After observing the archbishop's habits for a month, he recommended replacing his feather bedding. This worked, and won Cardano an influential supporter.

Cardano's later life was more difficult. In 1560 one of his sons was convicted of poisoning his wife, and executed. In 1570 Cardano was accused of heresy for computing Jesus' horoscope, and was tortured and imprisoned by the Inquisition. His influential friends got him released from prison after a few months. He then traveled to Rome where he was astrologer to the Pope. Cardano committed suicide in 1576. One report claimed that he predicted the exact date of his death, and made sure his prediction was correct.

Cardano published over 200 works on a wide variety of topics, including mathematics, science, medicine, astrology, music, philosophy, religion, and gambling.

Cardano, unlike Fiore, was a first-rate mathematician. Equipped with Tartaglia's poem, he and his student Ludovico Ferrari managed to obtain solutions and proofs for all of the cubic forms. Before publishing anything, he and Ferrari, hearing rumors of del Ferro's discovery, traveled to Bologna, where Fiore's successor allowed them to inspect del Ferro's work.

Cardano now considered himself free to publish Tartaglia's solution, since it was essentially the same as del Ferro's. In 1545 he produced his masterpiece, *Artis Magnae, Sive de Regulis Algebraicis* (*The Great Art*, or *The Rules of Algebra*), now simply called the *Ars Magna*, with solutions to the cubic. Tartaglia felt himself cheated, although Cardano had mentioned Tartaglia's role in the discovery. After various exchanges of public insults, another contest was held, although Tartaglia's opponent was not Cardano, but his student Ferrari. This time Tartaglia lost.

Let us take a closer look at the solution of a cubic of the form $x^3 + cx = d$, using our modern notation. Tartaglia's poem handled this case, telling the reader to find numbers u and v such that $u^3 - v^3 = d$ and $3uv = c$. If we can find such numbers, we let $x = u - v$. Using the identity $(r + s)^3 = r^3 + 3r^2s + 3rs^2 + s^3$, we get

$$
\begin{aligned}
x^3 + cx &= (u - v)^3 + c(u - v) \\
&= (u^3 - 3u^2v + 3uv^2 - v^3) + c(u - v) \\
&= [(u^3 - v^3) - 3uv(u - v)] + 3uv(u - v) \\
&= u^3 - v^3 \\
&= d,
\end{aligned}
$$

so x is a solution to the cubic.

How do we find u and v? Cardano gives the following.

$$
u = \sqrt[3]{\sqrt{\left(\frac{d}{2}\right)^2 + \left(\frac{c}{3}\right)^3} + \frac{d}{2}}
$$

$$
v = \sqrt[3]{\sqrt{\left(\frac{d}{2}\right)^2 + \left(\frac{c}{3}\right)^3} - \frac{d}{2}}
$$

Therefore

$$
x = \sqrt[3]{\sqrt{\left(\frac{d}{2}\right)^2 + \left(\frac{c}{3}\right)^3} + \frac{d}{2}} - \sqrt[3]{\sqrt{\left(\frac{d}{2}\right)^2 + \left(\frac{c}{3}\right)^3} - \frac{d}{2}}.
$$

As an example, Cardano solved $x^3 + 6x = 20$. In this case, the formula gives

$$
x = \sqrt[3]{\sqrt{108} + 10} - \sqrt[3]{\sqrt{108} - 10}.
$$

Cardano notes that this reduces to $x = 2$, but doesn't tell us how he knew that.

You may have noticed that the last cubic was not of the most general form; in particular, there was no x^2 term. A simple substitution, however, allows us to handle that case. For example, consider $x^3 + bx^2 + cx = d$. If we plug $x = y - b/3$ into this equation, we get a cubic in y that has no y^2 term, and it can be solved by Cardano's formula above. Details are left to the exercises.

Ludovico Ferrari (1522–1565)

Ludovico Ferrari was born into a poor family in Bologna, Italy. At the age of 14 or 15, he went to work in Milan as a servant of Cardano, who soon realized Ferrari's talent and taught him mathematics. At the age of 20, after defeating another job applicant in a debate, Ferrari assumed Cardano's old position as lecturer in geometry at the Piatti Foundation in Milan.

After he bested Tartaglia in debate in front of a huge audience, at the age of 26, Ferrari received many job offers. The one he accepted was as tax assessor to the governor of Milan. He became rich in this job, retired young, and moved back to Bologna where he became a mathematics professor at the university. He died the following year, of white arsenic poisoning. There has been some speculation that he was poisoned by his sister, who stood to inherit his wealth. This did not turn out well for her. She married two weeks after her brother's funeral, but her new husband took all of her money and left her in poverty.

After the cubic was conquered, it was natural to tackle the case of the quartic $ax^4 + bx^3 + cx^2 + dx + e = 0$. In fact, the quartic was solved by Ferrari, in time to be included in the *Ars Magna*.

After Ferrari's work, mathematicians naturally started studying the quintic (fifth degree) equation. This was rather more difficult, and wasn't handled successfully until the 19th century (see Section 3.3).

Cardano's *Ars Magna* was very influential, but was not easy for a student to read. In 1557 Rafael Bombelli began work on a textbook aimed at students, but included the new work on the cubic and quartic. He completed only the first three of the five planned parts of this work, called *Algebra*; they were published just before his death in 1572. In addition to the new algebra, Bombelli included many problems from Diophantus' *Arithmetica*, newly translated, and intended to include geometrical work on polynomials, similar to that of Al-Khayyāmī. Despite being incomplete, Bombelli's *Algebra* was the culmination of Italian Renaissance algebra.

Rafael Bombelli (1526–1572)

Rafael Bombelli was born in Bologna, Italy, the son of a wool merchant. He did not attend the university, instead being trained by the engineer and architect Pier Francesco Clementi. Bombelli became one of the leading engineers of his day, working to reclaim marshlands in several regions of Italy.

It was during a pause in one of his reclamation projects, in 1557, that he began his *Algebra*. He worked on his mathematics until the engineering project resumed in 1560. As mentioned above, he was never able to finish his text.

Bombelli did more than repeat his predecessors' work, however. In particular, he addressed a conundrum that had puzzled Cardano.

Recall the quadratic formula for solutions to $ax^2 + bx + c = 0$:

$$x = \frac{-b \pm \sqrt{b^2 - 4ac}}{2a}.$$

In the case where $a = 1$, $b = 2$, and $c = 2$, this simplifies to $-1 \pm \sqrt{-1}$. Before Bombelli, mathematicians had merely discarded such values, arguing that a negative number can have no square root. (Many rejected negative roots as well.)

What Cardano noticed was that, for $x^3 = 15x + 4$, his formula gave the root

$$x = \sqrt[3]{2 + \sqrt{-121}} + \sqrt[3]{2 - \sqrt{-121}}$$

instead of the obvious root $x = 4$. He naturally wanted to treat the expression above as nonsensical, but could not reject 4 as a root. He wrote to Tartaglia about this problem, but got no solution (Tartaglia misunderstood the problem).

Bombelli investigated this case, deciding to treat the square roots of negative numbers as actual numbers. In particular, he defined an arithmetic on numbers that included these square roots, arguing by analogy with regular square roots. For example,

$$\begin{aligned}
(1 + 2\sqrt{-1})(3 + 5\sqrt{-4}) &= (1 + 2\sqrt{-1})(3 + 5\sqrt{4}\sqrt{-1}) \\
&= (1 + 2\sqrt{-1})(3 + 5 \cdot 2\sqrt{-1}) \\
&= (1 + 2\sqrt{-1})(3 + 10\sqrt{-1}) \\
&= 3 + 10\sqrt{-1} + 6\sqrt{-1} + 20(\sqrt{-1})^2 \\
&= 3 + 10\sqrt{-1} + 6\sqrt{-1} + 20(-1) \\
&= -17 + 16\sqrt{-1}.
\end{aligned}$$

To address Cardano's problem, Bombelli used his new arithmetic to calculate

$$\sqrt[3]{2 + \sqrt{-121}} = 2 + \sqrt{-1} \text{ and } \sqrt[3]{2 - \sqrt{-121}} = 2 - \sqrt{-1}.$$

This gives the solution to $x^3 = 15x + 4$ as

$$x = \sqrt[3]{2 + \sqrt{-121}} + \sqrt[3]{2 - \sqrt{-121}} = 2 + \sqrt{-1} + 2 - \sqrt{-1} = 4.$$

Cardano's difficulty was resolved, at the cost of expanding the notion of number. In subsequent centuries these new numbers, which we now call *complex numbers*, have played a crucial role in the advance of many areas of mathematics.

Bombelli was the last great Italian algebraist of the Renaissance; the center of mathematical research had moved north. The most important figure of the late 16th century was François Viète in France.

François Viète (1540–1603)

François Viète was born in Fontenay-le-Comte in western France, the son of a lawyer. He graduated from the University of Poitiers in 1560. Like Bombelli, Viète was not a professional mathematician. Instead, he followed his father's career, working first in Fontenay, then in Paris. His law career was very successful. He worked on mathematics in his spare time.

In 1589–90, during a period of conflict between France and Spain, Viète was able to decipher coded messages sent from agents in France to the Spanish emperor Philip

II. One code he broke was so complex that Philip complained to the Pope that it must have been done by sorcery.

In 1593 a Belgian, Adriaan van Roomen, set a challenge to all European mathematicians to solve a particular polynomial of degree 45. The ambassador from the Netherlands (ruled by Spain at this time) reportedly claimed to the French king that there was no French mathematician capable of tackling the problem. The king thereupon summoned Viète, who produced two solutions within minutes. The following day, he presented the full set of twenty-three positive roots. (Negative roots were not then considered solutions.)

$\sim$

The Islamic mathematicians helped free algebra from its dependence on geometry, which ultimately led to the advances of the 16th century Italian mathematicians. It was Viète who started to free algebra from its dependence on arithmetic.

To begin, Viète employed an improved algebraic notation, although not the notation we now use. He designated unknown quantities (what we would call x, y, $\ldots$) by vowels, and constants (what we would call $a, b, \ldots$) by consonants.

Even though we have written formulas to explain the solution to various polynomials, before Viète solutions were always given as rules. Here, for example, is the beginning of a solution from Cardano.

> Let us divide 10 into equal parts and 5 will be its half. Multiplied by itself, this yields 25. From 25 subtract the product itself, that is 40, ...

Viète was the first to write general formulas for solving equations, instead of giving rules of solution. Thus he was the first to actually write down a version of the quadratic formula.

Viète made an important distinction between algebra and arithmetic. He considered algebraic equations to be separate entities, which could be manipulated symbolically without regard to any arithmetic interpretation. Thus, one could write $(x + 2)(x^2 - 1) = x^3 + 2x^2 - x - 2$ without insisting that x actually stand for anything.

This point of view led naturally to a study of the relationship of the coefficients of polynomials to their roots. For example, Viète considered the equation $bx - x^2 = c$, which has two roots. If we call the roots x_1 and x_2, we have $bx_1 - x_1^2 = bx_2 - x_2^2$ (they are both equal to c), so that $bx_1 - bx_2 = x_1^2 - x_2^2$. Factoring, we get $b(x_1 - x_2) = (x_1 + x_2)(x_1 - x_2)$. Dividing by $x_1 - x_2$ yields $b = x_1 + x_2$. Thus the coefficient b is the sum of the two roots. Substituting this result into $bx_1 - x_1^2 = c$ and solving tells us that the product of the two roots is c (details left to the exercises). Viète also wrote equations relating the coefficients of cubics to their roots.

EXERCISES

2.44 Abraham bar Hiyya derived the formula $A = (1/2)C(d/2)$ for the area of a circle, where C is the circumference and d the diameter of the circle. Show that this is equivalent to our usual $A = \pi r^2$. (Hint: π is defined to be C/d.)

2.45   One of Leonardo's problems is about two men with money, measured in *denarii*. The first says to the second: if you give me 7 *denarii*, I will have five times as much as you have left. The second says to the first: if you give me 5 *denarii*, I will have seven times as much as you have left. How much does each man have now? (Hint: the answer uses fractions.)

2.46   Find a number that, when divided by 7 has remainder 6, when divided by 8 has remainder 7, when divided by 9 has remainder 8, and when divided by 10 has remainder 9.

2.47   What numbers do the following Roman numerals represent?
   a) VI
   b) IX
   c) DCC
   d) MMCM

2.48   Write the following in Roman numerals.
   a) 14
   b) 326
   c) 1999
   d) 2014

2.49   What is XXII times IV?

2.50   What is the answer to the abacist's interest problem quoted in this section?

2.51   Use Alberti's method (and a ruler) to draw a 4×3 checkerboard in perspective.

2.52   Check that Cardano's values of u and v satisfy $u^3 - v^3 = d$ and $uv = c/3$.

2.53   a) Apply Cardano's formula to find a solution to $x^3 + 3x = 36$.
   b) Use a calculator to simplify your answer.

2.54   Plug $x = y - b/3$ into the equation $x^3 + bx^2 + cx = d$, and show that the result is a cubic in y with no y^2 term. You may wish to use the identities $(r + s)^3 = r^3 + 3r^2 s + 3rs^2 + s^3$ and $(r + s)^2 = r^2 + 2rs + s^2$.

2.55   a) Use the substitution of the last exercise on $x^3 + 3x^2 + 2x = 6$.
   b) Find a solution to $x^3 + 3x^2 + 2x = 6$.

2.56   Complete the proof in the text that the product of the two roots of $bx - x^2 = c$ is c.

2.57   Modern theory of equations tells us that if the two roots of the quadratic equation $ax^2 + bx + c = 0$ are x_1 and x_2, then we can rewrite the quadratic as $a(x - x_1)(x - x_2) = 0$. Use this form to show that the sum of the roots is $-b$, and their product is c.

CHAPTER 3

MODERN MATHEMATICS

Mathematics and Poetry are... the utterance of the same power of imagination, only that in the one case it is addressed to the head, in the other, to the heart.

THOMAS HILL (1818–1891)

This chapter tells the story of mathematics in a time when the way people live begins to undergo a massive shift, a shift in which mathematics plays a major role. One difficulty of this chapter is that the level of sophistication, and the sheer volume, of the new math prevents us from providing more than a sketch of the history. Nonetheless, we can at least touch on some of the major developments of this time, and make a few stops at some scenic attractions.

Mathematics for the Liberal Arts.
By Donald Bindner, Martin Erickson, Joe Hemmeter Copyright © 2012 John Wiley & Sons, Inc.

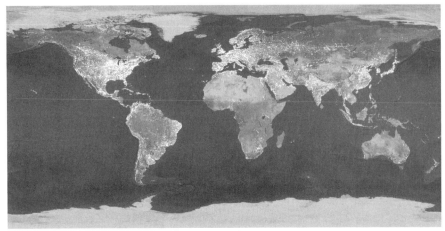

Earth.[1]

3.1   The 17th Century: Scientific Revolution

Introduction

The 17th century was one of the most creative periods in all of cultural history. Modern science was born, with mathematics at its heart. This Scientific Revolution, as it is usually called, combined with the Industrial Revolution of the following century, are largely responsible for creating our modern society.

Much of the first half of the century was dominated by religious wars, including the Eighty Years' War (1568–1648) in the Low Countries of Northwestern Europe, the English Civil Wars (1642–1651), and above all the Thirty Years War (1618–1648).

The Thirty Years War was the worst European war outside of the 20th century; millions died. It involved all the European powers, although most of the fighting took place in what is now Germany and the Czech Republic. In many of those regions, 25% or more of the population perished. The war ended with the Treaty of Westphalia (1648). Although devastated by the war, Germany did manage to produce the great mathematician and philosopher Gottfried Wilhelm Leibniz (1646–1716).

The 17th century was the Dutch Golden Age. The Dutch Republic, part of what is now the Netherlands, won its independence from the Hapsburg Empire, ruled at that time from Spain. Although not officially recognized by all powers until 1648, it was de facto independent from early in the century. It was at this time one of the great commercial centers of Europe, with Amsterdam probably the richest city in the world. The tolerant Dutch cities attracted Protestant refugees from the Southern Netherlands and Jewish refugees from Portugal and Spain. Dutch traders became

---

[1]"Earth at Night." C. Mayhew and R. Simmon (NASA/GSFC), NOAA/NGDC, DMSP Digital Archive.

dominant, displacing the Portuguese in Asia. Some historians credit the Dutch for establishing the first stock exchange, which financed the founding of the first multinational corporation, the Dutch East India Company, in 1602.

This was a period of brilliant artists such as Johannes Vermeer (1632–1675) and the great Rembrandt Harmenszoon van Rijn (1606–1669). Dutch universities, particularly the University of Leiden, were among the best. Among the prominent scientists were Anton van Leeuwenhoek (1632–1723), "the father of microbiology," who improved the microscope and discovered red blood cells and micro-organisms, and the astronomer, mathematician, and inventor of the pendulum clock, Christiaan Huygens (1629–1695).

France emerged from the Thirty Years War the most powerful nation in Europe. This was the century of the philosopher and mathematician René Descartes (1596–1650), the dramatist Molière (1622–1673), and the heyday of the absolute monarchy under Louis XIV, the Sun King (r. 1661–1715), who ruled from his sumptuous palace at Versailles. Paris was arguably the most important cultural center of the time.

In England the 16th century began in the time of Elizabeth I (1533–1603), William Shakespeare (1564–1616), and the philosopher (and Lord Chancellor) Francis Bacon (1561–1626), who promoted the "New Science" based on observation and experimentation. The middle of the century was difficult with the civil wars, the Great Plague of 1665–66, and the Great Fire of London in 1666. Nonetheless England emerged strongly from this period. London was rebuilt, led by the architect Christopher Wren (1632–1723). The Royal Society was founded in 1660, and hired the brilliant Robert Hooke (1635–1703) as curator of experiments. This was also the time of the incomparable Isaac Newton (1642–1727).

In Italy, by contrast, the most productive scientists of the century lived in the first half. The greatest of these was the man many consider to have been the first modern scientist, Galilei Galileo (1564–1642), who died in the year of Newton's birth. A brilliant polemicist as well as scientist, Galileo did much to promote the new astronomy of Nicolaus Copernicus (1473–1543). The Church was unwilling to accept this, and Galileo spent his last years under house arrest.

Above all, the 17th century saw the triumph of the new science, which was based on the primacy of experimentation and couched in the language of mathematics. Its first great success was in astronomy. Copernicus had published his heliocentric theory in 1543. Johannes Kepler (1571–1630), working from the careful observations of Tycho Brahe (1546–1601), studied planetary orbits, and discovered his three "planetary laws" in the first decade of the 17th century. In particular, he discovered that the planets travel around the Sun in elliptical orbits. In 1610 Galileo improved on the new telescope and was able to see through it a mini-solar system of moons orbiting Jupiter, as well as the phases of Venus, the mountains of the Moon, and sunspots, observations inconsistent with the old Ptolemaic astronomy. Finally, Isaac Newton, using the new calculus he helped to develop, formulated his laws of physics and from them derived Kepler's planetary laws. His new physics was published in his masterwork, *Philosophiæ Naturalis Principia Mathematica* (1687).

In mathematics, the 17th century is remembered for the invention of calculus, the "mathematics of change." There were other major advances as well, most notably the marriage of algebra and geometry—analytic geometry—with the invention of the Cartesian coordinate system. Also, computation was made easier with the invention of logarithms, and major work was done in probability theory and number theory.

The modern world of mathematical research is dominated by universities where most of the work is done, and by journals which disseminate the results. Neither institution was fully developed in the 17th century. Most mathematicians of the time were trained in universities, and some research was carried out there, but much was done outside them. In the early part of the century, especially, the patronage of rulers was essential. The Holy Roman Emperor Rudolph II hosted scientists including Johannes Kepler in Prague. Prince Maurice of Holland supported Simon Stevin (c. 1548–1620) and René Descartes. Prince Leopold de' Medici supported research in Florence.

Communication of research advances was still rudimentary at this time, although the custom of hoarding instead of publicizing one's work was beginning to break down. Most important in this respect were the new scientific societies, including the *Accademia dei Lincei* (1603–30) and *Accademia del Cimento* (Academy of experiments, 1657–67) in Italy, the *Academia Parisiensis* (c. 1635) and its successor, the *Académie des Sciences* (1666–present), in France, as well as the Royal Society in England. Mathematicians also corresponded by letter. The French monk Marin Mersenne (1588–1648), who organized the *Academia Parisiensis*, had a network of correspondents throughout Europe and the Middle East. He would have letters reproduced and sent to other researchers, spreading the latest discoveries.

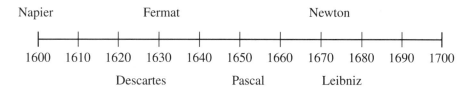

Seventeenth Century mathematicians.

Logarithms

The most detailed, laborious calculations at this time were done by astronomers. They used a number of techniques to simplify their work. For example, they utilized the following identity to multiply numbers.

$$2 \sin \alpha \sin \beta = \cos(\alpha - \beta) - \cos(\alpha + \beta)$$

To multiply two numbers, say $x = .3295$ and $y = .0827$, they would first use trig tables to find angles α and β such that $x = \sin \alpha$ and $y = \sin \beta$, in this case $\alpha \approx 19.238$ and $\beta \approx 4.744$. They would then look up the cosines on the right-hand side, subtract them

$$\cos(\alpha - \beta) - \cos(\alpha + \beta) \approx \cos(14.494) - \cos(23.982) \approx .9682 - .9173 = .0545$$

and divide by 2 to get $.3295 \times .0827 \approx .02725$.

The last calculation may seem a complicated way to multiply numbers. The point was to replace hand multiplication by table lookups, addition, and subtraction, which were easier. In general, multiplication and division were time-consuming and prone to error. That was why logarithms were invented in the early 17th century.

Logarithms were invented twice, by a Swiss mathematician named Joost Bürgi (1552–1632), and a Scotsman named John Napier. Napier's work was published first, in 1614 versus 1620 for Bürgi, and was much more influential.

John Napier (1550–1617)

John Napier was a Scottish aristocrat, the eighth laird of Merchiston. As a teenager, he studied at St. Andrews but did not get a degree. He probably traveled for a while in Europe, but lived most of his life as a landed aristocrat in Scotland. He had twelve children.

Napier was most famous in his time for his labors on behalf of religion. In particular, he wrote *A Plaine Discovery of the Whole Revelation of Saint John: Set Downe in Two Treatises*. In this work, he revealed that the Pope is the Antichrist, and that the world will probably end between 1688 and 1700. The book was quite popular, with twenty-one editions in English and numerous translations.

The following discussion uses the modern form of logarithms. Napier's original logarithms, although using the same essential ideas, were a bit different in detail.

What is a logarithm? In a word, it is a power. For example, since $100 = 10^2$, we call 2 the logarithm of 100, and write $\log 100 = 2$. Note that $\log 10 = 1$, since $10^1 = 10$, and $\log 1000 = 3$, since $10^3 = 1000$.

What happens to logarithms when we multiply numbers? Consider $10 \cdot 100 = 10^1 \cdot 10^2 = 10^3$. When we multiply numbers, their powers, that is their logarithms, are added to produce the new power. In other words, multiplication is reduced to addition. For any numbers m and n, we have

$$10^m \cdot 10^n = 10^{m+n}, \quad \text{so} \quad \log(10^m \cdot 10^n) = m + n = \log(10^m) + \log(10^n).$$

We have only talked about 10^n where n is a positive integer, but the notion of powers can be extended, so that we can talk about 10^x for any positive real number x. The full explication of this is beyond this text, but it starts with the following definitions, for n a positive integer.

$$10^{-n} = \frac{1}{10^n} \quad \text{and} \quad 10^{1/n} = \sqrt[n]{10}$$

With this extension, we have these important identities, for real numbers x and y:

$$\log(xy) = \log x + \log y \quad \text{and} \quad \log(x/y) = \log x - \log y.$$

Thus we replace multiplication by addition and division by subtraction.

How can we use this? Consider the problem of multiplying two numbers, say $x = 2.3456$ and $y = .827$. We can use a table of logarithms to look up $\log x \approx .3703$ and $\log y \approx -.0825$. (Numbers between 0 and 1 have negative logarithms.) We then add the logs to get $\log(xy) = \log x + \log y \approx .2878$. Finally, we look this log up in a table to get $xy \approx 1.9400$.

We also note some further properties of logarithms, for any positive numbers x and y and nonnegative integer n.

1. If $x < y$, then $\log x < \log y$.

2. $\log 1 = 0$

3. If $\log x \geq n$ and $\log x < n + 1$, then the number of digits to the left of the decimal point of x is $n + 1$.

4. $\log(x^y) = y \log x$

Napier published *Mirifici Logarithmorum Canonis Descriptio* (*Description of the Marvelous Canon of Logarithms*), with the first table of logarithms, in 1614. It had taken him twenty years to calculate the tables. (This text is also responsible for our use of the decimal point.) He explained the theory behind his logarithms in the work *Mirifici Logarithmorum Canonis Constructio* (*Construction of the Marvelous Canon of Logarithms*), which appeared in 1619, two years after his death.

Napier's logarithms spread like wildfire. Mathematicians and astronomers immediately appreciated their value in calculations. A number of people, including Napier himself, quickly refined his original ideas, and as early as 1628, Henry Briggs (1561–1630) and Adriaan Vlacq (1600–1667) published a table of logarithms of numbers from 1 to 100,000 to ten decimal places. This was the standard reference until the 20th century.

It may be difficult for someone in this age of computers to appreciate the value of logarithms. Two centuries after their introduction, however, the mathematician Pierre-Simon de Laplace (1749–1827) wrote that they "by shortening the labors, doubled the life of the astronomer."

Logarithms are still useful. Recall the Fibonacci sequence: $1, 1, 2, 3, 5, \ldots$, where each term is the sum of the preceding two. How would we calculate the millionth Fibonacci number? Even today, a calculator (or computer) is likely to choke on the calculation. Logarithms to the rescue!

One way to approximate the nth term f_n of the sequence is by using the *golden ratio*, $\frac{1+\sqrt{5}}{2}$.

$$f_n \approx \frac{1}{\sqrt{5}} \left(\frac{1 + \sqrt{5}}{2} \right)^n$$

The approximation becomes more precise as n grows. Here are the first few terms.

n	1	2	3	4	5	6	$\ldots$	20
f_n	1	1	2	3	5	8	$\ldots$	6765
approx	0.72	1.17	1.89	3.07	4.86	8.02	$\ldots$	6765.0000

Using the properties above, we can compute the log of the nth Fibonacci number:

$$\log(f_n) \approx \log\left[\frac{1}{\sqrt{5}}\left(\frac{1+\sqrt{5}}{2}\right)^n\right]$$

$$= \log\left(\frac{1}{\sqrt{5}}\right) + \log\left(\frac{1+\sqrt{5}}{2}\right)^n$$

$$= \log\left(\frac{1}{\sqrt{5}}\right) + n\log\left(\frac{1+\sqrt{5}}{2}\right).$$

We can compute the two logs above using a table (or calculator), giving

$$\log(f_n) \approx -.3494850022 + n \cdot 0.2089874025.$$

The millionth Fibonacci number has $n = 10^6$, so in this case

$$\log(f_n) \approx -.3494850022 + 10^6 \cdot 0.2089874025$$

$$\approx 208987.4025 - .3495$$

$$= 208987.0530.$$

Thus, the millionth Fibonacci number is about

$$10^{208987.0530} = 10^{0.0530} 10^{208987} \approx 1.1298 \times 10^{208987}.$$

In particular, it has $208{,}988$ digits.

Napier also invented ways of multiplying and dividing numbers using small rods that became known as Napier's bones. These were forerunners of one of the most important computational aids in history, the slide rule.

The slide rule (Figure 3.1) used a mechanical equivalent of logarithms to approximate multiplication and division. Over time, other computations were added. For about three hundred years, the slide rule was an essential tool for the scientist and engineer. Before the arrival of the personal computer, the quintessential nerd was often pictured with a slide rule in the shirt pocket.

The Cartesian Coordinate System

In the 17th century, the mathematician and philosopher René Descartes (1596–1650) introduced a method of solving geometric problems using rectangular coordinates. Today, we call the system of coordinates the *Cartesian coordinate system*. The *Cartesian plane* is shown in Figure 3.2.

The Cartesian plane is determined by two infinite lines at right angles, called the *axes*. The *x-axis* is horizontal and the *y-axis* is vertical. Each axis is a real number line, with the two axes crossing at 0. The crossing point is called the *origin* of the coordinate system.

Each point in the plane has two *coordinates*. The first coordinate gives the point's horizontal position. The second coordinate gives its vertical position. We write the

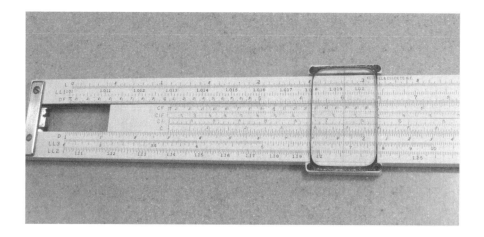

Figure 3.1 A slide rule.

coordinates as (x, y). For example, the point $(2, 1)$ is located 2 units to the right of the y-axis and 1 unit above the x-axis.

Geometry based on coordinates is called *analytic geometry*. In analytic geometry, algebraic equations are identified with sets of points in the plane (lines, circles, etc.).

René Descartes (1596–1650)

René Descartes was born in La Haye (later renamed La Haye-Descartes), France. He was the son of a lawyer, and himself obtained a law degree in 1616. After that, he traveled extensively throughout Europe, and participated for a while in the Thirty Years War, before settling in Paris in 1625. In 1628 Descartes moved to Holland, where he stayed for twenty years, although he moved eighteen times during this stay.

Descartes' fame resets as much on his philosophy as his mathematics. His most important works are considered to be *Discourse on the Method* (1637), *The Geometry* (1637), and *Meditations on First Philosophy* (1641). He is perhaps best known for his phrase "cogito ergo sum" ("I think, therefore I am."). Descartes believed in mind-body dualism and posited that the link between the two realms was the pineal gland at the base of the brain. He also subscribed to the vortex theory of gravitation, which has since been disproved.

Descartes was a colorful character. He said that a philosopher should not get out of bed before noon, and he claimed to have conceptualized the coordinate system while lying in bed and looking at a fly on the ceiling. He realized that the fly's position could be specified with three coordinates. Locating a point by coordinates is an elementary observation that others had made before, but Descartes gets the credit (and the name of the coordinate system) because he showed how to use it to solve

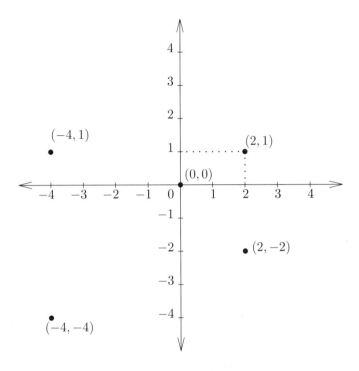

Figure 3.2   The Cartesian plane.

algebraic problems. The coordinate system reunified algebra and geometry after the schism that occurred in ancient Greece with the discovery of irrational numbers. Now, geometry and algebra are recognized as two aspects of the same mathematics. The coordinate system is the foundation for modern calculus and physics.

In 1649 Descartes was made philosopher at the court of Queen Christina in Sweden. Obligated to start work early in the morning, he died the first year of pneumonia.

Distance Formula   From the Pythagorean Theorem, we obtain a formula for the distance between any pair of points in the plane (Figure 3.3). Let $P(x_1, y_1)$ and $Q(x_2, y_2)$ be two points in the plane. Then the line segment PQ is the hypotenuse of a right triangle with side lengths $|x_1 - x_2|$ and $|y_1 - y_2|$.

From the Pythagorean Theorem, the distance between P and Q is

$$|PQ| = \sqrt{(x_1 - x_2)^2 + (y_1 - y_2)^2}.$$

Note that if $x_1 = x_2$ or $y_1 = y_2$, the triangle will have a leg of zero length, but the distance formula still holds.

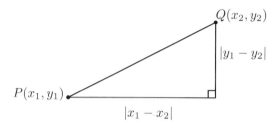

Figure 3.3   Two points determine a right triangle.

EXAMPLE 3.1

The distance between the points $(4, 6)$ and $(10, 10)$ is

$$\sqrt{(4-10)^2 + (6-10)^2} = \sqrt{36 + 16} = \sqrt{52}.$$

The three-dimensional Cartesian coordinate system is defined in an analogous way to the two-dimensional Cartesian coordinate system. Three mutually perpendicular number lines intersect at an origin. These lines are called the x-axis, y-axis, and z-axis. Each point in three-dimensional space is given by three coordinates, (x, y, z).

It is straightforward to extend the formula for the distance between two points in the plane to a formula for the distance between two points in three-dimensional space. Let $P = (x_1, y_1, z_1)$ and $Q = (x_2, y_2, z_2)$ be two points in three-dimensional space. Then the distance between P and Q is

$$|PQ| = \sqrt{(x_1 - x_2)^2 + (y_1 - y_2)^2 + (z_1 - z_2)^2}.$$

This formula can be proved using two applications of the Pythagorean Theorem. We leave it to you to draw a diagram and show the steps.

Pierre de Fermat (1601–1665)

Pierre de Fermat's father was a merchant, and his mother was from a family of lawyers. Pierre became a lawyer, and spent his entire career practicing law in France.

Fermat was a lawyer by vocation, but a mathematician by avocation. He tended not to publish his work, but instead disseminated his ideas through correspondence with other mathematicians. His work on analytic geometry predates Descartes' publications, but wasn't circulated as widely. He described a method for finding maxima and minima of functions and tangent lines to curves, which are basic problems of calculus (see Chapter 4). A number of subsequent French mathematicians gave him credit for inventing calculus, but most modern scholars disagree. Collaborating with Blaise Pascal, Fermat helped lay the foundations of probability theory.

In number theory, Fermat investigated certain Diophantine equations (equations whose solutions must be integers or rational numbers). He proved that there are no nonzero integer solutions to

$$x^4 + y^4 = z^4.$$

Fermat thought he had a proof that there are no nonzero integer solutions to

$$x^n + y^n = z^n,$$

for $n \geq 3$. (If $n = 2$ then the side lengths of Pythagorean triangles provide infinitely many integer solutions.) In 1637 he wrote of his discovery in the margin of his copy of Diophantus' *Arithmetica*, saying the margin was too narrow to contain the proof. However, no proof appears in his correspondence and it is likely that his proof was in error. His conjecture, which became known as Fermat's Last Theorem, generated much mathematical research over the following three centuries, until it was finally proved in 1995 by Andrew Wiles (see Section 5.16).

Pascal's Triangle

As described in Chapter 2, Jia Xian discovered the triangle of binomial coefficients in the 11th century. However, the triangle is named after Blaise Pascal (1623–1662), because he studied it extensively and used it to calculate odds in gambling games. He presented his findings in *Treatise on the Arithmetical Triangle* (published posthumously in 1665).

Recall that Pascal's triangle has a 1 at the top. Each successive row has 1s at the two ends. To find each other entry, add the two numbers above that entry. The triangle can be extended indefinitely.

$$
\begin{array}{c}
1 \\
1 \quad 1 \\
1 \quad 2 \quad 1 \\
1 \quad 3 \quad 3 \quad 1 \\
1 \quad 4 \quad 6 \quad 4 \quad 1 \\
1 \quad 5 \quad 10 \quad 10 \quad 5 \quad 1
\end{array}
$$

Pascal's triangle has many uses. It gives the coefficients when the binomial power $(x + y)^n$ is multiplied out. For example,

$$(x + y)^3 = x^3 + 3x^2y + 3xy^2 + y^3,$$

and we see that the coefficients, 1, 3, 3, 1, are the entries in the third row of Pascal's triangle. (We say "third row" instead of "fourth row" because the top row is considered to be row 0.)

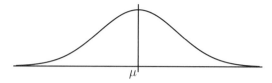

Figure 3.4   The normal distribution from probability theory.

Another use of Pascal's triangle, related to probability, is to find the number of combinations from a given set. For example, let S be the set $\{a, b, c, d, e\}$. The two-element combinations from S (order of elements unimportant) are

$$\{a, b\}, \ \{a, c\}, \ \{a, d\}, \ \{a, e\}, \ \{b, c\}, \ \{b, d\}, \ \{b, e\}, \ \{c, d\}, \ \{c, e\}, \ \{d, e\}.$$

We see that there are 10 two-element combinations from the five-element set S. The number 10 is the second entry of the fifth row of Pascal's triangle (as with row entries, column entries are numbered starting with 0). How many three-element combinations from S are there?

If each entry of a row of Pascal's triangle is divided by the sum of the entries in that row, there results the binomial distribution from probability theory. If the row number is large, this distribution approximates the *normal distribution* well known for its bell-shaped curve (Figure 3.4). Abraham de Moivre (1667–1754) was the first to study this probability distribution.

Blaise Pascal (1623–1662)

Blaise Pascal was born in Claremont Ferrand, France, the son of a judge. His mother died when he was three. His father homeschooled Blaise, and in 1635 introduced him to Marin Mersenne's circle in Paris.

Pascal was a mathematician, physicist, and philosopher. His interests were extremely varied. He invented a mechanical calculator when he was 16. He investigated conic sections, probability, and hydrodynamics, as well as theology. He corresponded with several great thinkers of his day, including Pierre de Fermat (1601–1665) and Christiaan Huygens (1629–1695). Pascal was able to work out some problems in calculus before the subject was fully developed by Isaac Newton (1642–1727) and Gottfried Leibniz (1646–1716).

Pascal had poor health and an irascible personality. He never married and died at the age of 39. This saying is attributed to him: "The more I see of men, the better I like my dog."

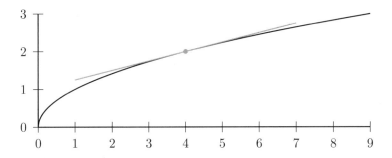

Figure 3.5   Tangent to $y = \sqrt{x}$ at the point $(4, 2)$.

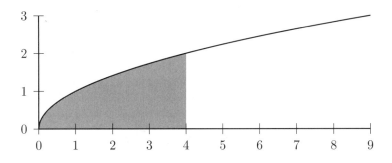

Figure 3.6   Area under $y = \sqrt{x}$ between 0 and 4.

Calculus

Calculus begins with two basic geometric questions. Given a curve and a point, can we find a tangent line to the curve at that point; and given a curve and an interval, can we find the area under the curve?

An example of the tangent line problem is illustrated in Figure 3.5. The curve is $y = \sqrt{x}$, and a tangent line is indicated through the point $(x, y) = (4, 2)$.

An example of the area problem is illustrated in Figure 3.6. Again, the curve is $y = \sqrt{x}$. The area of interest extends from $x = 0$ to $x = 4$.

These geometric questions had been studied for millennia, but they came to the fore in this century, in part because of the invention of analytic geometry. This greatly expanded the number of curves available to the mathematician: every algebraic expression in x and y was associated to a curve in the Cartesian plane.

A number of mathematicians studied the tangent line problem in the first half of the century, most notably Fermat and Descartes. As in the invention of analytic geometry, these two were rivals in developing methods to find tangent lines. Fermat also tied this problem to that of finding maximum and minimum values of a quantity.

The area problem had been studied by Archimedes and Kepler, among others. In the early 17th century, advances were made by Fermat, and by two students of

Galileo, Bonaventura Cavalieri (1598–1647) and Evangelista Torricelli (1608–1647). The latter made a remarkable discovery about what is now known as Torricelli's trumpet, the solid obtained by rotating the hyperbola $xy = k^2$ about the x-axis, and extending it from $x = a$ to $x = \infty$, where a is some positive number. Torricelli showed that this solid has finite volume but infinite surface area.

In the middle of the century, several mathematicians discovered connections between our two geometric problems. It turns out that they are in some sense inverses of one another. Important figures in this discovery were the Dutchman Hendrick van Heuraet (1634–1660?), the Scot James Gregory (1638–1675), and the Englishmen Isaac Barrow (1630–1677). The connection is now called the Fundamental Theorem of Calculus.

Although many of the major pieces of calculus were now in place, it remained to put them all together. Like logarithms, calculus can be said to have two fathers. It was first invented by Isaac Newton around 1665–66 at his country home in Grantham, England, where he retreated from an outbreak of plague that was running through Cambridge. At the same time that he was inventing calculus, he was conducting experiments into the nature of light and discovering the laws of motion that would ultimately become his paradigm-changing law of universal gravitation.

Sir Isaac Newton (1642–1727)

If I have seen further it is by standing on the shoulders of giants.

ISAAC NEWTON

Nature and nature's laws lay hid in Night.
God said, 'Let Newton be!' and all was light.

ALEXANDER POPE (1688–1744)

Isaac Newton was born on January 4, 1643 in Woolsthorpe, England (though by the English calendar in use at the time it was December 25, 1642). His father had died two months before he was born, and Newton spent much of his youth in the care of his grandparents. Although he was destined to become a great mathematician, he had little scholarly training in mathematics until 1661 when at age 18 he enrolled in Cambridge University's Trinity College.

Newton did some important work on infinite series, but he is remembered primarily for three discoveries that he made during the plague years of 1665–67. He discovered the particle nature of light, invented calculus, and discerned his law of universal gravitation.

In 1672 Newton published a paper called "New Theory about Light and Colours" in the *Philosophical Transactions of the Royal Society* in London. His paper described light as particles in which white light is a combination of different kinds of colored light. This controversial discovery contradicted the prevailing wave theory of light, and he experienced criticism over the work, especially by the influential Robert Hooke (of Hooke's law of elasticity).

The reaction to his theory of light left Newton reluctant to incur more criticism, and it wasn't until 1687 that Edmond Halley (for whom Halley's comet is named) finally persuaded Newton to publish his masterwork, *Philosophiæ Naturalis Principia Mathematica* (*Mathematical Principles of Natural Philosophy*). This contained his law of universal gravitation, which according to popular legend is said to have been inspired by an apple falling from a tree. Interestingly, although Newton used his calculus to derive his conclusions, he included only geometric proofs in the arguments of the *Principia*.

Newton himself did not publish his calculus until 1704, when he included it as an appendix to his book *Opticks*. This was nearly 40 years after he developed it, and well after calculus had been independently developed by Leibniz.

In addition to pursuing his research, Newton was twice elected to Parliament and, in 1705, was knighted by Queen Anne. He was appointed warden of the London mint in 1696 and was promoted to master of the mint in 1699, a post he held for decades. He died on March 20, 1726, and was buried, after a grand funeral ceremony, in Westminster Abbey.

$\sim$

Gottfried Wilhelm Leibniz (1646–1716)

Gottfried Wilhelm Leibniz was an attorney, mathematician, and philosopher from Leipzig. His father, a professor of ethics and vice chairman of the philosophy faculty at the University of Leipzig, died when he was six years old, leaving him an extensive library. By age eight, Leibniz was given access to his late father's library, and thus began a life of independent learning.

When he failed to achieve a doctorate in October of 1666 from the University of Leipzig, he transferred to the University of Altdorf and received a doctorate in February of 1667 for a dissertation on difficult legal cases. He worked as an attorney for the remainder of his life because he felt he could do more good for people as a lawyer than as a professor.

Even as an attorney, Leibniz continued to study philosophy, science, and mathematics. He published "Discourse on Metaphysics" in 1686, which defined his philosophy. His book *Theodicy* came out in 1710. In it he considered questions of church doctrines that required resolution for reunification of the church (something he worked toward much of his lifetime).

As a mathematician, Leibniz was mostly self-taught. He is most famous for his calculus, which he created around 1675 (about ten years after Newton). Although Newton gets credit for creating calculus first, Leibniz was responsible for *disseminating* calculus throughout Europe, beginning with his 1684 publication of *Nova Methodus Pro Maximis et Minimis* (*New Method for Maxima and Minima*). Moreover, Leibniz created a lucid notation for calculus, using symbols that are still recognized and popular today, such as dx for infinitesimal changes and the elongated S symbol $\int$ for integration.

Leibniz was a polymath. He could write in Latin, French, German, and other languages, though he spoke only poor English. He was a copious letter writer, and authored as many as 300 letters per year. Among his notable inventions were calculating machines that could add, subtract, multiply, and divide. One could multiply a 10-digit number by a 4-digit number with only four turns of a crank. A later model could operate on numbers up to 12 digits. He was an accomplished geologist who held that the earth had originally been molten, then covered with oceans.

Leibniz died November 14, 1716, and was buried inside the Neustäder church (something that was rare for a person without a title). Though at the time of his death his reputation had been tarnished by a dispute with Newton over the invention of calculus, by the end of the 18th century his accomplishments were again celebrated. In particular, he is recognized today as one of calculus' true and independent creators.

If calculus were merely about solving two isolated geometrical questions, it might be important but it probably would not have had the profound impact that it has had. Tangent lines and areas, however, can be used in a great variety of interesting problems. Leibniz' 1684 paper was about maximizing and minimizing quantities. Newton used the principles of calculus to answer physical questions about force and acceleration. It has since been used to derive statistical methods, principles of economics, and rules of engineering. It is hard to imagine an endeavor of mathematics or science that has not been enhanced somehow by the powerful tools of calculus.

Calculus will be taken up in detail in Chapter 4.

EXERCISES

Use this log table for the first five exercises.

x	2	3	1.211	4.116	5.122
$\log x$	0.30103	0.47712	0.08314	0.61448	0.70944

x	5.227	6.114	6.203	8.895
$\log x$	0.71825	0.78633	0.79260	0.94915

3.1   Use the log table to approximate $\log 6$. (Hint: $6 = 2 \cdot 3$.)

3.2   Use the log table to approximate $\log(2/3)$.

3.3   Use the log table to approximate $\log(2 \cdot 3^5)$.

3.4   Use the log table to approximate 1.211×5.122.

3.5   How many digits does 2^{1000} have?

3.6   Plot the points $(2, 3)$ and $(5, 7)$. Find the distance between the points.

3.7   Find five points at distance 3 from $(0, 4)$.

3.8   What is the slope of the line through $(0, -2)$ and $(-3, -4)$?

3.9   What is the slope of the line passing through the origin and the point $(-5, 4)$?

3.10   Find an equation of the circle with center $(3, 2)$ and radius 5.

3.11   Find the distance between $(1, 2, 3)$ and $(4, 5, 6)$.

3.12   Give an equation for the sphere of radius 5 with center $(0, 2, -3)$.

3.13   A cube has vertices $(0, 0, 0)$, $(0, 0, 1)$, $(0, 1, 0)$, $(0, 1, 1)$, $(1, 0, 0)$, $(1, 0, 1)$, $(1, 1, 0)$, and $(1, 1, 1)$. Let $P = (3/4, 0, 0)$, $Q = (0, 3/4, 0)$, $R = (1, 1/4, 1)$, and $S = (1/4, 1, 1)$. Show that $PQRS$ is a square and find its side length. A cube of this side length is called *Prince Rupert's cube*. It is the largest cube that can pass through a cube of side length 1 (the cube that we started with).

3.14   Use Pascal's triangle to find the expansion of $(x + y)^5$.

3.15   Use Pascal's triangle to find the number of three-element combinations from a set of six elements. List the combinations.

3.16   Divide each entry of the eighth row of Pascal's triangle by the sum of the entries in that row. Plot the nine values on a graph and connect the points by a curve. Do you recognize this curve?

3.17   The tangent line in Figure 3.5 has slope $m = 1/4$ and goes through the point $(4, 2)$.

    a) Use the point-slope form of a line,

$$y - y_0 = m(x - x_0),$$

    to write the equation of the line.

    b) Write the equation of the tangent line in the form $y = mx + b$.

3.18   For the curve in Figure 3.6, the area under the curve and above an interval $[0, x]$ is given by the formula $A = \dfrac{2x\sqrt{x}}{3}$.

    a) Find the area under the curve and above the interval $[0, 1]$.

    b) Find the area under the curve and above the interval $[0, 5]$.

    c) Find the area under the curve and above the interval $[1, 5]$ by subtracting your answer for (a) from your answer for (b).

3.19   Newton's universal law of gravitation says that the gravitational force between two point-masses m_1 and m_2 is given by

$$F = G\frac{m_1 m_2}{r^2},$$

where G is a constant and r is the distance between the point-masses. Explain how this law shows that the acceleration due to gravity of an object near the Earth's surface may be taken to be a constant (which we call g). (Hint: use Newton's law that the force on an object equals its mass times its acceleration.)

3.2   The 18th Century: Consolidation

Introduction

The 18th century saw the continuation of the Scientific Revolution of the previous century. Newtonian physics was triumphant, and scientists worked to apply the new scientific method to other fields. Equally important was the birth of the Industrial Revolution, which, together with the new science, was eventually to transform everyday life in ways more profound than anything since the advent of agriculture.

France was the dominant political and cultural center of 18th century Europe. It was the richest and second most populous (after Russia) country in Europe. Early in the century, the French had expanded their colonial holdings overseas, but lost most of them in the Seven Years' War (1756–63). This war, which pitted France, Austria, and Russia against Britain and Prussia, took place on several continents and killed about a million people. In India, the war was called the Third Carnatic War, and ended with the English East India Company establishing dominance over the other European traders. (The height of English colonialism in India was not until the 19th century.) In the American theater, the war is called the French and Indian War; it established English control of Canada.

In 1789 the French Revolution began, the central political conflict in the modernization of Europe, specifically the reduction in power of the royalty and the Church. On July 14 the revolutionaries attacked the prison and armory called the Bastille. A decade of political unrest followed. At the beginning, the king remained in place, but was forced to cede some power to the National Assembly. The members of the Assembly sat in two sections, the supporters of the king and church on the right, and the supporters of the revolution on the left. This was the origin of the custom of dividing political factions into the left and the right.

Unrest continued, eventually leading to the execution of the king in 1793. This was followed by the Reign of Terror (1793–94), during which many thousands were executed by guillotine. France was also engaged in wars with its neighbors, whose kings were not too happy with the ideas of the revolution. After some initial defeats, the French army was victorious. One of its generals, Napoleon Bonaparte, took control of France in 1799.

In Eastern Europe, this century saw the rise of Prussia and Russia, at the expense of Poland. The Kingdom of Prussia was formed in 1701 and transformed into a world power especially by Frederick the Great (r. 1740–86). Frederick was a great patron of science and the arts; he himself was a flutist and composer. He also was interested in philosophy and an admirer of the Enlightenment (see below).

Russia greatly expanded its empire under Peter the Great, who ruled 1682–1725 (jointly with his half-brother until 1696, when he was 24 years old). In addition to his conquests, he modernized Russia, adapting much from Western Europe, and founded St. Petersburg, modeled after the canal city of Venice. This city hosted the greatest of 18th century mathematicians, Leonhard Euler (1707–1783).

The other Eastern European power was the Hapsburg empire based in Austria. Although the Hapsburgs had their political difficulties in this century, their capital,

Vienna, was a major cultural center, notable especially for its musical luminaries, including Haydn, Mozart, and Beethoven.

The Industrial Revolution began in England. It was preceded and made possible by an "Agricultural Revolution," which increased agricultural efficiency and made available the labor to work in the new industrial factories. There were many aspects to the new agriculture, which were added gradually over the centuries. These included new crops from America, improved crop rotation, better transportation, increased availability of financing, and new mechanical inventions. In 1701 Jethro Tull introduced his horse-drawn seed drill, which placed seeds more efficiently in rows, at the right depth, and covered them with soil. Several improvements were made to the plow during this century, and there was increased use of iron plows, which replaced wooden plows. In 1784 Andrew Meikle invented the threshing machine, used to separate the grain from the rest of the plant.

The Industrial Revolution was powered by coal, of which England had an abundance. In 1709 Abraham Darby I built a blast furnace at Coalbrookdale that used coke, a fuel derived from coal, to produce pig iron. The coke replaced charcoal made from wood. This, and other technical innovations throughout the century, made iron increasingly available for the tools of the Industrial Revolution.

A central technology of the early Industrial Revolution was the steam engine, which used a coal fire to heat water into steam, which in turn drove a piston to do the work. The first commercially successful steam engine was introduced around 1712 by Thomas Newcomen. It was used primarily in coal mines. As the shallower coal seams were exhausted, miners needed to dig deeper to mine the coal. These mines often flooded; the new steam engines were used to pump out the water. In 1769 James Watt invented an improved steam engine. By 1800 his firm had produced 496 engines, mostly powering machinery in mills.

The textile industry underwent major developments in this century, turning it from a cottage industry into a factory industry. The spinning jenny, introduced in 1764 by James Hargreaves, made the process of producing yarn from wool more efficient. Richard Arkwright built the first cotton mill, a factory combining spinning and weaving machines. In 1794 the American Eli Whitney patented the cotton gin, which separated the seed from the cotton fiber.

The 18th century is also called the Age of Enlightenment. Enlightenment was a philosophy that emphasized the rational over the supernatural. It promoted science and opposed superstition. The Enlightenment had its origins in the 17th century, in the works of the philosophers Baruch Spinoza (1632–1677) and John Locke (1632–1704), as well as the new science. Its apex, however, was in the 18th century, throughout Europe and North America, but centered in France. Among its stars were the writer Voltaire (1694–1778), the philosopher Jean-Jacques Rousseau (1712–1778), and Denis Diderot (1713–1784), editor of the influential *Encyclopédie, ou dictionnaire raisonné des sciences, des arts et des métiers* (*Encyclopaedia or a Systematic Dictionary of the Sciences, Arts and Crafts*). This was a 35-volume work, first published between 1751 and 1772, which attempted to include all of the world's knowledge. Its many contributors included all the leading lights of the French Enlightenment.

The Enlightenment had a political aspect as well, one which rejected the divine right of kings and emphasized the consent of the governed, and the ideals of freedom and equality. These ideals were expressed in the American Declaration of Independence (1776) and the French Declaration of the Rights of Man and the Citizen (1789). They inspired the world's first modern liberal democracy, the United States of America.

The 18th century witnessed a remarkable collection of intellects in Scotland, beginning with the philosopher Francis Hutcheson (1694–1746). This group included the philosopher David Hume (1711–1776), the poet Robert Burns (1759–1796), the geologist James Hutton (1726–1797), the chemist Joseph Black (1728–1794), and the philosopher Adam Smith (1723–1790), whose 1776 work *An Inquiry into the Nature and Causes of the Wealth of Nations* is considered the beginning of modern economics. Also native to Scotland were the inventors James Watt (1736–1819) and Andrew Meikle (1719–1811).

In science, the 18th century saw the birth of modern chemistry with the work of Joseph Black (1728–1799), Joseph Priestley (1733–1804), and especially Antoine Lavoisier (1743–94) (who was executed in the Reign of Terror); they isolated oxygen, nitrogen, hydrogen, and carbon dioxide, and discovered the law of conservation of mass.

Modern geology was also born, inspired in part by mining. At the beginning of the century, the most popular theory traced geological origins to the biblical flood. During the century, an effort was made to establish a strictly scientific basis for geology. The year 1795 saw the publication of James Hutton's *An Investigation of the Principles of Knowledge and of the Progress of Reason, from Sense to Science and Philosophy*, whose 2138 pages developed theories of the origin of rocks, including sedimentary rocks formed in the sea and later raised up, concluding that the Earth was in fact much older than the biblical few thousand years, and still geologically active.

Research in electricity accelerated in the 18th century, helped considerably by the invention of the Leyden jar, a device for storing charges. The connection of electricity to lightning was discovered by Benjamin Franklin (1706–1790), and to frog muscles by Luigi Galvani (1737–1798). Charles-Augustin de Coulomb (1736–1806) discovered his famous law, a formula for the force between two charges at a given distance. The century ended with the invention of the first electric battery by Alessandro Volta (1745–1827).

In mathematics, much work was done to consolidate the advances of the previous century. This included reconciling the different versions of calculus presented by Newton and Leibniz. Scientists also continued refining and extending the Newtonian theory of mechanics, which describes how bodies move when acted upon by forces. This theory was as much mathematics as physics. The typical application used Newton's laws to arrive at a type of equation, called a *differential equation*. Then the hard work began, studying the differential equation to extract useful formulas giving the motion of the bodies. Such studies were carried out on planets and moons, on violin strings and drums.

In addition to differential equations, important advances were made in this century in number theory, especially by Euler. He and others also studied the complex numbers earlier investigated by Bombelli. In fact, complex numbers became one of the most important topics in mathematics right through the 19th century.

Finally, we note that this century was characterized by a major split between British and Continental mathematicians, which began with a priority dispute between Newton and Leibniz over the invention of the calculus. As a result, even though science was well represented in the British Isles, most of this century's mathematics was developed on the Continent.

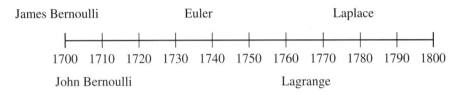

Eighteenth Century mathematicians.

The Budding of 18th Century Mathematics

The invention of Calculus in the previous century by Isaac Newton and Gottfried Leibniz ushered in a new way of approaching both mathematics and physics. Eminent in this endeavor were the illustrious Bernoulli family.

James (1654–1705) and John (1667–1748) Bernoulli

Probably no other single family has produced so many capable mathematicians as the Bernoulli family. As many as eight different members of this Swiss family were mathematicians of note, beginning with the brothers James (Jacob) and John (Johann). Many of them were widely known, as this story illustrates.

> Once, while traveling with a learned stranger who asked his name, he said, "I am Daniel Bernoulli." The stranger could not believe that his companion actually was that great celebrity, and replied [in jest], "I am Isaac Newton."[1]

Though their father wished James to study theology and John to study medicine, both eventually pursued mathematics, particularly the new calculus of Leibniz. In Basel, Switzerland, James was professor of mathematics from 1687 until his death in 1705. He was succeeded in that post by his younger brother John, who, until that time, had taught at Groningen, Holland.

---

[1] Florian Cajori, *A History of Mathematics*, fifth edition, Chelsea, New York, 1991.

Figure 3.7 A logarithmic spiral.

James Bernoulli was responsible for several notable mathematical achievements. He was perhaps the first mathematician to actively use polar coordinates to solve problems. He studied many special curves including the catenary (hanging chain) and the lemniscate (figure eight), but he was particularly interested in the logarithmic spiral, which has the polar equation $r(\theta) = ae^{b\theta}$ (see Figure 3.7). He even asked that the spiral be engraved on his tombstone, cleverly referring to the way the curve looks like a copy of itself when magnified via the epitaph, "I shall arise the same, though changed."[2]

James also had an interest in series. He proved that the harmonic series diverges (i.e. it has an infinite sum).

$$1 + \frac{1}{2} + \frac{1}{3} + \frac{1}{4} + \cdots = \infty$$

He also determined that the reciprocals of the squares have a finite sum, though it was Euler (a student of his brother John) who finally discovered the actual value to be $\pi^2/6$.

$$1 + \frac{1}{2^2} + \frac{1}{3^2} + \frac{1}{4^2} + \cdots < \infty$$

One of James Bernoulli's most famous results is the "Law of Large Numbers," which is sometimes referred to simply as Bernoulli's Theorem. It states:

If the probability of an event is p, and if n independent trials are made, with k successes, then $\dfrac{k}{n} \to p$ as $n \to \infty$.

An easy way to think about Bernoulli's Theorem is to consider a fair coin. Call flipping heads a "success." Although the probability of heads is 50%, if you flip a coin 10 times, you may get any number of heads (successes) and tails (failures), even 10 heads in a row. However, the theorem guarantees that no matter what has happened with any finite number of flips, as you flip more and more, the proportion of heads will always tend (eventually) toward 50%.

Together, James and John made many contributions to the early calculus, sometimes in collaboration and often in competition. The brothers were regular correspondents with Leibniz, and together the three men discovered what would today constitute much of a college course in calculus.

---

[2]David Eugene Smith, *History of Mathematics, Volume I*, Dover, New York, 1951.

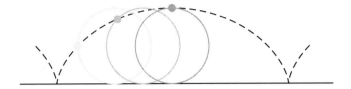

Figure 3.8   A cycloid.

In 1696 John Bernoulli proposed the brachistochrone problem as a calculus challenge to Newton. It was addressed, "to the shrewdest mathematicians in the world." The problem can be stated,

> Let two nails in a wall, one not directly above the other, be connected by a wire. Find the shape of the wire that allows a (frictionless) bead to fall from the higher point to the lower in the least amount of time.

John knew, via calculus, the answer to be a cycloid (actually, an upside-down cycloid), which is a curve traced by a fixed point on a rolling circle, as in Figure 3.8. The challenge was correctly solved by four other mathematicians: his brother James, Leibniz, Newton, and l'Hôpital.

John Bernoulli was also indirectly responsible for the content of the world's first calculus text, because as tutor to Guillaume de l'Hôpital (1661–1704), he contributed many of the ideas that were incorporated into l'Hôpital's book, *Analyse des infiniment petits*, published in 1696.

Complex Numbers

A recurring theme in the history of mathematics is the expansion of what we consider to be a number. We have seen this theme played out when the Pythagoreans discovered the existence of irrational numbers. The theme occurred again when zero was invented in China and India. As we saw at the end of Chapter 2, acceptance of square roots of negative numbers came about through the solution of cubic and quartic equations. Mathematicians defined a new number, i, such that $i^2 = -1$. Certainly, i is not a real number, since the square of a real number is always nonnegative. It is called an *imaginary number*. A *complex number* is a number of the form $a + bi$, where a and b are real numbers. Complex numbers, at first regarded with skepticism, were fully incorporated into mathematics once Caspar Wessel (1745–1818) and Jean-Robert Argand (1768–1822) showed that the new numbers are not mysterious at all. They can be plotted on a Cartesian coordinate plane, with the horizontal axis giving the real part of the number and the vertical axis the imaginary part. See Figure 3.9.

As if to close the deal, Carl Friedrich Gauss (1777–1855) proved that every polynomial of degree n has n complex roots (the Fundamental Theorem of Algebra). For

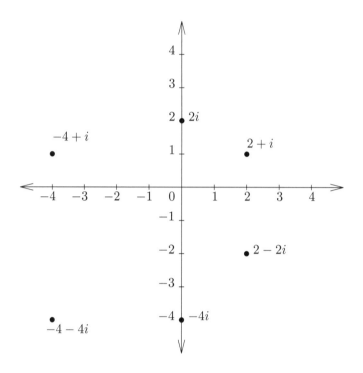

Figure 3.9   The complex plane.

example, the quartic polynomial

$$x^4 - 7x^3 + 18x^2 - 22x + 12$$

has four roots, two real roots, 2 and 3, and two complex roots, $1 + i$ and $1 - i$. Thus the quartic factors as

$$(x - 2)(x - 3)(x - (1 + i))(x - (1 - i)).$$

Complex numbers have elegant geometric properties. The addition of complex numbers is quite simple. To add two complex numbers, add their real parts and add their imaginary parts. This is a vector addition. Multiplication is more interesting. Take, for example, the case of multiplication by i. The product of 1 by i is i; the product of i by i is -1. This means that the unit vectors on the real and imaginary axes are rotated 90° counterclockwise. Indeed, multiplication by i rotates the entire plane 90° counterclockwise. Multiplication by a real number, on the other hand, stretches each complex number by that number. Multiplication by an arbitrary complex number stretches and rotates each complex number. Thus, we may think of multiplication as stretching and twisting the complex plane.

Complex numbers are instrumental in many branches of mathematics, as well as in applied fields such as electrical engineering. The imaginary has proved to be eminently practical.

Euler

> Read Euler, read Euler, he is the master of us all.
>
> PIERRE-SIMON DE LAPLACE (1749–1827)

Leonhard Euler (1707–1783) was the most prolific mathematician in history, and one of the best. He contributed to many areas of mathematics, and, although he never had a teaching job, his texts have been immensely influential in mathematics education.

Euler helped change the emphasis in calculus from studying curves to studying functions. He originated the now ubiquitous notation $f(x)$. He was the first to treat sine and cosine as functions, instead of just lengths. For example, we now can write $x(t) = \sin t$, where t is a number, not necessarily an angle. This notion is quite important in physics, where, for example, x may measure the position of an object at time t.

Leonhard Euler (1707–1783)

Leonhard Euler was born in Basel, Switzerland, the son of a Protestant minister. He was a child prodigy, graduating from the University of Basel at age fifteen. Although his father originally intended for him to become a minister, Leonhard showed an early interest in mathematics. This was nurtured by contact with a family friend, Johann Bernoulli. In 1726 Euler joined Nicolaus and Daniel Bernoulli, sons of Johann, at the St. Petersburg Academy of Sciences. He worked in Russia until 1741, when he joined the Berlin Academy of Sciences. In 1766 he returned to St. Petersburg, where he spent the rest of his life.

In addition to his mathematics, Euler produced important work in physics, engineering, and astronomy. He helped create improved lunar tables, which were used to solve the difficult problem of determining longitude at sea. He wrote a successful work of popular science. After he returned to Russia, he became almost completely blind, but was still able to perform involved calculations in his head, and continued his prodigious output of research until his death.

Euler spent much time studying logarithms and exponential functions, which were quite important in solving a variety of differential equations. (Exponential functions are of the form $f(x) = a^x$, where a is some constant, called the base.) Euler discovered a remarkable connection between these functions and trigonometric functions.

To state Euler's discovery, we need to introduce one of the most important mathematical constants: e. We consider a simple problem in compound interest. Suppose that we invest $1 at 100% interest. After 1 year, we might have $2, the original $1 plus $1 in interest. But suppose that our interest was figured twice a year. After six months, we would have earned fifty cents interest, so would have $1.50. In the next six months, we would earn 50% more, but on the full $1.50; in other words, we would earn another $.75, for a total of $2.25.

Suppose now that we figure ("compound") interest three times. After four months, we have $1 + 1/3$ dollars. After eight months, we have

$$(1 + 1/3) + (1 + 1/3)(1/3) = (1 + 1/3)^2$$

dollars. After one year, we have

$$(1 + 1/3)^2 + (1 + 1/3)^2(1/3) = (1 + 1/3)^3$$

dollars, approximately $2.37.

It is clear that the more times we compound the interest, the more money we have at the end, because we are earning interest on more interest. Does this mean that we can earn as much as we like, by compounding more often? Unfortunately, no; we get at most about $2.72. As we compound more and more often, the amount of money at the end gets closer and closer to our number e, which is approximately 2.718281828. Put another way, the limit of $(1 + 1/n)^n$, as n goes to infinity, is e. In passing, we note another nice formula.

$$e = 1 + \frac{1}{1} + \frac{1}{1 \cdot 2} + \frac{1}{1 \cdot 2 \cdot 3} + \frac{1}{1 \cdot 2 \cdot 3 \cdot 4} + \cdots$$

Euler was able to define an exponential function with e as the base. He could even deal with powers of e that are complex numbers. The connection to trigonometric functions that he found was this:

$$e^{ix} = \cos x + i \sin x,$$

where x is given in radians. For example, if $x = \pi$, then $\sin x = 0$ and $\cos x = -1$, so $e^{\pi i} = -1$, or

$$e^{\pi i} + 1 = 0.$$

This is called *Euler's Identity*. Remarkably, it ties together the five most important numbers in mathematics: 0, 1, π, e, and i.

Euler also solved a famous problem posed in 1644, the *Basel Problem*. The problem was to compute the sum

$$\frac{1}{1^2} + \frac{1}{2^2} + \frac{1}{3^2} + \cdots .$$

He showed that the sum is $\pi^2/6$. In fact, his method could be used to find the sum

$$\frac{1}{1^n} + \frac{1}{2^n} + \frac{1}{3^n} + \cdots$$

for any positive even integer n. He also tried to compute

$$\frac{1}{1^3} + \frac{1}{2^3} + \frac{1}{3^3} + \cdots$$

but was unsuccessful. Nobody knows what this sum is even today, although it has been proved to be an irrational number.

Finally, we mention an important formula in geometry discovered by Euler. This concerns polyhedra, solids with flat faces and straight edges. An example is a cube. Let us denote by V the number of vertices (corners) of a polyhedron, by E the number of edges, and by F the number of faces. So in the case of a cube, $V = 8$, $E = 12$, and $F = 6$. (Check!) What Euler discovered is the formula $V - E + F = 2$ for polyhedra with no holes. Although he published a proof of this, it was incomplete. The first complete proof was given by Adrien-Marie Legendre (1752–1833) in 1794.

This formula has a nice application to the classification of Platonic solids. Recall that the five Platonic solids (Figure 3.10) were constructed in Euclid's *Elements*. These polyhedra have the following two properties:

Each vertex is the endpoint of the same number of edges, which we denote by m.

Each face is bounded by the same number of edges, which we denote by n.

For a cube, $m = 3$ and $n = 4$. Figure 3.11 shows V, E, F, m, and n for the Platonic solids.

| Cube | Tetrahedron | Octahedron | Dodecahedron | Icosahedron |

Figure 3.10   The Platonic solids.

Solid	V	E	F	m	n
Tetrahedron	4	6	4	3	3
Cube	8	12	6	3	4
Octahedron	6	12	8	4	3
Dodecahedron	20	30	12	3	5
Icosahedron	12	30	20	5	3

Figure 3.11   Parameters for the Platonic solids.

We will use the two conditions above, and $V - E + F = 2$, to show that there can only be these five Platonic solids. First, note that each edge has two endpoints, so if we add up all edge-vertex pairs, we get $2E$. Since each vertex is the endpoint of m edges, that same sum is mV. In other words, $2E = mV$.

Similarly, each edge bounds two faces, so if we add up all edge-face pairs, we get $2E$. Since each face is bounded by n edges, that same sum is nF. In other words, $2E = nF$.

Let us use the last two results to substitute for F and V in $V - E + F = 2$.

$$\frac{2E}{m} - E + \frac{2E}{n} = 2$$

Divide by $2E$.

$$\frac{1}{m} - \frac{1}{2} + \frac{1}{n} = \frac{1}{E}$$

Add $1/2$ to both sides.

$$\frac{1}{m} + \frac{1}{n} = \frac{1}{E} + \frac{1}{2}$$

Since each vertex is the endpoint of at least three edges, $m \geq 3$. Since each face is bounded by at least three edges, $n \geq 3$. Suppose that m and n were both greater than 3. Then

$$\frac{1}{E} + \frac{1}{2} = \frac{1}{m} + \frac{1}{n} \leq \frac{1}{4} + \frac{1}{4} = \frac{1}{2}.$$

Since $1/E > 0$, this clearly is impossible. So we must have $m = 3$ or $n = 3$.

If $m = 3$, we must have $n < 6$, since $1/n + 1/3 > 1/2$. The three cases $n = 3, 4, 5$ yield the tetrahedron, cube, and dodecahedron, respectively.

If $n = 3$, we must have $m < 6$. The three cases $n = 3, 4, 5$ yield the tetrahedron, octahedron, and icosahedron, respectively. Therefore, these are the only possible Platonic solids.

Euler's influence on mathematics was huge. One last example: Euler introduced the notation π for the ratio of the circumference of the circle to its diameter.

Number Theory

Another area of mathematics that Euler advanced was number theory, the branch of mathematics that studies properties of the positive integers. Among other contributions, he provided proofs for a number of theorems that Fermat stated without proof. He also proved Fermat's Last Theorem for the case $n = 3$, i.e., that there is no nontrivial integer solution to $x^3 + y^3 = z^3$.

The other mathematician of this century who was in Euler's class was Joseph-Louis Lagrange. Like Euler, Lagrange contributed to many areas of mathematics. In number theory, he proved a number of important results, including several stated without proof by Fermat and others.

Joseph-Louis Lagrange (1736–1813)

Joseph-Louis Lagrange was born Giuseppe Luigi Lagrangia, in Turin, Italy. His family was French on his father's side, and he later adopted the French version of his name. Like Euler, Lagrange was a prodigy. He became a lecturer in mathematics at the Royal Artillery School of Turin at age nineteen.

In 1766, when Euler left Frederick the Great's Berlin Academy to return to St. Petersburg, Lagrange replaced him. He stayed in the Prussian capital until Frederick's death in 1787. He then accepted an invitation to Paris, where he was given apartments in the Louvre. It was in Paris that he finished his great work, *Mécanique Analytique*.

In *Mécanique Analytique* (*Analytic Mechanics*), published in 1788, Lagrange reformulated Newtonian mechanics. The new formulation made use of an extension of calculus called the calculus of variations, which Euler and Lagrange had developed. Although this was a work of mathematical physics, Lagrange himself viewed mechanics as a branch of pure mathematics. The *Mécanique Analytique* was the basis of all subsequent work in the area. It was so beautifully done that William Rowan Hamilton (1805–1865) called it a "kind of scientific poem."

Besides his work in mathematics and mechanics, Lagrange contributed to astronomy, and to education. Since he was nonpolitical, he survived the crazy years following the French Revolution, unlike his chemist friend Lavoisier, who was guillotined. Lagrange helped develop the metric system and reform science education in France.

One theorem of number theory proved by Lagrange was Wilson's theorem, discovered around 1770 by John Wilson (1741–1793), an English mathematics professor. Wilson did not realize that the theorem was known to Ibn al-Haytham around 1000, and even earlier to Bhaskara I in the 7th century. Such is the vagary of naming in mathematics. Wilson gets the credit, even though he wasn't the original discoverer, nor did he prove the theorem. It was Lagrange who proved it.

Wilson's theorem is a criterion by which we can tell whether an integer is a prime number. An integer n, greater than 1, is prime if and only if $(n-1)! + 1$ is a multiple of n. (Wilson only stated the "if" part of the theorem.) For example, 7 is prime, and $(7-1)! + 1 = 721$, a multiple of 7. On the other hand, 6 is not prime, and $(6-1)! + 1 = 121$, not a multiple of 6. The only shortcoming of Wilson's theorem is that the computation of $(n-1)!$ takes too long for the criterion to be useful. However, early in the 21st century, an effective procedure was found for determining when a positive integer is prime.

Also around 1770 Lagrange proved that every positive integer is a sum of four squares. For instance, $97 = 8^2 + 5^2 + 2^2 + 2^2$. Here is a challenge: How many ways can you write 10 as a sum of four squares, where the order of the terms matters, and the numbers being squared can be positive or negative? For example, $3^2 + 1^2 + 0^2 + 0^2$, $3^2 + 0^2 + 0^2 + 1^2$ and $(-3)^2 + 1^2 + 0^2 + 0^2$ are all considered different. See the next section for the answer.

One of the most important theorems of number theory was conjectured in 1796 by Adrien-Marie Legendre (1752–1833). It is called the Prime Number Theorem, and has to do with the distribution of prime numbers. It says that the number of prime numbers less than a given number n is approximately $n/\ln n$, where $\ln n$ is the natural logarithm of n. For example, the number of primes less than 10^6 (one million) is approximately $72,382$, whereas the actual number is $78,498$ (an error of about 8%). The relative error decreases as n increases. As we will report in Section 3.3, the Prime Number Theorem was proved about 100 years after it was conjectured. A new machinery was needed for the proof: complex variables.

Paris

Paris was the premier center of mathematics in this century. Among its illustrious mathematicians, in addition to Lagrange, were Alexis Clairaut (1713–1765), Jean Le Rond d'Alembert (1717–1783), Gaspard Monge (1746–1818), Pierre-Simon de Laplace (1749–1827), and Adrien-Marie Legendre (1752–1833).

Pierre-Simon de Laplace (1749–1827)

Pierre-Simon de Laplace was born at Beaumont-en-Auge, in Normandy, France, the son of a farmer. He studied at the University of Caen, then went to Paris in 1769, with a letter of recommendation to the mathematician Jean d'Alembert. With the help of d'Alembert, he obtained a professorship at the École Militaire in Paris.

Laplace's greatest accomplishments were in mathematical physics. In particular, he worked on the problem of the stability of the solar system. Newton had determined the larger motions of the solar system, due to the gravitational attraction of the Sun on the planets, but understood that the planets interacted among themselves. He despaired of solving this more general problem exactly, and in fact opined that divine intervention was necessary from time to time to maintain the system's stability. Laplace, in a series of papers, showed that the solar system was on average stable. (This was not the final word, as Poincaré demonstrated one hundred years later that the solar system is in fact chaotic.)

In the process of rewriting celestial mechanics, Laplace developed a number of new mathematical techniques. His influence is memorialized in terms such as the Laplace transform and Laplace's equation, which every undergraduate mathematics and physics major encounters. In addition, Laplace did groundbreaking work in probability and statistics. He also took time in 1780, in joint research with the chemist Antoine-Laurent Lavoisier, to demonstrate that respiration is a form of combustion.

The leading mathematicians of the day did not work at universities. The modern research university was a 19th century invention. The best education in science and mathematics in 18th century France was provided at military schools, which trained engineers. The best research was done at academies, especially at the Paris Academy of Sciences.

Many of the military schools and universities were closed during the French Revolution. The need for education did not disappear, however. In 1794 the École Centrale des Travaux Publiques was founded. Shortly after, it was renamed the École Polytechnique. Monge was heavily involved in developing the curriculum, and most of the leading French mathematicians of the time taught there, including Lagrange and Laplace. The École Polytechnique became a model for engineering schools worldwide.

The 1790s also saw the invention of the metric system. Lagrange was the president of the committee that established the metric system. Laplace and Monge also served on this committee.

EXERCISES

3.20 Find the sum and product of the complex numbers $2 + 3i$ and $3 + 2i$. Find the quotient $(2 + 3i)/(3 + 2i)$ in the form $a + bi$.

3.21 Explain what happens geometrically when the complex number $3 + i$ is multiplied by $-i$. Do the same for multiplication by $2i$.

3.22 Find all four roots of the quartic polynomial $x^4 - x^3 + x^2 - x$.

3.23 Can you give an example of a quartic polynomial with no real roots?

3.24 Give a quintic polynomial with roots 0, 1, -1, i, and $-i$.

3.25 What is $e^{\pi i/2}$?

3.26 What is $e^{4\pi i}$?

3.27 A soccer ball is a polyhedron. Find the values of V, E, and F for it.

3.28 Give an example of a polyhedron, other than the dodecahedron, such that $V = 20$, $E = 30$, and $F = 12$.

3.29 Give an example of a polyhedron, other than the icosahedron, such that $V = 12$, $E = 30$, and $F = 20$.

3.30 Use Euler's solution of the Basel Problem to find the sum of the infinite series

$$\frac{1}{1^2} - \frac{1}{2^2} + \frac{1}{3^2} - \frac{1}{4^2} + \frac{1}{5^2} - \frac{1}{6^2} + \cdots.$$

3.31 Verify Wilson's theorem for $n = 5$, 8, 9, 10, and 11.

3.32 How many ways can you write 4 as a sum of four squares, where the order of the terms matters, and the numbers being squared can be positive or negative? Can you answer the same problem for the number 10?

3.3   The 19th Century: Expansion

Introduction

The most important development of the 19th century was the continuation, indeed the acceleration, of the Industrial Revolution. Much of what we are today, we became in the 19th century.

In Europe, the early years of the century were dominated by the Napoleonic wars, which ended in 1815 with the defeat of France by a coalition of European powers. Great Britain emerged from these wars as the dominant power, a position it occupied through most of the century. This dominance was due to it being home to the Industrial Revolution. One estimate has the United Kingdom possessing about 20% of the *world's* manufacturing production in 1860. (It had about 2% of the world's population.) In 1900 London was the largest city in the world, with about 6.5 million people.

This was a century of empire. The biggest was of course the English empire, especially with its holdings in South Asia, crowned by India. But it also ruled over Canada, Australia, New Zealand and, later in the century, large parts of Africa. The other great European empire was the Russian, which ruled from Poland in the West to Alaska in the East. (The Russian Alaskan holdings were sold to the United States in 1867. There were also early Russian settlements in Hawaii and California.)

In Asia, this was a bad century for China. Its weakness was demonstrated by its defeat in the Opium Wars (1839–42 and 1856–60). China had tried to cut off the illegal importation of opium, controlled by the British, but were defeated and required to make major trade concessions. Partly as a consequence, China was convulsed by the Taiping Rebellion (1850–64), in which about 20 million people died.

Following the forcible opening of Japan to Western trade by the American Commodore Matthew Perry in 1853, Japan rapidly adopted Western technology and industry. By century's end, it had become a major imperial power, defeating China in a war in 1895 and Russia ten years later.

Not all empires prospered at this time. The Spanish and Portuguese empires in America collapsed in the early years of the century. Between 1810 and 1822, independence from Spain and Portugal was obtained by nearly the entire South American continent, as well as Mexico.

The second half of the 19th century saw the rise of two new powers: Germany and the United States. Germany was united from a number of smaller states by the Prussian Otto von Bismarck. By the time of the Franco-Prussian War (1870–71), Germany was the continent's most powerful country. Also at this time, America was emerging from its civil war, and rapidly industrializing.

Overshadowing these changes was the Industrial Revolution. This century saw the spread of the previous century's innovations. To these were added diverse new technologies. The telegraph, the Internet of its day, for the first time allowed people many miles apart to communicate almost instantaneously. The railroad revolutionized transportation. The electric motor was invented, followed by the electric grid and the electric light. The telephone changed the world of communication. World

trade expanded. The Suez Canal opened in 1869. Refrigeration was invented, allowing, for example, meat to be shipped from Argentina to Europe. The population of
Europe roughly doubled in this century, as it had in the previous one.

These changes in technology were accompanied by equally important cultural
and political changes. People started moving from farms into cities. Some parts of
the world, including Britain and America, saw a gradual expansion of democracy.
Slavery was abolished in the United States after the Civil War. Russia emancipated
its serfs in 1861.

The nineteenth was a great century for "isms." In addition to imperialism and
colonialism, Europe saw a surge in nationalism, including the unification of both
Germany and Italy. Socialism, communism, and anarchism all arose in response to
the problems created by the new industrialization, including appalling urban squalor
(see Dickens).

Supported by the new wealth, the arts flourished in this century. A (somewhat
arbitrary) hall of fame would include Johann Wolfgang von Goethe and Jane Austen,
Vincent van Gogh and Auguste Rodin, Ludwig van Beethoven and Antonin Dvo-
rak and Peter Ilyich Tchaikovsky, Richard Wagner and Giuseppe Verdi and Georges
Bizet, Charles Dickens and Fyodor Dostoyevsky and Mark Twain.

The word "scientist" originated in this century, and science produced many advances. Physicists achieved a remarkable unification of electric and magnetic phenomena. Michael Faraday (1791–1867) performed experiments that elucidated the
nature of electromagnetism, and invented the electric motor. James Clerk Maxwell
(1831–1879) discovered that light was an electromagnetic wave, and reduced all of
electricity and magnetism to four equations.

Also in physics, the study of heat and energy led to the development of thermo-
dynamics and statistical mechanics. Some of the important contributors were Sadi
Carnot (1796–1832), Rudolf Clausius (1822–1888), William Thomson, Lord Kelvin
(1824–1907), Ludwig Eduard Boltzmann (1844–1906), and Josiah Willard Gibbs
(1839–1903) (who worked in relative obscurity in America).

In chemistry, John Dalton (1766–1844) developed the modern theory of atoms.
Dmitry Ivanovich Mendeleev (1834–1907) constructed his periodic table of elements, and used it to predict three new elements—germanium, gallium, and scan-
dium—which were later discovered.

Biology saw the publication in 1859 of *On the Origin of Species* by Charles Dar-
win (1809–1882), which provided a theoretical foundation for biological evolution.
This theory was independently proposed by Alfred Russel Wallace (1823–1913), an
outgrowth of his herculean efforts as a professional collector of biological specimens
from around the world. (In his eight years exploring the East Indies, he collected,
cataloged, and returned to England 125,660 plant and animal specimens, including
83,000 beetles.) The germ theory of disease, with its profound implications for public health, was developed by Louis Pasteur (1822–1895), Joseph Lister (1827–1912),
and Robert Koch (1843–1910).

Mathematics thrived in the 19th century. There was progress on old problems,
such as which figures can be constructed using compass and straightedge, and solving fifth degree polynomials. The foundations of mathematics were elucidated, with

formal definitions of integers and real numbers. The analysis of complex numbers continued, with major applications to physics and to other areas of mathematics. Perhaps most impressive were the new topics: Fourier analysis, non-Euclidean geometry, vectors and matrices, logic and set theory, and abstract group theory.

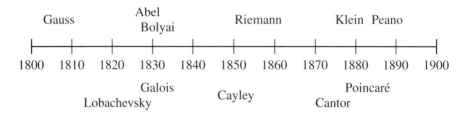

Nineteenth Century mathematicians.

Complex Variables

In the previous section, we mentioned the advent of complex numbers. Augustin-Louis Cauchy (1789–1857) and Georg Friedrich Bernhard Riemann (1826–1866) further developed this branch of mathematics by studying complex *functions*, which are functions from the complex plane to the complex plane. These are more difficult to picture than real functions, because both the domain and the range are 2-dimensional. However, complex functions have elegant properties that make them, in some respects, better behaved than real functions. For example, the real function $f(x) = x^{5/3}$ has a derivative easily obtained by the power rule: $f'(x) = (5/3)x^{2/3}$. However, using the power rule again, the second derivative, $f''(x) = (10/9)x^{-1/3}$, is not defined at 0. By contrast, when a complex function has a derivative, all higher order derivatives exist. A differentiable complex function is called an *analytic function*. Analytic functions can be expanded in terms of power series (infinite polynomials). For instance, the exponential function $f(z) = e^z$ has the power series expansion

$$f(z) = 1 + z + \frac{z^2}{2!} + \frac{z^3}{3!} + \frac{z^4}{4!} + \frac{z^5}{5!} + \cdots.$$

While the real exponential function is always nonnegative, the complex exponential function takes all complex values except 0.

Complex variables have proven to be a powerful tool in many branches of mathematics, including number theory.

Number Theory

In 1796 Gauss showed that a regular polygon with 17 sides is constructible using only a straightedge and compass. This result, first appearing in Gauss's *Disquisitiones Arithmeticae* (1801), extends Euclid's *Elements*, where constructions of an equilateral triangle and a regular pentagon are given. Gauss went on to show that a regular polygon with n sides is constructible if n is either a power of 2 or $n = 2^m p_1 p_2 \cdots p_r$, where $m \geq 0$ and $p_1, p_2, \ldots, p_r$ are distinct primes of the form $2^{2^k} + 1$, for $k \geq 0$. Some admissible values of n are 3, 5, $20 = 2^2 \cdot 5$, and $85 = 5 \cdot 17$, while some inadmissible values are 7, 13, and $25 = 5^2$. Primes of the required form are called *Fermat primes*. Indeed, $2^{2^k} + 1$ is a prime for $k = 0, 1, 2, 3$, and 4, but for no known higher values of k.

Carl Friedrich Gauss (1777–1855)

Carl Friedrich Gauss was born into a poor family in Brunswick, Germany. He showed strong mathematical ability at an early age. One story has it that at age ten he surprised his teacher by adding all the numbers from 1 to 100 instantly. He did this by pairing 1 and 99, 2 and 98, 3 and 97, etc. The forty-nine pairs that sum to 100 total 4900. Together with 50 and 100, this totals 5050. Thus, Gauss found the formula

$$1 + 2 + 3 + \cdots + n = \frac{n(n + 1)}{2}$$

which we encountered in Chapter 1.

Gauss' teachers recognized his genius, and, beginning in 1791, the Duke of Brunswick financed his education. Gauss studied at the University of Göttingen from 1795 to 1798, then obtained his doctorate at the University of Helmstedt. He joined the University of Göttingen as professor of astronomy in 1807, and remained there for the rest of his life.

Like Euler, Gauss was a mathematical titan who reinvigorated traditional areas of mathematics, inaugurated new fields, and made many discoveries that now bear his name. His research transformed nearly all branches of mathematics, including number theory, geometry, algebra, analysis, and probability, as well as physics.

Gauss became quite famous in 1801 for his rediscovery of the asteroid Ceres, which had been discovered earlier, but lost because its orbit was not known. This rediscovery was made possible by Gauss' development—at age eighteen—of a method to handle experimental errors, the method of least squares.

Also in 1801 Gauss published *Disquisitiones Arithmeticae (Research in Number Theory)*. This monograph encompasses discoveries by past mathematicians and Gauss's new contributions. In the preface, Gauss acknowledges four predecessors: Euler, Fermat, Lagrange, and Legendre. An important element of the work is Gauss's new congruence notation. (See Chapter 5.)

Gauss worked extensively in geometry, studying alternative geometries to Euclidean geometry (non-Euclidean geometries). He also laid the foundations for differential geometry, which uses calculus to study geometrical curves and surfaces.

Gauss is renowned in physics as well, for many discoveries. In 1833 he and Wilhelm Weber invented an early electric telegraph, able to communicate over a distance of 1200 meters.

Gauss gave four proofs of the Fundamental Theorem of Algebra, probably because he considered the theorem so important. He gave the final proof when he was in his 70s.

$$\sim$$

The most important development in number theory in the 19th century was cross-fertilization with other branches of mathematics, such as complex variables and group theory. Recall that Lagrange had proved that every positive integer is the sum of four squares. In 1834 Carl Gustav Jakob Jacobi (1804–1851) extended Lagrange's theorem by finding a formula for the number of ways that an integer n can be written as the sum of four squares. The formula depends on whether n is odd or even. If n is odd, then the number of ways that n can be written as a sum of four squares is 8 times the sum of the positive divisors of n. If n is even, it is 24 times the sum of the odd positive divisors of n. We posed the question, how many ways can you write 10 as a sum of four squares (where the order of the terms and the sign of the numbers being squared is important)? According to Jacobi's formula, the number of ways is

$$24(1 + 5) = 144.$$

The proof of Jacobi's theorem uses the theory of complex variables.

The crowning achievement in number theory in the 19th century was the proof of the Prime Number Theorem in 1896, by Jacques Hadamard (1865–1963) and Charles Jean de la Vallée-Poussin (1866–1962). The theorem, described in the previous section, was conjectured by Gauss and Legendre. Its proof uses functions of complex variables.

The Quintic

In Chapter 2 we discussed the fact that cubic and quartic equations are solvable using the four arithmetic operations (addition, subtraction, multiplication, and division) and radicals (square roots, cube roots, etc.). After the discovery of these solutions, mathematicians sought to solve the quintic equation,

$$ax^5 + bx^4 + cx^3 + dx^2 + ex + f = 0.$$

Lagrange had set the stage for understanding solutions of equations by introducing one of the most important ideas: permutations of roots of a polynomial. A *permutation* is a rearrangement. For example, shuffling a deck of cards is a permutation of the 52 cards. Lagrange's concept of permutations of polynomial roots led to breakthroughs in polynomial theory by Abel and Galois, and to the development of group theory, with implications throughout mathematics.

Niels Abel (1802–1829) proved that there are quintics that cannot be solved by radicals, showing in particular that $x^5 - 10x + 2$ is not solvable. Évariste Galois (1811–1832) later provided a more general theory for which quintics are solvable. He had the key insight that a polynomial can be solved in radicals if and only if its associated Galois group (a set of permutations of its roots) is "solvable." (A solvable group can be built up from commutative groups—groups in which $xy = yx$ for all elements x and y.) The Galois group of Abel's example consists of all permutations of the five roots. This group is unsolvable.

Niels Henrik Abel (1802–1829)

Niels Henrik Abel was born on the island of Finøy, in southwestern Norway, but spent most of his life in southeastern Norway. The family was poor, and the situation grew worse when Abel's father, a Lutheran minister, died in 1820. Abel's mathematics teacher raised funds for his education, however, and he was able to attend the University of Christiania in Oslo. He never was able to find a university position, and died of tuberculosis at age 26.

Abel is perhaps most famous for proving the impossibility of solving the quintic by radicals (mentioned above), but contributed to many areas of mathematics. His contributions are memorialized in terms including abelian groups, abelian functions and abelian integrals. In 2002 the Abel Prize, a prestigious mathematical award, was created in his honor.

Évariste Galois (1811–1832)

Évariste Galois was born in Bourg-la-Reine, a town near Paris. His education was spotty: he twice failed the entrance examination for the École Polytechnique, instead attending the École Normale, a school for training teachers. Galois then became involved in the chaotic politics of the time, and was expelled from that school.

Galois submitted several papers on polynomial equations to the Academy of Sciences. A couple of times they lost his manuscripts. Augustin-Louis Cauchy did review two of his papers, but did not publish them.

Galois died at age 20, of injuries sustained in a duel. He stayed up the night before, writing a letter to a friend, including several manuscripts with some annotations. The letter concluded: "Ask Jacobi or Gauss publicly to give their opinion, not as to the truth, but as to the importance of these theorems. Later there will be, I hope, some people who will find it to their advantage to decipher all this mess." His manuscripts were finally published in 1846, but not fully appreciated for decades after that.

In his short, tragic life, Galois had monumental mathematical insights. He was the first to use the word *group* ("groupe," in French) to refer to a collection of permutations. The concept of a group has since become enormously important in mathematics. Galois was also the first to describe and investigate finite fields (see Chapter 5). The great Hermann Weyl (1885–1955) wrote of Galois' last testament: "This letter, if judged by the novelty and profundity of ideas it contains, is perhaps the most substantial piece of writing in the whole literature of mankind."

The Abstract Concept of a Group

The groups studied by Galois and Abel were groups of permutations of roots of polynomials. Cauchy studied more general permutation groups. Arthur Cayley (1821–1895) gave the first abstract definition of a group in 1854: A *group* is a set together with an operation on pairs of elements of the set. Suppose that we indicate the operation on x and y by xy. (The operation might not be commutative, that is, xy doesn't necessarily equal yx.) The operation must be associative: $(xy)z = x(yz)$, for all elements x, y, and z. There must be an identity element, e, such that $xe = ex = x$, for all elements x. Finally, each element x must have an inverse, x^{-1}, such that $xx^{-1} = x^{-1}x = e$. An example of a group is the set of positive real numbers, with the operation multiplication. The identity element is 1. Each positive real number x has a multiplicative inverse $1/x$.

Arthur Cayley (1821–1895)

Arthur Cayley was born in England, but spent most of his first seven years in Russia, where his merchant father moved the family, before returning to England. He showed early talent in mathematics, and entered the University of Cambridge at the age of 17. He stayed some years at Cambridge as student and fellow, but left in 1846 to study law. He practiced law for 14 years, while producing over 300 mathematics papers in his spare time.

At the age of 42, Cayley gave up his lucrative legal career to become a professor of mathematics at Cambridge. He stayed at Cambridge the rest of his life, continuing to produce research at a prodigious rate. He was also an early supporter of university education for women, and served in the 1880s as chairman of the council of Newnham College, one of the two women's colleges at Cambridge.

Cayley wrote over 900 mathematical papers in the areas of algebra, including linear algebra and abstract algebra, geometry, differential equations, and combinatorics. His conception of matrix algebra would prove to be important in quantum mechanics in the next century.

$\sim$

Non-Euclidean Geometry

Recall from Chapter 1 that Euclid's *Elements* contains five postulates, the fifth of which is more complicated than the other four. The fifth postulate is equivalent to the following statement about the existence of parallel lines:

> Given a line $\mathcal{L}$ and a point P not on $\mathcal{L}$, there exists a unique line $\mathcal{L}'$ passing through P which does not intersect $\mathcal{L}$.

We say that $\mathcal{L}'$ is *parallel* to $\mathcal{L}$ (Figure 3.12).

Over the next two thousand years, mathematicians attempted to prove Euclid's fifth postulate from the other four. Some thought they had succeeded, but there was

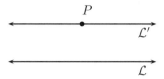

Figure 3.12   A parallel line given by Euclid's fifth postulate.

always a mistake in the reasoning. Finally, it became clear that the fifth postulate cannot be proved from the first four. One can assume that the postulate is false and derive consistent geometry that is quite different from Euclidean geometry. A geometry in which Euclid's fifth postulate does not hold is called a *non-Euclidean geometry*.

If the fifth postulate is false, there are two possibilities: (1) there are no lines parallel to a given line and passing through a given point; or (2) there is more than one such parallel line. The first case is called *elliptic geometry*; the second case is called *hyperbolic geometry*. (These terms were introduced by Felix Klein.)

In the early 1800s, the Hungarian mathematician János Bolyai (1802–1860) and the Russian mathematician Nikolai Ivanovich Lobachevsky (1792–1856) formulated versions of hyperbolic geometry. Gauss had discovered hyperbolic geometry earlier but didn't publish his findings. Later, Bernhard Riemann (1826–1866) formulated elliptic geometry.

Nikolay Ivanovich Lobachevsky (1792–1856)

Nikolay Ivanovich Lobachevsky's father died when he was young, and his mother moved to Kazan, a Russian city about 700 kilometers east of Moscow, just west of Siberia. Lobachevsky studied at the Kazan State University, where most of his teachers were German. He later joined the faculty there, and was promoted to full professor in 1822. He served as rector of Kazan University from 1827 to 1846, when he retired. The university prospered in his time as rector, and he worked to improve education throughout the entire Kazan district.

Lobachevsky's first paper on non-Euclidean geometry was written in 1823, but not published until much later. His first work on the subject to make it to print appeared in an obscure Kazan journal in 1829. A later text of his was praised by Gauss in 1842, but not everyone appreciated the new geometry. Lobachevsky's work did not receive full recognition until after his death, with the work of Riemann and Klein.

Lobachevsky worked in other areas of mathematics as well. There is a method still used to approximate solutions of algebraic equations, called the Lobachevsky–Gräffe or Dandelin–Gräffe method, discovered independently by Lobachevsky, Germinal Pierre Dandelin (1794–1847), and Karl Heinrich Gräffe (1799–1873).

János Bolyai (1802–1860)

János Bolyai was the son of a Hungarian mathematician, Farkas Bolyai. János was a prodigy, but his family could not afford to send him to a first-rate university. Instead, he studied at the Royal Engineering College in Vienna, completing the seven year course in four years. He then served in the Austro-Hungarian Imperial Army engineering corps from 1822 to 1833. He was a renowned swordsman, as well as an accomplished violinist, and spoke at least nine languages.

Bolyai's father encouraged him in his mathematics, but tried to dissuade him from studying Euclid's fifth postulate, having studied it himself to no avail. János ignored the warning, and instead produced a consistent geometry that did not obey the fifth postulate. In 1832 this work appeared as an "Appendix Explaining the Absolutely True Science of Space" in his father's textbook *An Attempt to Introduce Studious Youth to the Elements of Pure Mathematics.*

Carl Friedrich Gauss had discovered a non-Euclidean geometry earlier, but had not published his work. When he read Bolyai's appendix, he wrote to a friend, "I regard this young geometer Bolyai as a genius of the first order." Gauss's letter to his friend Farkas Bolyai on his son's work was less than tactful, however: "To praise it would amount to praising myself. For the entire content of the work... coincides almost exactly with my own meditations which have occupied my mind for the past thirty or thirty-five years." This response was a blow to János. A further blow was his discovery, in 1848, that Lobachevsky had published a similar geometry in 1829. As was the case with Lobachevsky, Bolyai's work was not fully appreciated until after his death.

In Euclidean geometry, the sum of the angles of a triangle is $180°$. In elliptic geometry, the sum of the angles is greater than $180°$, while in hyperbolic geometry, the sum is less than $180°$. In hyperbolic geometry, all similar triangles have the same area.

Besides providing new mathematical worlds to explore, non-Euclidean geometries point to a more axiomatic—and less intuitive—trend in mathematics. It had been taken for granted that Euclidean geometry was *the* geometry, but now we realize that this geometry is only a logical extension of the axioms we assume, and these axioms do not necessarily represent logical or physical truth. In fact, it is not known what geometry provides the best model of the physical universe. More and more, mathematicians would emphasize the logical foundations of the various branches of their discipline. This would ultimately lead to a reevaluation of mathematical logic itself in the 20th century.

Bernhard Riemann (1826–1866)

Georg Friedrich Bernhard Riemann was born in the village of Breselenz, in northern Germany, the son of a Lutheran pastor. The family was relatively poor, but found

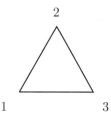

Figure 3.13   An equilateral triangle has six symmetries.

enough money to send Bernhard to the University of Göttingen, to study theology. He soon switched to mathematics, moved to the University of Berlin for two years, then returned to Göttingen, where he completed his dissertation under Gauss's supervision. He spent most of the rest of his career at Göttingen, eventually being appointed to the mathematics professorship in 1859.

Riemann contributed to many areas of mathematics, and to physics. He is best known for his work in geometry. Riemannian geometry, as it is now called, is the mathematical foundation of Einstein's general relativity. Riemann's work in geometry also inspired the work of another mathematician, Charles Lutwidge Dodgson (aka Lewis Carroll). The looking glass connecting England to Wonderland in Carroll's *Through the Looking-Glass* was an example of a Riemann cut, the original of what is now called a wormhole.

Riemann published only one paper in number theory, eight pages long. In it, he applied new ideas in complex analysis, and produced probably the leading unsolved problem in mathematics today, the Riemann Conjecture. The solution of the Riemann Conjecture would yield important results on the distribution of prime numbers, and will eventually win someone one million dollars.

Riemann also did important work on the foundations of calculus. See, for example, Sections 4.14 and 4.15.

In 1862 Riemann fell ill with tuberculosis. In the next few years, he made several trips to Italy to treat the disease, but eventually succumbed to it. He died in Selasca, Italy, at the age of 39.

$\sim$

Groups and Geometry

Recall that a group is a set with an operation satisfying certain properties. Groups can describe the symmetries of an object.

Here is a small example. Consider the equilateral triangle in Figure 3.13, whose vertices are numbered 1, 2, and 3.

An equilateral triangle can be rotated or flipped over and occupy the same space. There are six symmetries, including all rotations and flipping over the triangle. The group consists of these six symmetries, where the operation is performing one symmetry after another (this is called composition). The group contains a smaller group of three symmetries, namely, the three rotations. (One of the rotations is the *identity element*, which leaves the triangle unmoved.) The smaller group is a *subgroup* of the larger group.

The five Platonic solids (tetrahedron, cube, octahedron, dodecahedron, and icosahedron) have symmetry groups. Let's consider the dodecahedron (Figure 3.14). It has twelve faces which are regular pentagons. The dodecahedron can be picked up and set down on any of its twelve faces. Then it can be rotated into any one of five positions. After these maneuvers, the dodecahedron occupies its original space. Since there are twelve choices for which face goes on the bottom, and five choices for the rotation, there are altogether $12 \cdot 5 = 60$ symmetries of the dodecahedron. This 60-element group is "simple," meaning that it cannot be broken down into smaller symmetry groups.

Figure 3.14   The dodecahedron has 60 symmetries.

We will give another example of a symmetry group. Figure 3.15 shows a *finite geometry* with seven points and seven lines. The points are labeled 1, 2, 3, 4, 5, 6, 7. The lines are 123, 154, 264, 176, 374, 275, 356. Points are collinear if they lie on the same line. For example, 1, 4 and 5 are collinear, while 1, 4, and 7 are not collinear. In finite geometry, the only thing that matters is the relationship between points and lines. The points can be placed anywhere and the lines can be drawn curved or straight. Notice that the line 356 looks like a circle in our diagram. This finite geometry has one of the properties of Euclidean geometry: for every two points there is a unique line containing them. But every two lines intersect; there are no parallel lines, so this is a non-Euclidean geometry.

A symmetry of this finite geometry is a way of relabeling the points so that all the collinearity relationships are preserved. For instance, if we simply switched labels 1 and 2, we would mess up the collinearity, since 1, 5, and 4 would no longer be collinear (for example). In a symmetry, lines have to stay lines. How many symmetries does the finite geometry have? We know that there are 7! possible permutations of the seven point labels, but as we have seen, not all permutations are symmetries. All seven points are interchangeable, since they are all on the same number of lines

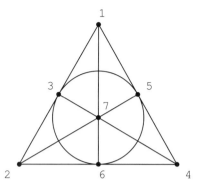

Figure 3.15 A finite geometry with 168 symmetries.

(three). We are therefore free to relabel point 1 as any of the seven points (including leaving it as 1). Say we label it $1'$. Once this decision is made, we can relabel 2 with any label other than $1'$. There are six choices. So far, we have been free to make $7 \cdot 6$ choices. Once we relabel 1 and 2, we are forced (by collinearity) to choose for 3 the label of the point collinear with the $1'$ and $2'$. However, now that we have relabeled 1, 2, and 3, we can relabel 4 with any label not already used (four choices). Once we relabel 4, collinearity forces all of the other labeling decisions. Altogether, there are $7 \cdot 6 \cdot 4 = 168$ ways to relabel the finite geometry, and that is the number of symmetries. Since the symmetries form a group, this establishes a group of 168 elements (one element for each hour of the week!). Like the symmetry group of the dodecahedron, this group is simple.

Let's bring the discussion back to Euclidean geometry. When we say that two figures are congruent in Euclidean geometry, we mean that there is a motion that brings one figure into the other. Such a motion must be "rigid" in that it preserves all lengths of line segments and angles. We call these motions *isometries* of the plane.

There are four types of isometries of the plane, as shown in Figure 3.16. The first two, translations and rotations, preserve *orientation* of the plane. If you translate or rotate a picture of a left-handed glove, it will stay left-handed. The other two isometries, reflections and glide-reflections, reverse orientation. They give a mirror image of a figure. A left-handed glove would be transformed into a right-handed glove. The collection of isometries, under composition, forms a group.

Felix Klein (1849–1925), while at the university at Erlangen, Germany, proposed his *Erlangen Program* (1872), in which he emphasized that geometry should be studied according to its symmetries. Thus, all the properties of Euclidean geometry are inherent in the group of isometries. Other geometries have different notions of equivalence and therefore different symmetry groups. The wider the definition of equivalence, the larger the corresponding symmetry group.

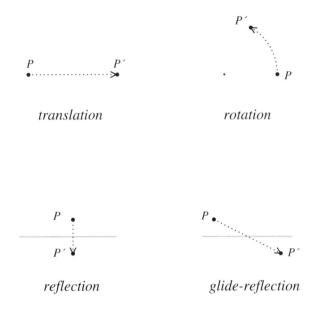

Figure 3.16   The four types of isometries of the plane.

Klein showed that the geometries known in his time—Euclidean, non-Euclidean, and projective—can all be viewed as instances of projective geometry. (Examples of projective geometry are perspective geometry, discussed in Chapter 2, and the finite geometry of Figure 3.15.) Projective geometry is therefore very general; there are more symmetries than in the special cases of Euclidean and non-Euclidean geometry; correspondingly, there are fewer "invariants," properties retained under symmetry. For example, there are no angles in perspective geometry. (When you look at a square from a slant-wise perspective, its right angles become slanted.)

Felix Klein (1849–1925)

Christian Felix Klein was born in Düsseldorf, Prussia (now in western Germany), the son of a government official. He studied physics and mathematics at the University of Bonn.

In 1872, at the age of 23, he was appointed professor at the University of Erlangen. It was on the occasion of this appointment that he set out his Erlangen Program. He subsequently worked at the Institute of Technology in Munich (1875–80), the University of Leipzig (1880–86), and finally at the University of Göttingen (1886–1913).

Klein built Göttingen into one of the leading research centers of its day and, as editor, turned the *Mathematische Annalen* into one of the leading mathematical research journals. He supervised the dissertation of Grace Chisholm Young, who in

Figure 3.17   A polyhedron with a hole.

1895 was the first woman to earn a doctorate in Germany, in any field. Klein also worked to improve the teaching of mathematics in secondary schools.

From Galois' concept of a group of permutations of the roots of a polynomial, to the abstract definition of a group, to the use of groups in geometry, the rise of group theory has been a vital and increasingly important part of modern mathematics.

Topology

In Euclidean geometry, two figures are "the same" if they congruent. In Chapter 2 we saw that in perspective geometry figures need not be congruent to be considered "the same." Any two quadrilaterals are "the same" in perspective geometry.

Topology is a type of geometry in which shapes can twist, bend, and stretch, but not break. Let us use the term *equivalent* for "the same." In topology, two figures are equivalent if we can bend, stretch, and twist one figure to make it identical to the other.

An easy-to-understand example of topology is the equivalence between a doughnut with one hole and a coffee cup with a handle. The two shapes are equivalent because one can be deformed into the other (if they are malleable) by bending, stretching, and twisting. The common shape is called a *torus*. We will see a construction of a torus in the forthcoming discussion of graph theory.

Let's take a look at the relevance of topology to a formula of Euler that we examined earlier. Euler's formula states that $V - E + F = 2$, for a polyhedron with V vertices, E edges, and F faces. We stipulated that the polyhedron has no holes. Figure 3.17 shows a polyhedron with a hole. The shape is topologically equivalent to a torus.

For this polyhedron, $V = 16$, $E = 32$, and $F = 16$. Notice that Euler's formula does not hold. However, Euler discovered a generalization of his formula, later further generalized by Henri Poincaré, that relates V, E, and F to the number of holes,

g, in the polyhedron:
$$V - E + F = 2 - 2g.$$
If $g = 0$ (no holes), we are back to Euler's original formula. For the polyhedron in Figure 3.17, we have $g = 1$ and $V - E + F = 2 - 2 \cdot 1 = 0$.

A polyhedron with g holes is topologically equivalent to a smooth surface with g holes. We call g the *genus* of the surface. These kinds of surfaces are all *orientable*, meaning that if you put a clock face on the surface and move it around, it will always keep time in a clockwise fashion. Other surfaces are nonorientable. Whether orientable or nonorientable, surfaces are called (two-dimensional) *manifolds*. A manifold is a geometry that is locally like Euclidean space. A person who inhabited a small region of a manifold wouldn't know the difference between it and the plane. Three-dimensional manifolds are more complicated, and they are still not fully understood.

Discoveries in topology may tell us something about the shape of the universe. It is not known what the true topology of the universe is.

Henri Poincaré (1854–1912)

Henri Poincaré was born in Nancy, France, the son of a medical professor. The family was prominent; a cousin was President of France during World War I.

Poincaré obtained his doctorate from the École Polytechnique in 1879, studying differential equations. While working on his mathematics doctorate, he also got a degree in engineering. He taught at the University of Caen for two years, then joined the University of Paris, where he remained for the rest of his life.

In 1886 King Oscar II of Sweden offered a prize to anyone who could establish whether the solar system is stable, so that planets won't fly off into space or crash into one another. Poincaré began his study of the problem by considering a simpler one, a system with three objects, the 3-body problem. (The 2-body problem had already been solved.) He found this to be difficult enough, but his proof that 3-body systems are stable won the prize. After the prize was awarded, and his paper printed, Poincaré discovered an error. He rewrote the paper and paid for it to be reprinted (costing him more than he had won), in the process proving that the solution of the 3-body problem cannot be written as a formula, and that the system is chaotic. Small changes in the initial positions of the bodies can lead to large differences later. This was the origin of chaos theory, which became an important area of mathematics in the second half of the 20th century.

Poincaré did other important work in physics and astronomy, including relativity theory, and in many areas of mathematics. But he is known best for his work in topology, and is often called the father of algebraic topology. He formulated what has become known as the Poincaré Conjecture, which concerns three-dimensional manifolds. This conjecture attracted the attention of some of the best mathematicians, leading to Fields Medals (the mathematical equivalent of Nobel Prize) in 1966, 1986, and 2006, when the conjecture was finally proved.

Mathematical Analysis

When calculus was unleashed upon the world in the 1600s, it was an instant success. Mathematicians and physicists realized that it provided a mathematical framework in which major problems could be expressed and solved. Calculus explained Kepler's planetary laws. Calculus explained the tidal effects of the Moon. In our age, calculus has taken astronauts to the Moon and back to Earth.

However, the foundations of calculus were a little shaky from the start. Newton wrote his *Philosophiæ Naturalis Principia Mathematica* in the language of geometry. When he needed to justify a limit, he used complicated geometrical reasoning. Leibniz, on the other hand, offered a new mathematical object—the infinitesimal. The derivative (rate of change) of the function y with respect to the variable x is dy/dx, where the "differentials" dy and dx represent infinitesimal changes in y and x, respectively. In a later publication, Newton introduced the terms "fluent" for function and "fluxion" for derivative, an appeal to physical intuition that didn't help to clear up matters.

The philosopher George Berkeley (1685–1753), not liking infinitesimals or fluxions, gave a biting critique of calculus in a book titled *The Analyst* (1734):

> And what are these Fluxions? The Velocities of evanescent Increments? And what are these same evanescent Increments? They are neither finite Quantities nor Quantities infinitely small, nor yet nothing. May we not call them the ghosts of departed quantities?

Although calculus was practical and popular, some shoring up of its foundations was in order.

The two main operations of calculus are differentiation and integration. Both are defined in terms of limits. To give calculus a firm foundation, limits must be defined carefully. This was first done in 1817 by Bernard Bolzano (1781–1848). The "epsilon-delta" definition quantifies the relationships among the function, the independent variable, and the limit. For the record, the definition is that the *limit* of the function $f(x)$, as x approaches a, is L, or in symbols,

$$\lim_{x \to a} f(x) = L,$$

if for every positive number ϵ, there exists a positive number δ, such that for all x satisfying $0 < |x - a| < \delta$, we have $|f(x) - L| < \epsilon$. This is a precise way of saying that $f(x)$ gets as close as we like to L when x is sufficiently close to a.

With the definition of limit, we define a function f to be *continuous* at a point a if $\lim_{x \to a} f(x) = f(a)$. This rigorously captures the commonplace idea of a continuous function being one that can be drawn without lifting one's pen.

Having a rigorous definition of limit puts calculus on a sound basis, but it raises other issues. Once the definition is accepted, it is tested against various kinds of "pathological functions," which are functions with strange properties. Mathematicians delighted in finding functions that are nowhere continuous, or continuous only at one point, or everywhere continuous and nowhere differentiable. Giuseppe Peano (1858–1932) found, in 1890, a one-dimensional curve that fills the entire plane. He was motivated by Georg Cantor's theory of infinite sets, which establishes a one-to-one correspondence between the points on a line and the points in the plane, so that the two sets have the same cardinality (size).

Pathological functions have changed the landscape of mathematics. Any theory must account for them. An irony of pathological functions is that they are actually the "normal functions." Most functions (in a precise mathematical sense) are pathological. The functions that we ordinarily encounter (polynomial, rational, trigonometric, exponential, logarithmic, etc.) are atypical.

Giuseppe Peano (1858–1932)

Giuseppe Peano was born on a farm in northwestern Italy. He was educated in nearby Turin, and taught at the University of Turin from 1880 until the day he died.

Besides the space-filling curve mentioned above, he is best known for the Peano Axioms. These axioms provide a rigorous foundation for the positive integers, from which true statements about the positive integers (theorems) can be deduced. Peano also introduced several symbols of set theory that are commonly used today, e.g., union ($\cup$), intersection ($\cap$), "element of" ($\in$), and "subset of" ($\subset$).

Peano wrote textbooks for mathematicians at all levels, including for secondary school teachers, and was the creator and advocate of a universal language based on Latin, French, German and English.

$\sim$

Mathematical Logic

Much of the mathematical theory behind modern computers was invented in the 1800s. Charles Babbage (1791–1871) created a plan for what he called an "Analytical Engine," a machine capable of arithmetic operations and memory. Although it hasn't been built yet (a project to do so is now underway), its design envisioned all of the concepts necessary for modern computing. Ada Lovelace (1815–1852), often called the first computer programmer, wrote the first computer algorithm for this machine.

The mathematical underpinnings of computing were developed at the same time as a new understanding of logic was achieved, and these advances went hand-in-hand. One of the major mathematical contributions was *propositional calculus*. A *proposition* is a statement that is true or false. A calculus is a system of calculations. So, a propositional calculus is a way of calculating with statements.

Statements can be simple, such as "The cat is in the snow," or fantastic. Lewis Carroll (1832–1898) offered many delightfully convoluted propositions in his *Symbolic Logic* (1897). Some examples:

"Animals, that do not kick, are always unexcitable."

"No plum-pudding, that has not been boiled in a cloth, can be distinguished from soup."

"A man, who has lost money and does not eat pork-chops for supper, had better take to cab-driving, unless he gets up at 5 a.m."

The point is that no matter how ridiculous the statements, they may be deemed either true or false.

Statements can be combined with connectives to make new statements. The simplest connectives are 'or' and 'and'. If p and q are statements, we write

$$p \vee q \quad \text{and} \quad p \wedge q$$

for 'or' and 'and', respectively. By definition, $p \vee q$ is true when p or q is true; $p \wedge q$ is true when both p and q are true. The operation 'not' specifies the negation of a statement. The notation $\sim p$ represents 'not' p. Thus, $\sim p$ is true when p is false.

A statement that is always true is called a *tautology*. An example is $p \vee \sim p$, because, for every statement p, either p or its negation is true. An example in words: "It is raining carrots or it is not raining carrots."

George Boole (1815–1864), in his book *Laws of Thought* (1854), showed that combinations of statements can be manipulated algebraically. If p is a proposition, we introduce a variable P which equals 0 or 1 depending on whether p is true or false. If p is true, then $P = 1$. If p is false, then $P = 0$. The propositional calculus with these variables is called *Boolean algebra*. In Boolean algebra, the 'or' operation is usually written with a '$+$' symbol, while the 'and' operation is written with a '$\cdot$' symbol. Thus $P + Q = 1$ when $P = 1$ or $Q = 1$, while $P \cdot Q = 1$ when $P = 1$ and $Q = 1$. Negation is indicated by an overline bar. Thus $\overline{P} = 1$ if and only if $P = 0$.

The tautologies of propositional calculus are called *identities* in Boolean algebra. Consider the tautology

$$\sim (p \vee q) \Leftrightarrow (\sim p) \wedge (\sim q).$$

The $\Leftrightarrow$ symbol means that the two sides are logically equivalent, both true or both false. The corresponding statement in Boolean algebra is the identity

$$\overline{P + Q} = \overline{P} \cdot \overline{Q}.$$

This identity is due to Augustus De Morgan (1806–1871), who expanded upon Boole's work.

What does propositional calculus have to do with computing? In electronics, the voltage in a circuit may be high (represented by '1') or low (represented by '0'). Circuits called *logic gates* can emulate the logical operations $\vee$, $\wedge$, and $\sim$ (or their Boolean equivalents). Arithmetic operations can thus be carried out via logic gates.

In computers, numbers are represented in binary (base-2) notation. Here are the numbers 1 through 10 in binary:

$$1, \ 10, \ 11, \ 100, \ 101, \ 110, \ 111, \ 1000, \ 1001, \ 1010.$$

Notice that 2 is written in binary as 10. This is because the 1 in the second place from the right represents 2 just as it represents 10 in our usual base-10 system. Similarly, a 1 in the third place from the right represents 2^2, or 4, and a 1 in the fourth place from the right represents 2^3, or 8.

The places in decimal numbers are called digits, and the places in binary numbers are called *binary digits*, or *bits*. A computer can add two binary numbers by performing an operation on the corresponding bits one pair at a time. Let's see how this works in the simplest case, adding two one-bit binary numbers X and Y. As these numbers can be 0 or 1, when we add them, we will get a sum of 0, 1, or 10 (which is 2 in binary). In each case, we have a sum bit, S, and a carry bit, C. We want the carry to be 1 precisely when both X and Y are 1, that is,

$$C = X \cdot Y.$$

We want S to be 1 if either X or Y is 1, but not both. Thus

$$S = (X + Y) \cdot \overline{(X \cdot Y)}.$$

Since the logical operations can be built with logic gates, a circuit can be built to add two one-bit binary numbers. In a similar fashion, all arithmetic operations can be embodied by circuits, and indeed, all features of a computer, including memory, can be created.

Mathematical Foundations

As mathematics expanded outward in the 19th century, there was also attention inward, to the foundations of mathematics. The discovery of so-called pathological functions (e.g., functions that are continuous at no point), and new number systems (e.g., the complex numbers) led mathematicians to question the basis of their science. What is a function? What is a number? These are by no means trivial questions, and they demand rigorous definitions. In turn, the definitions give rise to new, surprising, counter-intuitive examples.

We are familiar with several kinds of numbers, such as integers, rational numbers, and real numbers. Rational numbers are simply quotients of integers. Real numbers are more subtle. What is a real number? We know that not all real numbers are rational, since the Pythagoreans proved that $\sqrt{2}$ is irrational. But $\sqrt{2}$ is rather special, as it satisfies a polynomial equation with integer coefficients, namely, $x^2 = 2$. It is tempting to say that a real number is a root of a polynomial with integer coefficients. But real numbers are harder to pin down than that. The Archimedean constant, π, is not the root of any such polynomial. A better definition of real numbers is that they are all possible decimal expressions, such as

$$0.324234234234234444342342343244444343242342342... .$$

This gets at the intuitive idea that a real number can be approximated more and more closely by a decimal. However, the definition is still unsatisfactory because it is circular. We are saying that a real number can be written as a decimal, and a decimal approximates a real number.

The mathematician Richard Dedekind (1831–1916) provided a definition of real numbers in terms of rational numbers. Let's take the case of $\sqrt{2}$. He defined this number as the set of all positive rational numbers whose square is less than 2, that is,

$$\{r : r \text{ is rational and } r^2 < 2\}.$$

This may seem a strange definition, as we are defining a single real number in terms of an infinite set of rational numbers. However, the definition works, and from it the real numbers have all the expected properties: addition, multiplication, distributive law, etc.

The notion of defining a number in terms of a set is a useful tool in establishing the foundations of mathematics. It was a goal in the 1800s to reduce all of mathematics to set theory, and ultimately to logic. So far, we have indicated how real numbers are defined in terms of rational numbers, and rational numbers in terms of integers. Integers can be defined in terms of sets, thereby reducing all number systems to sets. We start with the empty set, the unique set with no elements, denoted $\emptyset$. We identify this set with the integer 0. Then the integer 1 is identified with the set containing 0, i.e., $\{0\}$. Note that this set is different from the empty set, since it contains one element. Next, the integer 2 is identified with the set containing 1, i.e., $\{1\}$. Continuing in this manner, all the nonnegative integers are defined. It only remains to define negative integers. These are defined in terms of sets of ordered pairs of nonnegative integers.

We glossed over the definition of rational numbers, saying that they are quotients of integers. Actually, they too are defined as sets. For instance, each rational number is defined as a set of pairs of integers x and y. Intuitively, we think of x and y as being the numerator and denominator, respectively, of the rational number. Since a rational number can be written in different forms (by multiplying numerator and denominator by any nonzero integer), each rational number is represented by infinitely many different pairs of integers. The usual operations of addition, subtraction, multiplication, and division can be defined in terms of pairs of integers. For example, the sum of a rational number represented by x and y and a rational number represented by x' and y' is the rational number represented by $xy' + x'y$ and yy'. (Think about the way two fractions are added using a common denominator.)

According to the procedure we have described, all real numbers are ultimately defined in terms of sets, and indeed in terms of the simplest set, the empty set. It was natural, then, for mathematicians to try to put set theory on a solid foundation. The most eminent mathematician in this quest was Georg Cantor (1845–1918).

Set Theory

Cantor's set theory was controversial, for he proved the existence of different sizes of infinity. At first, this may sound absurd. Most people think of infinity as being a process that goes on forever, or as an all-encompassing concept. How can anything be greater than infinity?

Cantor started out with a simple premise. Two sets are the same size if there is a one-to-one correspondence between their elements. This makes perfect sense. The number of socks you are wearing is the same as the number of shoes you are wearing, if your left shoe contains one sock and your right shoe contains one sock.

The genius of Cantor's principle is that it applies to infinite sets as well. The set of all even integers is the same size as the set of all odd integers. Simply pair each even integer with the integer one greater (an odd integer). This one-to-one correspondence means that the two sets are the same size. So far, nothing surprising has happened. We have shown that two infinite sets, the set of even integers and the set of odd integers, are the same size. In fact, each of these sets is the same size as the set of all integers, for we can easily pair each even integer with an integer: pair the even integer $2n$ with the integer n. It follows that there are as many even integers as there are integers, which isn't too surprising as both are infinite sets, although it does show that a proper subset of the integers can be as large as the set itself.

The big surprise is that there are larger infinite sets. A natural one is the set of real numbers. It turns out that there is no way to pair each real number with an integer. It follows that the set of real numbers is of a larger order of infinity than the set of integers. Moreover, Cantor showed that for every infinite set, there is a larger infinite set. There is no largest infinite set.

Cantor's proof that each infinite set gives rise to a larger infinite set is really quite simple. For any given set S, he considered the set T containing all subsets of S. Suppose that there is a one-to-one correspondence between the elements of S and the elements of T. In this correspondence, an element of S may correspond to a subset of S which contains that element, or not. Let X be the subset of S containing all elements of S which correspond to subsets of S which do not contain them. Then X, being a subset of S, must correspond to some element of S. Call it x. Is x an element of X? If not, then x is an element of X, by the definition of X. This is a contradiction, so x cannot be an element of X. But then, by definition, x is an element of X, a contradiction. This illogical situation means that the assumption that there is a correspondence between the elements of S and the elements of T is false. By Cantor's criterion, the two sets are not the same size, Clearly, T is at least as large as S, since it contains all the one-element subsets of S. The only possible conclusion is that T is larger than S. If S is an infinite set, then T is a larger infinite set.

If we start with the set of integers, then Cantor's method yields a larger infinite set equivalent to the set of real numbers. The question arises as to whether there might be an infinite set of intermediate size, larger than the integers but smaller than the real numbers. The assumption that there is no intermediate infinite set is known as the Continuum Hypothesis (the real numbers are called the continuum). This hypothesis was put forth by Cantor in 1878 and would become the first of David Hilbert's famous problems for the 20th century (to be discussed shortly).

It was shown in the 20th century, by Kurt Gödel and Paul Cohen, that the Continuum Hypothesis cannot be proved one way or the other from the standard axioms of set theory. Consistent mathematics can be obtained either by assuming that there exists such an intermediate set, or that there does not.

Georg Cantor (1845–1918)

Born in Russia of Danish parents, Georg Cantor lived most of his life in Germany. He earned his doctorate at the University of Berlin. He taught at the University of Halle from 1869 until his retirement in 1913.

Cantor started out studying number theory, but in the 1870s turned his attention to infinite sets. After defining the notion of the size of an infinite set, he defined an arithmetic on these *transfinite numbers*. These ideas encountered considerable resistance. Some thinkers could not accept the existence of infinite numbers. Leopold Kronecker (1823–1891), in particular, fought this notion, and succeeded in blocking Cantor's appointment to the faculty of the University of Berlin.

Unfortunately, the last part of Cantor's life was beset by mental problems, exacerbated by the uncertain reception of his work. Today, Cantor's theory of infinite sets is regarded as one of the greatest achievements in mathematics and in all of human intellectual thought. Set theory is now an important part of the foundations of mathematics. It has had an effect in education, as well, forming an important component of the "New Math" revolution of the 1960s. [You may enjoy the song *New Math*, by comedian and mathematician Tom Lehrer (b. 1928). Look it up.]

EXERCISES

3.33 Use the power series for e^z and a calculator to approximate the value of e^i.

3.34 Calculate the Fermat primes $2^{2^3} + 1$ and $2^{2^4} + 1$.

3.35 Look up a method for constructing a regular pentagon using a straightedge and compass. Carry out the method.

3.36 The "golden ratio" is the number $(1 + \sqrt{5})/2$. It is intimately connected to the regular pentagon. Can you find some of the connections?

3.37 Write each of the numbers $1, 2, 3, \ldots, 20$ as the sum of four squares.

3.38 How many ways can 5 be written as a sum of four squares, where the order of the terms and the signs of the numbers being squared matters? List the ways.

3.39 Find all of the permutations of the three symbols a, b, and c. (Hint: there are six.)

3.40 Find all of the permutations of the four symbols a, b, c, and d. How many are there?

3.41 How many permutations are there on n symbols?

3.42 Is the set of all real numbers, with the operation of multiplication, a group?

3.43   A *great circle* on a sphere is a circle of largest circumference, such as the equator on the Earth. Suppose that a geometry consists of all points on a sphere, and the "lines" are all great circles. Does the parallel postulate hold for this geometry? What kind of geometry is this?

3.44   We found the number of symmetries of the regular dodecahedron. Do the same for the other four Platonic solids.

3.45   The symmetry group of the regular dodecahedron is the same as that of one of the other Platonic solids. Which one? Two other Platonic solids have the same symmetry group. Which ones?

3.46   What is the result of reflecting the Cartesian plane with respect to the x-axis and then with respect to the y-axis? What happens to an arbitrary point (x, y)?

3.47   In Euclidean geometry, what is the result of reflecting the plane with respect to two parallel lines? What is the result of reflecting the plane with respect to two intersecting lines?

3.48   In Euclidean geometry, if the plane is rotated 30 degrees counterclockwise about a point P and then rotated 60 degrees counterclockwise about a point Q, at distance 1 from P, what is the net result?

3.49   A polyhedron on an orientable surface has 24 vertices, 52 edges, and 26 faces. How many holes does the surface have? Can you draw such a polyhedron? (Hint: put a square hole through the top face of the polyhedron of Figure 3.17.)

3.50   Look up George Berkeley's *The Analyst* (1734). What is the subtitle of this work? What do you suppose the subtitle means?

3.51   Look up Peano's space-filling curve. Draw the first few steps used in creating the curve in the plane. Do you see how the final curve will fill up two-dimensional space?

3.52   Write the propositional calculus tautology

$$\sim (p \lor q) \Leftrightarrow (\sim p) \land (\sim q)$$

in the language of Boolean algebra.

3.53   Is the statement $\sim (p \land q) \Leftrightarrow (\sim p) \lor (\sim q)$ a tautology?

3.54   Explain why the statements $p \Rightarrow q$ and $(\sim p) \lor q$ are logically equivalent.

3.55   Let $S = \{1, 4, 9, 16, 25, 36, \ldots\}$, the set of perfect squares. Explain why S has as many elements as the set of all positive integers.

3.56   Let $S = \{1, 10, 10^2, 10^3, 10^4, \ldots\}$, the set of powers of 10. Does S have as many elements as the set of all positive integers?

3.57   Is e a transcendental number? (Hint: the Internet is a great resource.)

3.4   The 20th and 21st Centuries: Explosion

Introduction

The first half of the 20th century was dominated by the two world wars. World War I, 1914–1918, pitted Germany, Austria-Hungary, and Turkey against France, Great Britain, Russia, Italy, and Japan. In 1917 the United States entered the war and Russia left it, after the Bolshevik Revolution. The battles were characterized by trench warfare, made necessary by the new weapons, especially the machine gun and rapid-fire artillery. The war ended in defeat for Germany and its allies. An estimated 20 million people died.

Two decades later, after the Great Depression and the rise of Hitler in Germany, the second world war commenced. In this war, lasting from 1939 to 1945, the Axis powers—Germany, Italy, and Japan—battled the Allies—France, Great Britain, the Soviet Union, and the United States. The war was fought across most of the world. New technologies, led by tanks and airplanes, again increased the death toll. This was the deadliest war in history, with estimates of the number killed ranging from 35,000,000 to 60,000,000.

The United States emerged from the war as the dominant military and economic power, although the Soviet armies ruled Eastern Europe. Soon after, these two powers were locked in the Cold War, a time of great tension and many proxy wars, but fortunately no nuclear conflicts. Western Europe, which had been devastated by World War II, began rebuilding, helped by the United States' reconstruction program known as the Marshall Plan.

The U.S. continued to dominate the world's economy through the 1950s and 1960s. Gradually other nations grew stronger, led by Japan and Germany, joined later in the 20th century by the newly industrial "Asian Tigers": Hong Kong, Singapore, South Korea, and Taiwan. The early part of the 21st century has seen the spectacular rise of China, as well as the other BRIC (Brazil, Russia, India, and China) countries.

The Industrial Revolution continued and spread through much of the world. Automobiles became common. Airplanes were invented, and radio, and television. There were great advances in chemistry, including nylon and detergents; in medicine, especially with antibiotics; in labor-saving devices for the home, such as the vacuum cleaner and washing machine. The Space Age began in 1957, when the Soviet Union launched the first artificial satellite of Earth, Sputnik 1; men walked on the Moon in 1969.

The transistor, invented in 1947, led to a revolution in electronics. In the late 1950s came the integrated circuit, in which as many as ten transistors were put on a single silicon chip. Today, there are commercially available chips containing billions of transistors each.

The first digital electronic computers were built in the 1940s. Mathematicians, including Alan Turing (1912–1954), John von Neumann (1903–1957), and Donald Knuth (b. 1938), played an important part in the invention of computers and computer science. In the 1970s the Internet grew out of a network called ARPANET, which was created in 1969 by the Advanced Research Projects Agency (ARPA) of the U.S. Department of Defense. The World Wide Web was invented by the Englishman Tim Berners-Lee and colleagues at CERN in Switzerland, in the years 1989–92.

As the Industrial Revolution intensified and spread, the way people live has continued to change. The proportion of the United States population living in rural areas went from 60% in 1900 to 19% in 2000. As of 2010 some 23 cities in the world had metropolitan populations of more than 10 million people.

Science continued its dizzying growth in this century. Early in the century, Albert Einstein (1879–1955) introduced his theory of relativity. The study of atoms led physicists, including Max Planck (1858–1947), Einstein, Niels Bohr (1885–1962), Werner Heisenberg (1901–1976), and Erwin Schrödinger (1887–1961), to quantum mechanics. This theory has had far-reaching consequences in chemistry, technology (e.g., lasers, semiconductors), even philosophy, not to mention the bomb.

Our understanding of astronomy evolved dramatically since the turn of the 20th century. On 26 April 1920 the Great Debate took place at the Smithsonian Museum of Natural History. The issue was whether the fuzzy patches visible in telescopic photographs, known as spiral nebulae, were inside our galaxy, the Milky Way. By the end of that decade, the work of Edwin Hubble (1889–1953) demonstrated not only that these nebulae were galaxies in their own right, but that the galaxies were receding from each other. In 1965 Arno A. Penzias (b. 1933) and Robert W. Wilson (b. 1936) discovered a faint "noise" of radio waves pervading the universe, and confirmed the theory of the Big Bang, that the universe exploded some 13–14 billion years ago.

The work of astronomers Fritz Zwicky (1898–1974) and Vera Rubin (b. 1928), among others, revealed that much of the mass of the universe is not actually visible. This is called dark matter. In 1998 observations by Saul Perlmutter (b. 1959), Brian P. Schmidt (b. 1967), and Adam Riess (b. 1969) showed that the expansion of the universe discovered by Hubble is not being slowed down by gravity, as everyone had thought, but is in fact accelerating. This led to the hypothesis of a mysterious "dark energy," which comprises roughly 73% of all of the universe's energy. Most of the rest is dark matter. The part of the universe comprised of protons, neutrons, and atoms, i.e., the part of the universe that we understand, is less than 5%.

On Earth, the theory of continental drift was formulated by Alfred Wegener (1880–1930) and, after some initial setbacks, was confirmed, and expanded into the theory of plate tectonics by Harry H. Hess (1906–1969), J. Tuzo Wilson (1908–1993), and others. (Important to the theory is a theorem in spherical geometry due to Euler.) This theory is central to much of modern geology, explaining features from mid-ocean mountain chains, to the islands of Hawaii, to the Himalayas.

The central actor in modern biology is DNA (deoxyribonucleic acid). In 1944 Oswald Avery (1877–1955) discovered that DNA was the stuff of heredity. Remarkably, he never won the Nobel Prize, perhaps because he was sixty-seven years old when he published the paper. In 1953 the structure of DNA was described by James Watson (b. 1928) and Francis Crick (1916–2004).

In mathematics, a large majority of all of the research done ever has been done since 1900.

There are currently more than 500 regularly published mathematical journals, the majority of which are devoted to reporting new mathematical discoveries. The next time someone claims that "math never changes" you can point out this fact. The explosion of mathematical research reached a dizzying level in the 20th century. New mathematical fields were invented, new applications were discovered, and some major open problems were solved.

A combination of factors caused the vast mathematical boom in these times. One of these was the expansion of universities, including the funding of research by governments. Another factor was the Cold War, which stimulated (through fear) an immense expenditure on mathematics in the U.S. Yet another reason was the increasing role of computers and the Internet in mathematical research. Today, it is common for mathematicians to collaborate with colleagues whom they have never met in person. Mathematicians can investigate ideas quickly and easily using mathematical exploration software, they can write up their findings using mathematical typesetting software, and they can disseminate their results rapidly over the World Wide Web.

In this section we give an overview of some of the main events in mathematics in the last hundred years. It is impossible to cover all topics, or even to list all topics. All we can do is briefly describe some of them.

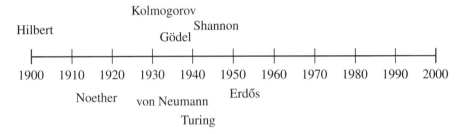

Some Twentieth Century mathematicians.

Hilbert's Problems

The 20th century opened with the mathematician David Hilbert (1862–1943) presenting a list of 23 important unsolved mathematical problems. He introduced his list in a speech to the International Congress of Mathematicians in 1900, adding to the collection of open problems in a later publication. Hilbert's problems helped to shape mathematical research in the 20th century, and a few of the original problems remain unsolved in the 21st century.

Some of Hilbert's problems are technical and would require lengthy explanations. Others are easier to explain, such as the tenth problem on the list. This problem is concerned with equations whose solutions are positive integers. Such equations are called *Diophantine*, after the Greek mathematician Diophantus (c. 200–294), who discovered methods for solving certain types of them. An example is the equation

$$x^2 + y^2 = z^2.$$

You are probably familiar with this as it is the formula relating the three sides of a right triangle given by the Pythagorean Theorem. You may know some integer solutions, such as

$$3^2 + 4^2 = 5^2$$

and

$$5^2 + 12^2 = 13^2.$$

What happens if we change the exponent in the equation from 2 to 3? Are there positive integer solutions to the equation

$$x^3 + y^3 = z^3 ?$$

Now the problem is much more difficult. Try as we may, we cannot find three positive integers x, y, z which satisfy the equation. It can be proved that this equation has no positive integer solutions (Euler showed this).[1]

Hilbert asked whether there is a method that would tell whether *any* given polynomial equation has positive integer solutions. For instance, we would be able to say whether the equation

$$x^3 y + y^2 z = x + 2y + 5$$

has a solution in positive integers x, y, and z. Such a method, if it existed, would be very useful. However, in 1970 Yuri Matiyasevich proved that there does not exist such a procedure. This is a stronger statement than merely saying that it is difficult to tell whether certain equations have integer solutions. There is *no* procedure that will tell, for any given polynomial equation, whether it has integer solutions. Thus, Hilbert's tenth problem was resolved in the negative.

It may seem strange that someone could prove that a general method cannot exist. However, many such negative results were established in 20th century mathematics, and some of the most important theorems have this negative character.

David Hilbert (1862–1943)

David Hilbert was born near Königsberg in Prussia (now Kaliningrad, Russia). He got his doctorate from the University of Königsberg in 1885, and taught there until 1895, when he became a professor at the University of Göttingen. He was a major reason why Göttingen was the best center of mathematics in the world throughout the first decades of the 20th century.

Hilbert was one of the most influential mathematicians in history. His famous list of unsolved problems, known as Hilbert's problems, was a catalyst for research in the 20th century. As the list shows, Hilbert was interested in all areas of mathematics of his day, and consequently he is recognized as perhaps the last mathematician who

---

[1] The proposition that the equation $x^n + y^n = z^n$ has no nontrivial solution for $n \geq 3$ is Fermat's Last Theorem (more about that in Chapter 5).

was conversant with all contemporary mathematics. After Hilbert, mathematics has expanded too much for anyone to comprehend it all.

Hilbert investigated the foundations of geometry, finding some holes in the proofs of Euclid and adding axioms to gain precision. In his study of mathematical analysis, Hilbert contributed a new mathematical object called Hilbert space (a generalization of Euclidean space).

Hilbert endorsed Cantor's theory of infinite sets, describing one of the results in an intuitive, yet mind-boggling, way. His thought-experiment is called the Grand Hotel. Suppose that a hotel has infinitely many rooms, numbered $1, 2, 3, \ldots$. All the rooms are occupied when a new guest arrives. However, the manager merely moves the guest in room 1 to room 2, the guest in room 2 to room 3, the guest in room 3 to room 4, etc. The new guest then moves into room 1. This explains why the infinity of positive integers is the same as the infinity of positive integers plus one.

Next, a coach carrying infinitely many people arrives at the hotel. Each person wants a room. The manager moves the guest in room 1 to room 2, the guest in room 2 to room 4, the guest in room 3 to room 6, etc. The guest in room n moves to room $2n$. This frees up all the odd-numbered rooms for the infinitely many new guests. This explains why infinity plus infinity is the same size as infinity.

Now infinitely many coaches arrive, each carrying infinitely many people who want rooms. Can you see how all the new guests can be accommodated? The conclusion is that infinity times infinity is the same size as infinity.

Hilbert knew and collaborated with many of the great mathematicians of the turn of the century, including Hermann Minkowski, Adolf Hurwitz, Carl Lindemann, Hermann Weyl, and Ernst Zermelo. He was anguished when the Nazis forced Jewish mathematicians out of the University of Göttingen, where he was Chair.

Hilbert is remembered for his focus on the unsolved problems of mathematics and for his optimistic spirit toward the resolution of these problems. His motto was "We must know. We will know."

Unsolvable Problems

The oldest posed problems which have turned out to be unsolvable are the three famous Greek geometric construction problems: trisection of an arbitrary angle, duplication of the cube, and quadrature of the circle. These constructions are to be done using only a straightedge and compass. Each of these tasks has been mathematically proved to be impossible. This is not to say that they are difficult, or that no one has discovered how to carry them out, but rather that they can never be accomplished using only the requisite tools.

One of the most important negative results in the history of mathematics, and one that continues to have a large impact on mathematical research, is a 1931 theorem of the logician Kurt Gödel (1906–1978), the *incompleteness theorem*. Gödel's theorem is concerned with limitations on what can be proved about the positive integers. The positive integers are a rich area of mathematics that has fascinated mathematicians for centuries. Gödel's shocking theorem is that there are true statements about positive integers that cannot be proved starting with the usual axioms about positive integers. The "usual axioms," known technically as the Peano Axioms, include such suppositions as the fact that every positive integer has a successor (a next larger positive integer). Gödel's theorem furthermore states that even if we added more axioms to the system of positive integers, there would still be true statements that cannot be proved within the system.

Kurt Gödel (1906–1978)

Kurt Gödel was born in Brünn, Austria-Hungary, now Brno, Czech Republic. He earned his doctorate at the University of Vienna, where he remained on the faculty until World War II started. He then fled east, taking the trans-Siberian railway across Russia to the Pacific, a ship to America, and another train to New Jersey, where he joined his friend Albert Einstein at the Institute for Advanced Study, from which he retired in 1976.

In 1931 Gödel published his incompleteness theorem in a landmark paper, "On Formally Undecidable Propositions of *Principia Mathematica* and Related Systems." The *Principia Mathematica* (1910), a three-volume book by Bertrand Russell (1872–1970) and Alfred North Whitehead (1861–1947), was a monumental effort to rewrite the foundations of mathematics as a branch of logic. The reasoning is slow and meticulous, with the proposition "$1 + 1 = 2$" not appearing until page 379.

Gödel was also interested in philosophy and physics. In 1949, he showed that relativity theory allowed for the possibility of time travel.

Gödel fought mental illness throughout his life. He became paranoid, convinced he was being poisoned. He ate food only after his wife had tasted it. When she became ill and had to be hospitalized, he starved himself to death.

The proof of Gödel's theorem owes a debt to a venerable logical conundrum known as the Liar's Paradox. Consider the statement "This is a false statement." Is this statement true or false? There is a dilemma, because if the statement were true, then it would be false, and if it were false, then it would be true. This is considered a paradox but not a serious mathematical one. What Gödel did was to make this paradox mathematically precise. He was able to formulate, within the system of positive integers, the following statement.

Statement X: "Statement X is not provable within the system of positive integers."

Statement X must be true, because if it were false, then statement X would be provable and hence we could prove a false statement (which isn't allowed). Since statement X is true, what it asserts is true, namely, that it is not provable. Therefore, statement X is a true statement that is not provable within the system of positive integers.

Throughout the 20th century, Gödel's ideas on the limitations of formal mathematical systems propagated into other areas of mathematics. An important offshoot

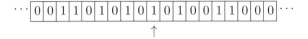

Figure 3.18   Turing machine.

is the analysis of algorithms proposed by the logician Alan Turing (1912–1954). An algorithm is a step-by-step method for solving a problem. In the 1930s (before modern computers were invented) Turing studied the question of what an algorithm can do, and concluded that all algorithms could be carried out on a theoretical computer now called a Turing machine (Figure 3.18). A Turing machine consists of a linear tape with square cells containing the symbols 0 or 1, and a pointer (↑) that points to one cell at a time. The machine is able to read the symbol indicated by the pointer, and decide whether to change the symbol and/or move the pointer.

Although a Turing machine is very simple, it is very powerful, as it can calculate anything that can be calculated. It can do computations, produce and store data, and make decisions about what steps to perform next. However, Turing conceived of a task that a Turing machine cannot carry out, and which is therefore impossible for any machine to carry out. The task is called the Halting Problem. He considered the following question: Is there an algorithm which will check other algorithms to see whether or not they contain infinite loops? (An infinite loop is a set of steps that repeats forever, not allowing the program to halt.) If you program computers you will realize that this would be a valuable program to have. When you finish writing a program you could call up this wonderful master program—call it the Halting Program—and the Halting Program would tell you whether your program has an infinite loop, that is, whether it will halt. Turing proved mathematically that the Halting Program cannot be written—it is a logical impossibility. The solution to the Halting Problem implies a logical contradiction, so the Halting Problem is unsolvable. Thus, computers cannot be used to solve every problem.

Alan Turing (1912–1954)

Alan Turing was born in London. His father worked in the Indian Civil Service, traveling between England and India, but Alan was educated in England. He did his undergraduate work at Cambridge, after which he was elected to a fellowship there. He submitted his famous paper on the Turing machine in 1936. The same year, he traveled to America, and he earned a PhD at Princeton University in 1938.

During World War II, Turing worked at Bletchley Park back in England and was a major participant in the successful effort to break German codes. After the war, in addition to foundational work in the theory of computing, he wrote a seminal paper on artificial intelligence. In the paper, he proposed what is known as the Turing test: if a human operator communicates remotely with a computer and another human, and can't tell which is which, then we must say that the computer has human-like intelligence. All of Turing's theoretical work in computer science was done before modern computers were in existence.

Turing was homosexual in an intolerant society. He was prosecuted, lost his security clearance, and given the choice of prison or chemical castration. He chose the latter. Two years later, Turing died of cyanide poisoning, probably a suicide, at age 41.

$$\sim$$

Computation

Our consideration of Turing's analysis of algorithms leads to a consideration of the role of computers in mathematics.

Computation is about solving problems. Which problems are computable and which are not computable? We have already described some problems, such as the Halting Problem, that cannot be solved at all. This is an extreme situation. For other problems, solutions of a qualified flavor exist. In some cases, the idea of a problem solution is fairly straightforward but the solution is impractical because of the heavy computations involved. In other cases, clever methods allow for fast solutions of seemingly difficult problems.

Some problems have only been solved using computers. One of the most prominent is the solution of the famous Four Color Conjecture, which was proposed in the 1800s and finally solved in 1976 by Kenneth Appel and Wolfgang Haken from the University of Illinois. The problem is simply stated: Can one color the regions of every map with four colors so that no two adjacent regions have the same color? Most people believed that you can color every map in such a way. The difficulty was to prove it for all possible maps. When Appel and Haken confirmed this using a computer, the state of Illinois issued a postal seal declaring "Four Colors Suffice." Appel and Haken solved the problem by examining some 1500 irreducible maps (a special term they had) on a computer. Many mathematicians are skeptical about computer proofs because they can't be verified directly by humans, only by other computers.

Some problems are solvable in theory but intractable even using a computer. The NP[1] problems have the property that there is a known algorithm for their solution but the algorithm requires exponential time on a computer, so that even to find a solution for a small case might take billions of years on the fastest available machine. For instance, in the Traveling Salesman Problem, one is given a list of cities and the distances between cities, and is asked to find an optimal route (one with least total distance) that visits all the cities. This is a practical problem, as it can model shipping or telecommunications situations.

In contrast, it has recently been shown that the problem of determining whether a given positive integer is a prime number is in the class P of deterministic polynomial algorithms. These algorithms can be executed relatively quickly.

[1] NP stands for "nondeterministic polynomial."

Some problems have been solved only by a large group of researchers working over a long period of time. A prime example of this (pun intended) is the classification of finite simple groups (to be described shortly).

The security of many types of transactions may ultimately depend on the intractability of prime factorization. The problem is to factor a number (positive integer) into its prime factors. A number is prime if it has only one and itself as divisors. The fastest computers today are unable to factor a reasonably large number (about 1000 digits) in a reasonable length of time (about one year). People working in the national security field use codes that rely on the factorization of large numbers. The banking system has been revolutionized by this idea. But a caveat: there are advances in factoring algorithms and computer speed every year. No code is unbreakable.

To natural philosophers of the 1700s, such as Lagrange and Laplace, the unsolvability of a problem would have been unimaginable. They believed that any problem that you can state precisely must have a solution. They believed, for example, that if you could specify the positions and velocities of every particle in the universe then you could determine the universe's past and ultimate future. We now know that even the 3-body problem has no mathematical solution. One cannot take a system of three particles and determine with certainty their future states or their past states.

Mathematical Abstraction

Recall the group, the mathematical structure that became so important in the 19th century. The notion of the group developed only gradually throughout the century. It first appeared in concrete settings, then in more general settings. Various mathematicians realized that they could consider an *abstract* group, one whose properties were reduced to the minimum necessary to capture the essence of the concept. This abstract group could then be studied, and any theorems one could prove in this setting would apply to all of its concrete realizations.

As an example, we will prove an elementary theorem about groups, using only the definition. Recall that a group consists of a set G and an operation that associates to any two elements x and y of G a third element xy of G, with these properties.

1. The operation is associative: for any x, y, and z in G, we have the equality $(xy)z = x(yz)$.

2. There exists an identity element, e, such that, for any x in G, we have $xe = x$ and $ex = x$.

3. Each element x in G must have an inverse, an element y in G such that $xy = e$ and $yx = e$.

Note that the last property asserts only the existence of inverses. Is it possible that an element could have more than one inverse? No.

Theorem. Each element of a group has only one inverse.

Proof. Suppose that x has two inverses, say, y and z. Then we have $xy = e$ and $yx = e$, and also $xz = e$ and $zx = e$. Using our properties,

$$\begin{aligned}
y &= ye \quad \text{(definition of identity)} \\
&= y(xz) \quad (z \text{ is an inverse of } x) \\
&= (yx)z \quad \text{(associativity)} \\
&= ez \quad (y \text{ is an inverse of } x) \\
&= z \quad \text{(definition of identity)}.
\end{aligned}$$

Therefore, $y = z$, which is what we needed to prove.   ∎

(The little black box on the right above means that the proof is done. Some people use the abbreviation *Q.E.D.* to mark the end of a proof. In Euclid's great book, the *Elements*, he concluded the proof of each proposition with "which was to be proved." This has come down to us, through translations, as the Latin *quod erat demonstrandum*, or *Q.E.D.* for short.)

Although the importance of groups was realized in the 19th century, the 20th century saw the enshrinement of abstraction as the hallmark of modern algebra, and many other abstract algebraic structures have been studied. We will mention (without formal definitions) two of them. The first is the *ring*. This notion captures many, although not all, of the features of the integers. In particular, it adds another operation. The two operations, naturally enough, are called addition and multiplication.

An abstraction is of limited usefulness if there is only one concrete realization. In fact, other rings of interest arose in number theory. Ernst Kummer (1810–1893), as a result of attempting to prove Fermat's Last Theorem, developed a new set of numbers, those of the form

$$a + b\sqrt{5}i$$

where a and b are integers. This set of numbers forms a ring under the usual addition and multiplication. This ring is interesting because there is no equivalent to the Fundamental Theorem of Arithmetic—unique factorization into primes. For example, $6 = 2 \cdot 3$ and $6 = (1 + \sqrt{5}i)(1 - \sqrt{5}i)$ are two different factorizations of 6 into the equivalent of primes, called irreducibles. (They cannot be factored in the ring.) By comparison, the ring consisting of numbers $a + bi$, where a and b are integers, does enjoy unique factorization. These numbers are called the *Gaussian integers*.

Another algebraic abstraction is the *field*. Rings do not have division (e.g., dividing two integers does not always result in an integer). Of course, our familiar real numbers do have division, as long as you don't divide by 0. A field is an abstract concept that captures the arithmetic of real numbers, including division.

Here is a different example of a field. The elements of this field are the numbers $0, 1, 2, 3, 4$. Addition and multiplication are the usual addition and multiplication, except that at the end we take the remainder upon division by 5. For example, $2 + 4 = 1$, since the remainder of 6 is 1. Similarly, we can write $2 \cdot 4 = 3$. One can show that this structure is a field. In particular, we can define division.

What is division? We write $a/b = c$ if $a = bc$. In our field of five elements, we can write $3/2 = 4$ since $3 = 2 \cdot 4$. Of course, it takes some work to show that this division obeys all of the usual field properties. More on this example and others can be found in Chapter 5.

Rings, fields, and many other abstract mathematical structures were codified in the early decades of the 20th century. The most important researcher in this development was Emmy Noether.

Emmy Noether (1882–1935)

Amalie Emmy Noether was born in Erlangen, Germany, the daughter of a mathematics professor. She originally studied French and English, and prepared to be a teacher. Although she passed the examination to become a teacher, she had in the meantime become more interested in mathematics. She audited classes at the University of Erlangen until women were officially allowed as students in 1904. She obtained her doctorate there in 1907. In the years 1907–1915 she taught at Erlangen, but was not paid.

In 1915 Noether was invited to the University of Göttingen by David Hilbert and Felix Klein to work on Einstein's new theory of general relativity. They wanted her to be hired as a privatdozent ("private lecturer," a sort of associate professor), but other faculty blocked the appointment because of her sex. Hilbert's indignant reaction: "I do not see that the sex of the candidate is an argument against her admission as privatdozent. After all, we are a university, not a bath house." For several years, she taught at Göttingen without position and without pay. Eventually, she did obtain an official position, and after 1922, was even paid.

In the late 1920s and early 1930s, Noether gathered a strong group of students around her. She was known as a generous teacher, freely sharing her great insights. Her students were themselves successful.

In 1933, with the ascent of the Nazis, many Jewish scholars lost their positions, including Noether. She accepted a position at Bryn Mawr College in a suburb of Philadelphia. There she was conveniently near the Institute of Advanced Study in New Jersey, which was home to fellow refugees Albert Einstein and Hermann Weyl, among many other leading scientists. Noether herself lectured at the Institute in 1934. The nearby Princeton University was not so welcoming; she described it as the "men's university, where nothing female is admitted."

In addition to her mathematics, Noether contributed to physics. In 1915 she proved Noether's Theorem, one of the most important theorems in modern physics. This theorem connects symmetries with conservation laws. For example, the fact that physical laws don't change over time is related to the conservation of energy.

Emmy Noether died in 1935, after surgery to remove a tumor.

Mathematical abstraction is not restricted to algebra. Another example comes from probability. This field had long been plagued by uncertainty in its foundations, which led to various paradoxes. These difficulties were overcome in 1933 by the Russian mathematician Andrey Nikolayevich Kolmogorov. He formulated a set of axioms which define a *probability space*. This concept is the bedrock on which modern probability theory has been built.

Andrey Nikolayevich Kolmogorov (1903–1987)

Andrey Nikolayevich Kolmogorov was born in Tambov, Russia. His mother died giving him birth. He was raised by his sister and aunt, who moved to Moscow when Andrey was seven years old.

As a student Kolmogorov had many interests. In 1920 he was enrolled simultaneously at the Moscow State University and the Chemistry Technological Institute. His first published paper was on history. He eventually settled on mathematics, and obtained his doctorate from Moscow State University in 1929, having already published a number of important papers. He then worked at that university, becoming professor in 1931. He held positions at a number of Moscow research institutions, but remained associated with Moscow State University his entire career.

Although Kolmogorov is best known for his work in probability theory, he contributed to many fields, including chemistry, ecology, physics, and computer science. He did major work on the theory of turbulence. In computer science, he was a founder of algorithmic information theory, where people study "Kolmogorov complexity."

Kolmogorov was important as well for his work in education. Many of his students obtained PhDs, and a number of them are quite famous for their own work. He wrote influential textbooks, and worked as well to improve education at secondary schools, especially for gifted students.

Finally, we mention a remarkable development of the late 20th century in group theory. It was inspired by a work published in 1870 by Camille Jordan (1838–1922), who found a way of decomposing the structure of a group that was analogous to how we represent integers as products of primes. The role of prime was played by what we call a *simple group*. Just as a prime is an integer that in some sense cannot be decomposed, a simple group is a group that cannot be broken down further. We have discussed some examples of simple groups: the symmetry group of the dodecahedron and the symmetry group of a seven-point geometry. All groups with a prime number of elements are simple. Simple groups are the building blocks that make up all symmetry in math and physics, and perhaps in art and music as well. After Jordan's work, it was natural to try to classify the simple groups, i.e., to find all of them.

In 1986 an announcement was made that the classification of all finite simple groups had been achieved. What made this so remarkable is the degree of collaboration required. It was a mammoth project that engaged more than 100 scientists from 10 nations, working over 50 years, writing about 10,000 pages of journal articles. It is doubtful that a single person could have done it. Perhaps joint effort will be the norm in the future. There may well be problems in medicine, biology, and other sciences whose solutions will take a concentrated and orchestrated effort over hundreds of years.

The difficulty of this project was underscored when, after the initial announcement, it was discovered that the classification of finite simple groups wasn't fully proved, as some omissions in the classification were found. In 2004 the completion of the classification was announced. These days, mathematicians are trying to simplify the hugely complex proof.

Game Theory

Game theory concerns mathematical questions about conflict. As a science, game theory had tentative beginnings in 1921 with the work of Émile Borel (1871–1956), who defined games of strategy, but began in earnest with the 1928 publication of the article "Zur Theorie der Gesellschaftsspiele" ("On the Theory of Parlor Games") by John von Neumann (1903–1957).

As used in game theory, the word "game" is more precise and focused than in general language. For example, common turn-based board games—such as tic-tac-toe, checkers, and chess—are usually not considered games in game theory. Most people who play tic-tac-toe for a period of time learn that there is little point in playing. Two experienced players will always achieve a tie. Board games such as checkers and chess share the same underlying characteristic: there is a single optimal (pre-scripted) way to play the game. It may not be feasible to explore every possible scenario of chess, but there *is* some optimal first move, second move, and so on. Games in game theory generally cannot be solved by exhaustive search.

Zero-Sum Games Although it is not a strict requirement, game play often occurs simultaneously rather than in turns. For example, a simple game between two players might work like this. Alice and Bob sit at a poker table. In each round of play, both players hide their hands under the table with a single poker chip, which they may or may not conceal in a fist. They bring their fist above the table and simultaneously open their hands to show. The outcome of a round in the game will be determined by these rules:

- If both Alice and Bob have a chip, then Alice will pay Bob $3.

- If both Alice and Bob have empty hands, then Alice will pay Bob $1.

- If one has a chip and the other is empty, then Bob will pay Alice $2.

It is common to describe games of this type in matrix form, and in fact they are often called matrix games. As a matrix, this game might look like:

		Alice	
		Chip	Empty
Bob	Chip	$(3, -3)$	$(-2, 2)$
	Empty	$(-2, 2)$	$(1, -1)$

Each ordered pair describes one possible outcome of the game, with the first number indicating the payoff for Bob and the second indicating the payoff for Alice. The pair $(3, -3)$, for example, indicates that Bob has gained $3 while Alice has lost $3; that is, Alice pays Bob $3 when they both have a chip. This game is called a *zero-sum* game, because in each individual outcome the sum of the payoffs is always 0.

Von Neumann's 1928 paper contained the "minimax theorem," which states that every 2-player zero-sum game has an equilibrium solution. Each player has a strategy that completely determines the long term outcome of the game. For example, in our game between Alice and Bob, Alice has this equilibrium strategy: she can randomly choose to play a chip $3/8$ of the time and to play an empty hand $5/8$ of the time.

No matter how Bob plays against Alice's equilibrium strategy, he can expect a long term payoff of $-0.125 per round. If he plays a chip, he has probability $3/8$ of winning $3, and probability $5/8$ of losing $2, for an expectation of $E = 3(3/8) - 2(5/8) = -0.125$. If he plays empty, his expectation is instead $E = 1(3/8) - 2(5/8)$, which is also -0.125. Notice that since the game is zero-sum, she must win $0.125 per play on average to balance Bob's loss, so Alice's equilibrium strategy is a winning strategy for her.

The minimax theorem guarantees that Bob also has an equilibrium strategy, and for this game it is to play a chip $3/8$ of the time and to play empty $5/8$ of the time (note that though this game is symmetric, it won't happen in every game that both players have identical strategies). If Bob follows this strategy, then no matter how Alice plays, her long term expectation is to win $0.125 per round.

It is not a coincidence that Bob's equilibrium strategy and Alice's equilibrium strategy lead to the same long term outcome. This always happens, and the long term value is said to be the *value of the game*. In this case, the value of the game is $-0.125 to Bob (or equivalently $+0.125 to Alice). Even for zero-sum games, game theory can't necessarily guarantee a winning outcome in an inherently unfair game. In this game, Bob at best hopes to lose "only" $0.125 per round on average.

Non-Zero-Sum Games Not every game in game theory has the property that the losses of one player exactly balance the gains of the other. Some of the most interesting games are ones where both players may lose or both may win. Games like this are referred to as non-zero-sum games. The most famous of these games is the "prisoner's dilemma," which was introduced by Merrill Flood and Marvin Dresher in 1950, and has come to be described by a story:

> Alice and Bob are arrested for committing some crime. The police don't have enough evidence to convict them on the full offense, and each defendant is facing a one year prison term (conviction on the full offense carries a five year prison term). Alice and Bob are individually offered an opportunity to testify against the other in return for a one-year reduction in sentence.

Both prisoners face the choice of cooperating with (by keeping silent) or defecting from (by testifying against) their partner in crime. As a matrix, the game looks like this:

		Alice	
		Cooperate	Defect
Bob	Cooperate	$(-1,-1)$	$(-5,0)$
	Defect	$(0,-5)$	$(-4,-4)$

The prisoner's dilemma is called a dilemma game because individual self interest leads to a poor result for both players. Consider the game from Bob's perspective. If both players keep silent, the outcome of the game is $(-1,-1)$ and he can anticipate one year in prison (as can Alice). If he believes that Alice will keep quiet, he can defect and testify against her to reduce his prison term to zero. But what if Bob knew that Alice would testify against him? He would face a five year prison term, and he should still defect in order to reduce the term to four years.

No matter what Alice does, it is in Bob's best interested to defect. Of course, Alice's point of view is similar, and the same reasoning requires that she should defect. However, the value of the game is $(-4,-4)$ if they both defect. Each, by independently acting in their best interest, has helped guarantee a worse outcome for both. If they had only cooperated, they would both be better off.

Like zero-sum games, non-zero-sum games have equilibrium solutions. They are called Nash equilibriums after the mathematician John Forbes Nash, Jr., who worked in game theory and whose life inspired the movie *A Beautiful Mind*. In the prisoner's dilemma, the Nash equilibrium is mutual defection.

The prisoner's dilemma is particularly interesting for its similarity to other social dilemmas, such as the tragedy of the commons, the free rider problem, and the (nuclear) arms race. It also has applications to evolutionary biology and psychology.

John von Neumann (1903–1957)

John von Neumann, born in Budapest in 1903, was the oldest son of a successful banker. He was raised in an environment of learning and from an early age showed an aptitude for languages and for mathematics. He was fluent in numerous languages, including Hungarian, German, Classical Greek, and English. He had a photographic memory, able to remember anything he read, and as a boy would memorize pages of the telephone directory as a party trick.

In 1921 von Neumann enrolled simultaneously at the University of Budapest and the University of Berlin. Although he wanted to pursue mathematics, he studied chemistry as a compromise with his father. By 1925 he had a degree in chemical engineering from the Swiss Federal Institute of Technology, and a year later he had earned a PhD in mathematics from the University of Budapest (with minors in physics and chemistry). He taught in Germany until 1930 when he came to Princeton.

Von Neumann did most of his pure mathematical research from 1925 to 1940. It was during this period that he essentially founded the area of game theory. He is said to have contributed in some significant way to most branches of mathematics, including set theory, logic, abstract algebra, and operator theory.

In 1943 von Neumann began work on the Manhattan Project. He was considered so important that he was one of the few scientists who had full knowledge of the entire project. His calculations were particularly instrumental in the success of the implosive design of the "Fat Man" nuclear bomb that was dropped on Nagasaki, Japan August 9, 1945.

Though he had a brilliant calculating mind, von Neumann was also a sociable, likeable person. He enjoyed throwing large parties, telling bawdy jokes, and playing pranks on his friends. Apparently his friends were not above returning the favor, according to an anecdote related in William Poundstone's *The Prisoner's Dilemma*.

> [Once] a young scientist at the Aberdeen Proving Ground worked out the details of a mathematical problem beforehand and came up to von Neumann at a party. The scientist presented the problem as one that had been stumping him. Von Neumann gazed into the middle distance and began to calculate. Just as he was about to arrive at each intermediate result, the scientist interrupted, "It comes to this, doesn't it?" Of course he was right each time. Finally the young scientist beat von Neumann to the answer. Von Neumann was shaken until he found out it was a setup.

Von Neumann's work after the war was distinctly applied. He helped the United States develop the hydrogen (fusion) bomb ahead of the Soviet Union. He served as consultant to defense agencies, the CIA, and several corporations.

One of his most lasting contributions was to the design of the modern computer, which is known in computer science as the "Von Neumann architecture." Among his ideas were that software programs should be stored on the computer as data (as opposed to being hard-wired separately), that the computer should work in a digital fashion (using discrete values like 0 and 1 rather than a continuous range of voltages), and that the computer should employ binary as its internal number system (rather than base 10 arithmetic). All popular modern computers share these very characteristics.

Von Neumann died from bone cancer in 1957, approximately one year after receiving the Medal of Freedom from President Dwight D. Eisenhower.

$\sim$

Graph Theory

In graph theory, a *graph* consists of a set of vertices, with some pairs of vertices joined by edges. Where the vertices are placed, and how the edges are drawn— straight or curved—doesn't matter.

Graph theory was introduced by Leonhard Euler in the 1700s, and then further developed by mathematicians in the 1800s. Research in graph theory accelerated in the 1900s and is going strong in the 2000s. Although Euler used graphs to solve the Königsberg Bridge Problem in 1735, the first graph theory textbook did not appear until two centuries later, in 1936: *Theorie der endlichen und unendlichen Graphen* (*Theory of finite and infinite graphs*) by Dénes König (1884–1944). See Section 5.12 for more on the Königsberg Bridge Problem.

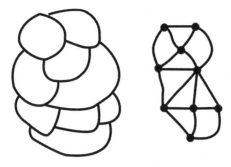

Figure 3.19   A map and its graph.

We mentioned the Four Color Conjecture (now Theorem) earlier. This is really a theorem of graph theory in disguise. We can replace any map with a graph (see Figure 3.19), where each region is represented by a vertex, and two vertices are joined by an edge when the two regions share a boundary. In the language of graph theory, the Four Color Theorem says that the vertices of every planar graph (one that can be drawn in the plane without edge crossings) can be colored with four colors so that no two adjacent vertices are the same color.

Do you see how the vertices of the graph in the diagram can be colored with *three* colors so that no two adjacent vertices are the same color?

We have talked about planar graphs. Let's take a look at a nonplanar graph, one that cannot be drawn in the plane without edge crossings. Our graph is the complete graph on five vertices, known as K_5 (Figure 3.20). In a complete graph, every two vertices are joined by an edge. Try as we may, we cannot draw this graph on a flat sheet of paper so that no edges cross.

Figure 3.20   The complete graph on five vertices, K_5.

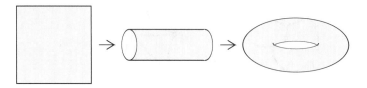

Figure 3.21   Construction of a torus from a square.

However, we can draw this graph without edge crossings on a different surface. The surface is known technically as a *torus*, which is just a fancy way of saying a closed surface with a hole in it. This is the third shape in Figure 3.21.

In the square model of the torus, points on opposite edges are identified. Thus, we can wrap the square around and glue together its top and bottom edges. This produces the cylinder in the middle diagram. The cylinder is then wrapped around and the two circular ends are glued together to form the torus.

Figure 3.22 shows K_5 on a torus without edge crossings. Each vertex is joined to every other vertex. Remember that lines going to the edge of the square come around to the other side.

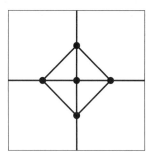

Figure 3.22   A K_5 on a torus.

A surface may have more than one hole. The number of holes is called the *genus* of the surface. In 1968 Gerhard Ringel (1919–2008) and J. W. T. Youngs (1910–1970) proved that the minimum genus for a surface on which one can draw the complete graph K_n without edge crossings is

$$\left\lceil \frac{(n-3)(n-4)}{12} \right\rceil, \quad \text{for } n \geq 3.$$

The notation $\lceil x \rceil$ is the "ceiling function." It outputs the least integer at least as large as x. Try the formula for $n = 3$, 4, and 5, to see that it gives the correct values.

Surfaces are studied in the branch of mathematics called *topology*. Topology is a generalization of ordinary geometry in which shapes can bend and stretch but not break.

Our time in history has been called the "Information Age," and with good reason. People are interconnected more than ever by the Internet and telephone communication. The World Wide Web can be modeled by a graph in which the edges are directed. Jon Kleinberg (b. 1971) is a contemporary researcher in this area of information theory. His work has shown that it is useful to think of the Web's structure as consisting of "authorities" and "hubs." Authorities have information to disseminate, while hubs link to many authorities. The mathematics of directed graph networks is being developed by today's mathematicians.

Paul Erdős (1913–1996)

Paul Erdős was born in Budapest, Hungary, the son of two high school mathematics teachers. His two older sisters died of scarlet fever the day he was born. He had a strange childhood. His father was captured in World War I, and spent six years in a Russian prisoner-of-war camp. His mother, after losing her daughters, was overly protective, not allowing Paul to go to school until age ten. He was quite a prodigy, being able to multiply three-digit numbers in his head at age three.

In 1934, at the age of 21, Erdős was awarded a doctorate in mathematics at Péter Pázmány University in Budapest. After graduation, with anti-Semitism increasing in Hungary, he moved to England, accepting a position at the University of Manchester. In 1938 he went for a year to the Institute for Advanced Study in New Jersey. During this year, he cofounded probabilistic number theory, one of many fields he started.

Beginning in the 1940s, Erdős became an itinerant world-traveler with no fixed address and no employment, living out of two half-filled suitcases. He collaborated with mathematicians in many countries, staying a few days to several months in each location, announcing upon his arrival that "My brain is open."

Much of Erdős' early work was in number theory. His most famous accomplishment there was his proof, with Atle Selberg, of the Prime Number Theorem. This had been proven in the 19th century, but only by using the sophisticated machinery of complex analysis. Erdős and Selberg found an "elementary" proof, one that did not require complex analysis.

During his later career, Erdős, while continuing to work in many areas, spent more time on discrete mathematics, pioneering several new branches, such as random graph theory (where the edges of a graph appear or don't appear according to a probabilistic model).

Erdős collaborated with so many mathematicians that the term "Erdős number" was coined to describe how close any mathematician was to Erdős. Erdős himself has Erdős number 0. Those who wrote a mathematical paper with Erdős have Erdős number 1. Those who wrote a paper with someone having Erdős number 1 have Erdős number 2, and so on. There are 511 mathematicians with Erdős number 1. Einstein's Erdős number is 2.

Erdős was a colorful character, and many stories are told of him. One of his notions was *The Book*, a transfinite work in which God keeps every theorem, with its most elegant proof. Although whimsical, *The Book* reflected Erdős' heartfelt Platonism, in which mathematics was discovered, not invented.

Erdős was the second most prolific mathematician in history, after Euler. Partly powered by amphetamines, he continued to produce mathematics into his eighties, dying of a heart attack at a mathematics conference in Warsaw. In 1993 director George Csicsery made the documentary *N Is a Number: A Portrait of Paul Erdős*, a fascinating window into the world of mathematical research, and the men and women who inhabit it.

Information Theory

Claude E. Shannon (1916–2001) was a pioneer in the field of information theory. He introduced the principle of *entropy*, which is randomness in an information channel. (Entropy is a concept in physics that Shannon transferred to information theory.) Shannon's investigations were motivated by practical problems: transmitting over noisy phone lines or detecting and correcting errors in calculations on the primitive computers of the 1940s.

Information can be encoded as a stream of 0s and 1s. For example, the stream

$$1\ 0\ 1\ 1\ 0\ 1\ 0\ 1\ 1\ 1\ 0\ 1\ 1\ 1\ 0\ 1\ 1\ 0\ 1\ 0\ 1\ 1\ 1\ 0\ 1\ 1\ 0\ 1\ 0\ 1 \ldots$$

might represent the words of a novel. If this information is stored and later retrieved, the person retrieving it could decode the stream and recover the words. In information theory, we are interested in the quantity of "information" conveyed by the symbol stream. If the two symbols, 0 and 1, are equally likely to appear, and the current symbol has no dependence on the previous symbols, then each new symbol conveys one *bit* of information. ("Bit" stands for "binary digit.")

However, in the source output above, it seems that the symbol 1 is more prevalent than the symbol 0, although we see only a finite number of symbols. Twenty 1s and ten 0s are shown. Hence, we may conjecture that 1 is twice as likely to occur as 0. Let's assume this for the sake of argument. Thus, the probability that a 1 occurs is $2/3$ and the probability that a 0 occurs is $1/3$. We also assume that the currently output symbol has no dependence on any preceding symbols. This information stream is different from one in which the two symbols are equally likely. It is less random, since we can guess with some measure of certainty what the next symbol will be (bet on a 1). Correspondingly, the information source conveys less information than one in which 0 and 1 are equally likely. When a symbol is revealed, it doesn't convey as much information, since we knew in advance that the symbol was rather likely to be a 1. In information theory, randomness and information are synonymous terms.

How much information is conveyed in a stream? If the two symbols occur with probabilities p and $1 - p$, then the entropy of the stream is defined as

$$-p \log_2 p - (1 - p) \log_2(1 - p).$$

Notice that the logarithms are base 2. (To compute logarithms base 2, use the formula $\log_2 x = \ln x / \ln 2$.) For our example stream, the entropy is

$$-(2/3) \log_2(2/3) - (1/3) \log_2(1/3) \approx 0.918296 \text{ bits.}$$

Each symbol in the stream reveals approximately 0.918296 bits of information.

One of Shannon's theorems of information theory tells us how economically we can encode information. It says that the average number of 0s and 1s needed to encode a symbol is equal to the entropy (randomness) of the stream. Let's apply this concept to our example. Perhaps we can devise a code that represents the same stream, but using fewer symbols. The standard way to accomplish this is to encode pairs of consecutive symbols. There are four possible pairs: 11, 10, 01, and 00. Given the probabilities of 1 and 0 occurring, the pairs occur with probabilities 4/9, 2/9, 2/9, and 1/9, respectively. The table shows an economical way to encode the pairs.

pair	probability	code
11	4/9	1
10	2/9	01
01	2/9	001
00	1/9	000

Notice that no code string is the prefix (beginning) of another code string. This allows us to decode a stream as it is output. Try encoding the above stream and note that the encoded stream requires fewer than thirty symbols.

Our encoding scheme requires, on average,

$$\frac{4}{9}(1) + \frac{2}{9}(2) + \frac{2}{9}(3) + \frac{1}{9}(3) = \frac{17}{9} \text{ bits.}$$

(The numbers in parentheses are the lengths of the code strings.) Since this code represents pairs of symbols, we divide the average number of bits by 2 to get the average number of bits necessary to encode each single symbol of the information stream: $17/18 \approx 0.944444$ bits. This is a reduction from 1 bit.

Shannon proved that the data stream can be encoded so that the average number of bits per source symbol is as close to the entropy as we like, but never below it. Thus, entropy is a measure of the compressibility of the information stream.

Perhaps the greatest applications of Shannon's theory is to information conveyed over a channel with "noise" (distortion of information). If information is sent over an information channel, such as a computer connection, noise may result in loss of data. One of Shannon's theorems states that, using an encoding procedure, information can be sent over a noisy information channel in a reliable way, without loss of information.

Claude E. Shannon (1916–2001)

Claude E. Shannon, the "father of information theory," was born in Michigan and did his undergraduate work at the University of Michigan, earning degrees in mathematics and electrical engineering. He did his graduate work at the Massachusetts Institute of Technology (MIT). In perhaps the most famous master's thesis ever, he applied Boolean algebra to establish the mathematical theory of digital circuits. (The Russian Victor Ivanovich Shestakov (1907–1987) independently applied Boolean algebra to the problem, but published somewhat later.) Shannon's PhD dissertation was *An Algebra for Theoretical Genetics.*

Shannon joined American Telephone and Telegraph's Bell Laboratories in 1941. This famous industrial research institution gave us, in addition to Shannon's information theory, transistors and the digital camera. Work done at Bell Labs has won seven Nobel prizes in physics. During the war, Shannon also worked on cryptography and on antiaircraft control systems.

Shannon published his groundbreaking paper on information theory in 1948. It was an immediate hit with engineers. It found wide applications. Shannon himself applied it to linguistics and to more lucrative pursuits, making a lot of money at blackjack and in the stock market.

In 1956 Shannon joined MIT, although he retained a connection with Bell Labs for many years after this. He retired in 1978 and died in 2001, a victim of Alzheimer's disease

In addition to theory-building, Shannon enjoyed building strange machines. One of them was a mechanical/electronic device that solves Rubik's Cube. Another is called the Ultimate Machine. You can probably find a video of this machine in action on YouTube.

$\sim$

The Millennium Prize Problems

Since David Hilbert gave his speech in 1900 outlining 23 important unsolved problems in mathematics, most of these problems have been solved. In 2000 the Clay Mathematics Institute announced a list of seven unsolved problems, known as the Millennium Prize Problems, whose solutions carry a prize of $1 million each. Since then, one of the problems, called the Poincaré Conjecture has been solved. The solver, Grigori Perelman, declined the prize.

Andrew Wiles, who proved Fermat's Last Theorem in 1995, gave a vivid description of what it is like to be a mathematician trying to solve unsolved problems.

> Perhaps I can best describe my experience of doing mathematics in terms of a journey through a dark unexplored mansion. You enter the first room of the mansion and it's completely dark. You stumble around bumping into the furniture but gradually you learn where each piece of furniture is. Finally, after six months or so, you find the light switch, you turn it on, and suddenly it's all illuminated. You can see exactly where you were. Then you move into the next room and spend another six months in the dark. So each of these breakthroughs, while sometimes they're momentary, sometimes over a period of a day or two, they are the culmination of, and couldn't exist without, the many months of stumbling around in the dark that precede them.

ANDREW WILES (B. 1953), FROM THE NOVA SPECIAL, *The Proof*, 1997

Conclusion

Mathematics is not written in stone. Unsolved problems abound and new techniques await discovery. Mathematics is a living subject. Every mathematical fact was discovered by a person or a team and is part of human history. Future mathematical facts will also be discovered by humans. (Mathematicians are sometimes aided in their work by computers, both in the discovery and the proofs of results.) This section has given an indication of some exciting new developments and open vistas in mathematics. Knowing something about the problems helps us appreciate the ongoing nature and challenge of mathematical work. It also helps us understand the extent to which mathematics affects our daily lives.

EXERCISES

3.58   Find an estimate for the number of mathematical research articles published per year.

3.59   Look up Hilbert's list of 23 problems. Report on one of them. What is the problem about? Is it solved? If so, who solved it?

3.60   Does the Diophantine equation $x^2 + xy + y^2 = 3$ have integer solutions?

3.61   Does the Diophantine equation $4x^2 + 4y^2 = 18$ have integer solutions?

3.62   At Hilbert's Grand Hotel, when infinitely many coaches arrive, each carrying infinitely many people, how can each person be given a room?

3.63   Show that a group has only one identity element.

3.64   This exercise concerns the field with elements $\{0, 1, 2, 3, 4\}$ described in the text.

a) Below is a partial table of addition in this field. For example, $0 + 2 = 2$ and $3 + 3 = 1$. Complete the table.

+	0	1	2	3	4
0			2		
1					
2					
3				1	
4					

b) Construct a multiplication table for the field.
c) Construct a subtraction table for the field. (Hint: $a - b = c$ means $a = b + c$.)
d) Construct a division table for the field.

3.65   Look up and explain one or more of the common dilemma games or social dilemmas:

- chicken dilemma

- stag hunt dilemma

- tragedy of the commons

- free rider problem

3.66   Investigate the "tit for tat" solution to the iterated prisoner's dilemma. Why does the method succeed?

3.67   Find a formula for the number of edges in the complete graph K_n.

3.68   Can K_6 be drawn on a torus without edge crossings? How about K_7 and K_8?

3.69   The graph $K_{3,3}$ consists of two sets of three vertices, and nine edges joining all pairs of vertices from the two sets. Draw $K_{3,3}$ on a torus without edge crossings.

3.70   Create a new 2-dimensional surface by starting with a square and identifying a pair of opposite sides, but twisting the square before joining them. The surface is called a *Möbius strip*. What unusual properties does this surface have?

3.71   Give an example of a map on a torus which requires five colors to be properly colored.

3.72   Find an encoding scheme for the information source shown on p. 212 that requires, on average, less than $17/18$ bits per symbol.

3.73   Prove (using calculus) that the entropy of an information source with two symbols is maximized when the two sources occur with equal likelihood.

3.74   Look up the Millennium Prize Problems. Report on one of them. What is the problem about? Is it solved? If so, who solved it?

3.75   Investigate a branch of mathematics not listed in this chapter. What is the branch of mathematics about? What are some of its applications? Are there any important unsolved problems?

3.5 The Future

This too shall pass.

ANONYMOUS

Mathematics has never been more vibrant than now. There are thousands of research mathematicians the world over, publishing original mathematics in hundreds of journals. A meeting of the main American mathematical societies in 2012 had almost 7200 registered participants. The International Conference of Mathematicians, held every four years, had representatives of more than 100 countries at its 2010 conference in Hyderabad, India. What might the future hold?

Predicting what mathematics will be developed is difficult for lesser mortals than David Hilbert. We can ask ourselves, however, in light of our knowledge of the history of the subject, whether and in what form the current golden age might continue.

First, there are several reasons to believe that it *will* continue, at least in the coming decades. One is the sheer number of mathematicians. The second is geographical distribution. The Fields Medal, the highest honor bestowed on researchers, has been awarded to 52 mathematicians, representing institutions in 10 countries. This widespread distribution suggests that, if support for mathematics declines in one or two major countries, there will be a haven for scholars in others, much as the United States was a haven for so many scholars fleeing Hitler.

The institutional base for mathematics has never been stronger. In first place among the institutions are the universities. But there are other important homes to mathematical research, including institutions such as the Institute for Advanced Study in Princeton, New Jersey, and corporate research labs such as Bell Labs in New Jersey or IBM Thomas J. Watson Research Center in New York. The largest employer of mathematicians in the United States is the National Security Agency.

The broad institutional support is a reflection of the breadth of mathematical applications in modern society, from traditional applications in the physical sciences, to more recent ones in communications, computing, statistics, biology and elsewhere. As long as mathematics is seen as essential to applications, the support is likely to continue.

Mathematics is healthy, but history has taught us that local conditions can and do often change. The center of research mathematics moved from Europe to America in the 20th century, although Europeans are still very active. Recently, the rise of Asia has been remarkable. The first Fields medalist based in Asia, Shigefumi Mori at the

University of Kyoto, won the award in 1990.[1] There is no reason not to expect that such geographical changes will continue.

Predicting the status of any cultural institution centuries in the future is futile. Physical conditions can change. As Will Durant wrote: "Civilization exists by geological consent, subject to change without notice." In addition, human societies are always changing, not always for the better. In this time of anti-evolution crusades and professional global warming skeptics, it is natural to reflect on the fragility of society's support of research. The linguist Nicholas Ostler put it this way. (He was primarily interested in the effect of science on the spread of English.)

> Dispassionate enquiry has never been an activity that appeals to a majority, however widely education is made available. Serious research remains a minority activity, which because it is disinterested will always need patronage from others who have accumulated power or wealth. But those political, military, business or religious elites cannot be trusted, especially if it seems that the results of enquiry are telling against their own power, or failing to buttress it: they will then often adjudicate in favor of tradition, or popular ignorance. It is easy to forget how much the ongoing popularity of science depends on its continuing to offer new golden eggs, or new golden bombs. When the flow of goodies slackens, as one day it may, the pursuit of science will be widely seen as an expensive indulgence by its paymasters, in industry and government.

NICHOLAS OSTLER (B. 1952)

---

[1] Mori was not the first Asian-born Fields recipient, however. There have been several others, starting with Kunihiko Kodaira in 1950, who worked at the Institute for Advanced Study. As always, centers of scholarship are magnets for immigrants.

TWO PILLARS OF
MATHEMATICS

CHAPTER 4

CALCULUS

To create a good philosophy you should renounce metaphysics but be a good mathematician.

BERTRAND RUSSELL (1872–1970)

We have taken an overview of mathematics in history. In this part of the book, we turn to two major subjects of mathematics, calculus and number theory, to be studied in detail. The first of these, calculus, is the focus of this chapter. Calculus has a rich historical tradition and is widely used in scientific endeavors today.

4.1 What Is Calculus?

Calculus is the mathematics of change. It deals with rates of change of all kinds. We can use calculus to compute physical rates, such as the rate at which a rocket rises or a bomb falls, or the rate at which a radioactive substance decays. We can compute rates in biology, such as the rate at which a bacteria colony grows, the rate at which solutions diffuse across a membrane, or the rate at which a disease spreads through a population. It has applications in economics, such as marginal cost, marginal profit,

Mathematics for the Liberal Arts. **221**
By Donald Bindner, Martin Erickson, Joe Hemmeter Copyright © 2012 John Wiley & Sons, Inc.

and elasticity of demand. We use calculus to solve engineering problems, like how to make a roller coaster exciting but not dangerous.

Using calculus, we can maximize and minimize quantities. How can we produce a can of green beans in the most efficient way possible? Calculus is part of the answer. How can we maximize the profit of a company, or an industry, or an economy? Calculus plays a part in the models that answer these questions.

If something changes, and nearly everything interesting changes, calculus may have a role in describing it or modeling it.

4.2 Average and Instantaneous Velocity

A good way to begin learning calculus is to think about falling objects. Consider the simple question: A pencil falls from a four foot high counter; how fast does it hit the floor?

To answer this question, we might create an experiment where we carefully time a pencil as it falls. If we did that, we would find that it takes the pencil almost exactly one-half second to fall four feet. Knowing that rate is distance divided by time, we might then conclude that the speed of the pencil is

$$\frac{4 \text{ ft}}{0.5 \text{ s}} = 8 \text{ ft/s}.$$

In practice, we try to be careful to measure distance by subtracting earlier positions from later positions. Since the height of the pencil at time zero was 4 feet and the height of the pencil after one-half second was 0 feet, we actually get the speed of the pencil to be

$$\frac{0 \text{ ft} - 4 \text{ ft}}{0.5 \text{ s}} = -8 \text{ ft/s}.$$

We use the word velocity to describe speeds when direction is important. In this case, the velocity is negative to indicate that the pencil is falling (i.e., moving in a downward direction).

A little more thought will convince us that this is not the right answer. Dividing total distance by total time gives us the *average* rate, the average velocity of the pencil. The pencil is barely moving when it first starts falling and then goes faster and faster under the influence of gravity. The average velocity (being somewhere in the middle) will be faster than the very slow speed at the beginning of the pencil's descent and too slow to be the speed at which the pencil hits the floor.

Perhaps, if we had a quick eye, we could see where the pencil was after 1/4 seconds. If we had a very quick eye, we would know that after 1/4 seconds the pencil was still about three feet above the floor. Since it falls the last 3 feet in only 1/4 seconds, we could redo our calculation to get

$$\frac{-3 \text{ ft}}{0.25 \text{ s}} = -12 \text{ ft/s}.$$

This is better, but it is still too slow for the same reason that our first estimate was too slow (because we want the velocity at the bottom, not somewhere in the middle of the descent).

What if we could continue making estimates over shorter and shorter time intervals? Over a very short interval, there isn't much time for the pencil to accelerate, so the average velocity and the velocity we want should be almost exactly the same.

It would take a fast eye to see where the pencil is $1/16$ or $1/32$ of a second before the pencil hits the floor, but a video camera takes pictures at a frame rate of one frame every $1/29.97 \approx 0.03337$ seconds. If we dropped our pencil on camera, we could flip through frames and discover these data points:

time (s)	height (ft)	time (s)	height (ft)
0.000	4.00	0.267	2.85
0.033	3.98	0.300	2.55
0.067	3.93	0.334	2.21
0.100	3.84	0.367	1.83
0.133	3.71	0.400	1.42
0.167	3.55	0.434	0.97
0.200	3.36	0.467	0.49
0.234	3.12	0.501	0.00

Graphically, the data look like Figure 4.1.

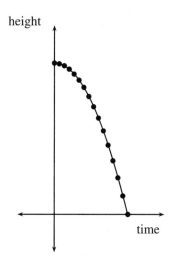

Figure 4.1   A graph of pencil heights over time.

In 1638 Galileo asserted that the distance fallen is proportional to the square of time. Thus, the distance an object falls has an equation of the form $d = kt^2$, where

k is a constant. Because our data points show how high the pencil is (starting from a 4 foot counter), adapting Galileo's idea, there should be a formula for our data that looks like $y = 4 - kt^2$. Indeed, you can check by hand that these points are very closely predicted by the formula

$$y = 4 - 16t^2.$$

This is the formula for the curve shown connecting the dots on the previous plot.

Once we know this formula, we can be as precise as we like about estimating the speed the pencil hits the floor. For example, the average speed during the last $1/16$ seconds would be

$$\frac{\text{distance}}{\text{time}} = \frac{y(0.5) - y(0.4375)}{0.5 - 0.4375} = \frac{-.9375}{.0625} = -15 \text{ ft/s}.$$

We could continue to use shorter and shorter intervals, but that would mean repeating essentially the same computation over and over. We can save ourselves a lot of tedium by introducing a small bit of algebra. Let h be the length of the time interval we want to use (it could be the last $1/4$ seconds, the last $1/16$ seconds, or even smaller). Then the average speed of the pencil from time $0.5 - h$ to time 0.5 seconds is

$$\begin{aligned}
\frac{\text{distance}}{\text{time}} &= \frac{y(0.5) - y(0.5 - h)}{0.5 - (0.5 - h)} \\
&= \frac{0 - \left(4 - 16(0.5 - h)^2\right)}{h} \\
&= \frac{-4 + 16(0.25 - h + h^2)}{h} \\
&= \frac{-4 + 4 - 16h + 16h^2}{h} \\
&= \frac{h(-16 + 16h)}{h} \\
&= -16 + 16h.
\end{aligned}$$

It is now quick work to calculate as many estimates of the pencil's collision speed as we like.

h	estimate
0.5 s	-8 ft/s
0.25 s	-12 ft/s
0.0625 s	-15 ft/s
0.01 s	-15.84 ft/s
0.001 s	-15.984 ft/s

As h gets smaller and smaller, it is apparent that $-16 + 16h$ gets closer and closer to -16. This gives us the true answer to our question. Over any time interval, the *average* velocity of the pencil will be some number larger (i.e., less negative) than -16 ft/s. But as the time intervals become shorter and shorter, the *average* velocities get closer to -16 ft/s, the speed of the pencil the *instant* it hits the floor.

There was nothing particular about the time $t = 0.5$ except that it happened to be the time when the pencil hit the floor. We could just as easily have estimated the speed of the pencil at any other instant during the descent. For example, let's check the speed of the pencil the moment it starts to fall, at time $t = 0$, by finding the average speed over the time interval $[0, h]$.

$$\frac{\text{distance}}{\text{time}} = \frac{y(h) - y(0)}{h - 0} = \frac{(4 - 16h^2) - 4}{h} = -16h.$$

In this case, as h gets smaller and smaller (closer and closer to zero) the average speed goes to 0. This agrees with our intuition. Just for an *instant* when the pencil starts falling, it isn't moving at all. Its *instantaneous velocity* at time 0 is 0 ft/s.

EXERCISES

4.1 Find the instantaneous velocity of a pencil dropped from a height of 4 ft when $t = 0.25$ s, i.e., the moment its height is 3 ft.

4.2 An object dropped from a height of 16 ft takes approximately one second to strike the ground, and it has a height function of $y = 16 - 16t^2$.

   a) Determine the velocity of the object as it hits the ground.
   b) Determine the velocity of the object at time $t = 0.5$ seconds.
   c) Determine the velocity of the object at time $t = 0.25$ seconds.
   d) Determine the velocity of the object at time $t = 0$ seconds.

4.3 A more precise height function for a pencil dropped 4 feet is $y = 4 - 16.1t^2$. It only takes about 0.49844 seconds (not a full half second) for the pencil to reach the floor.

   a) Find the velocity at which the pencil strikes the floor using this height function and impact time.
   b) Find the velocity of the pencil at time $t = 0$ seconds.

4.4 If your home were on the Moon, and a pencil were to drop 4 ft to the floor, its height function in feet would be $y = 4 - 2.65t^2$.

   a) Set $y = 0$ and solve for t to determine that it requires approximately 1.2286 seconds for the pencil to hit the floor.
   b) Determine the velocity the pencil strikes the floor.
   c) Does a pencil on the Moon strike the floor faster, slower, or the same velocity as a pencil on Earth?

4.5 A person shoots an arrow vertically into the air at 200 ft/s. Neglecting air resistance, the height of the arrow after t seconds is given by the formula $y = 6 + 200t - 16.1t^2$ (approximately).

    a) How high is the arrow when the bow is fired?

    b) Find the instantaneous velocity when $t = 0$ s.

    c) What is the instantaneous velocity of the arrow at time $t = 6.21$ s? What can you conclude about the maximum height the arrow reaches?

    d) At what time t will the arrow hit the ground?

    e) At what velocity does the arrow hit the ground?

    f) Compare the velocity of the arrow striking the ground with the velocity of the arrow leaving the bow, and comment on the safety of firing an arrow into the sky.

4.3 Tangent Line to a Curve

If we think about the process we developed in Section 4.2, each time we chose a time interval and computed the average velocity of the pencil over the interval, we were finding a value that looked something like

$$\text{velocity} = \frac{\text{distance fallen}}{\text{time elapsed}} = \frac{y_2 - y_1}{t_2 - t_1}.$$

Considering the problem geometrically (looking at the graph of what we are doing), we can see that the key to the whole process is slopes. Since slope is the ratio of the change in the y-coordinates (the rise) to the change in the t-coordinates (the run), we can write slope as $\Delta y / \Delta t$. (The Greek letter delta, Δ, stands for change.) Each of these average velocities is the slope of some line intersecting the height function at two points. See Figure 4.2.

We saw that finding the instantaneous velocity amounted to a limiting problem where Δt, which we called by the name h, was allowed to get closer and closer to zero. In the picture, we interpret this as a question about the slopes of lines that intersect a function when the points of intersection are brought closer and closer together. *Each average velocity for the pencil is the slope of some line through the curve, so if we want to know about the (instantaneous) velocity of the pencil, we can focus our attention on slopes of lines.*

A *tangent line* to a curve is a line that touches the curve in just one point, and closely approximates the curve near that one point. In Figure 4.3, a tangent line is drawn touching the curve where $t = 1/2$. Notice that the tangent line looks very much like *the limit* of the lines in Figure 4.2 when Δt approaches zero.

Figure 4.4 shows a general curve and a tangent line to the curve at a point P on the curve.

To find tangent lines in general, we use our experience with instantaneous velocities for inspiration. To find the tangent line at P in Figure 4.5, we begin by putting a second point Q on the curve somewhere nearby. We can put Q pretty much any-

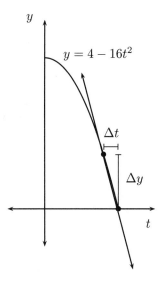

Figure 4.2   Average velocities are slopes of lines.

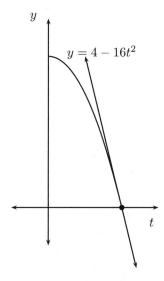

Figure 4.3   A tangent line is the key to instantaneous velocity.

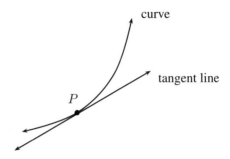

Figure 4.4   A curve and its tangent line.

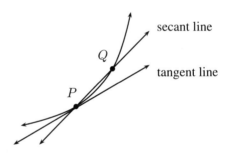

Figure 4.5   Tangent lines are derived from secant lines.

where on the curve to start out, though often you'll pick somewhere near P. The line through P and Q is called a *secant line*.

Now, if Q gets closer and closer to P (which is what happened when we observed our pencil over shorter and shorter time intervals), then the secant line through P and Q becomes a better and better approximation to the tangent line at P. If we can make Q coincide with P, by using a limiting process, then the secant line will "become" the tangent line at P. This is good for us, since the slope of the tangent P is going to tell us how fast our pencil hits the floor.

In the following examples, we will refer to P (the point where the tangent line touches the curve) as the base point. We will refer to Q as the second point.

The method of using a limiting process of secant lines to find a tangent line was discovered by Pierre de Fermat (1601–1665). It was expanded upon by the two discoverers of calculus, Isaac Newton (1642–1727) and Gottfried Wilhelm von Leibniz (1646–1716).

■ EXAMPLE 4.1

We will find an equation of the tangent line to the parabola $y = x^2$ at the point $(-1, 1)$. The base point is $(-1, 1)$. We take the second point to be $(-1 + h, (-1 + h)^2)$, where h is an arbitrary nonzero number. In Figure 4.6, h is positive, so the second point is to the right of the first point. But h could just as well be negative, with the second point to the left of the base point. The calculations work the same either way.

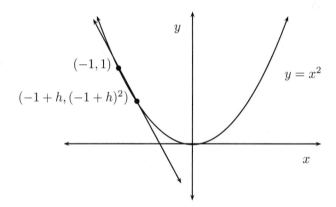

Figure 4.6   Tangent to $y = x^2$.

$$\text{slope of secant line} = \frac{\Delta y}{\Delta x}$$
$$= \frac{(-1 + h)^2 - 1}{h}$$
$$= \frac{1 - 2h + h^2 - 1}{h}$$
$$= \frac{-2h + h^2}{h}$$
$$= -2 + h.$$

When h diminishes to 0, we obtain the slope of the tangent line:

$$\text{slope of tangent line} = -2.$$

We have a slope, -2, and a point, $(-1, 2)$, so we can use the point-slope form of a line, $y - y_0 = m(x - x_0)$, to describe the tangent. Thus, an equation of the tangent line to $y = x^2$ at the point $(-1, 1)$ is

$$y - 1 = -2(x + 1).$$

If we want to solve this equation for y to put the line in the more familiar $y = mx + b$ form, we can:

$$y = -2x - 1.$$

EXAMPLE 4.2

Let's find an equation of the tangent line to the parabola $y = x^2$ at the point $\left(\frac{5}{8}, \frac{25}{64}\right)$ (see Figure 4.7).

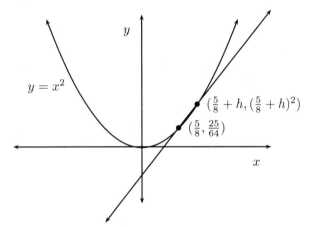

Figure 4.7   Tangent to $y = x^2$ at $\left(\frac{5}{8}, \frac{25}{64}\right)$.

$$
\begin{aligned}
\text{slope of secant line} &= \frac{\Delta y}{\Delta x} \\
&= \frac{\left(\frac{5}{8} + h\right)^2 - \frac{25}{64}}{h} \\
&= \frac{\frac{25}{64} + \frac{5}{4}h + h^2 - \frac{25}{64}}{h} \\
&= \frac{\frac{5}{4}h + h^2}{h} \\
&= \frac{5}{4} + h
\end{aligned}
$$

When h diminishes to 0, we obtain the slope of the tangent line:

$$\text{slope of tangent line} = \frac{5}{4}.$$

Using the point-slope form again, an equation of the tangent line is

$$y - \frac{25}{64} = \frac{5}{4}\left(x - \frac{5}{8}\right).$$

Looking over the last two examples, we see that the slope of the tangent line to the curve $y = x^2$ at the point $(-1, 1)$ is -2, and the slope of the tangent line to the same curve at the point $\left(\frac{5}{8}, \frac{25}{64}\right)$ is $\frac{5}{4}$. In both cases, the slope of the tangent line is double the value of the x coordinate. Let's use the next example to show that this is always the case for the curve $y = x^2$.

■ EXAMPLE 4.3

Let's find the slope of the tangent line to the parabola $y = x^2$ at the point (a, a^2), where a is an arbitrary number (Figure 4.8). Based on our previous work, we expect that our answer is going to be $2a$.

This time, for variety, we choose to write Δx for h.

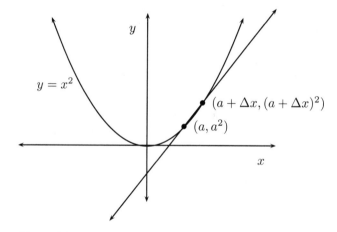

Figure 4.8 Finding the slope at an arbitrary value a.

$$\begin{aligned}
\text{slope of secant line} &= \frac{\Delta y}{\Delta x} \\
&= \frac{(a + \Delta x)^2 - a^2}{\Delta x} \\
&= \frac{a^2 + 2a\Delta x + (\Delta x)^2 - a^2}{\Delta x} \\
&= \frac{2a\Delta x + (\Delta x)^2}{\Delta x} \\
&= 2a + \Delta x
\end{aligned}$$

When Δx diminishes to 0, we obtain the slope of the tangent line:

$$\text{slope of tangent line} = 2a.$$

Notice that if we let $a = -1$, then we find the slope of the tangent line at $(-1, 1)$, and we get the same answer, -2, as we got in Example 4.1. If we let $a = 5/8$, we get the same answer, $5/4$, that we got in Example 4.2.

EXERCISES

4.6 Use the secant line method to find an equation of the tangent line to the parabola $y = x^2$ at the point $(3, 9)$.

4.7 Use the secant line method to find an equation of the tangent line to the parabola $y = x^2$ at the point $(-4, 16)$.

4.8 Use the secant line method to find an equation of the tangent line to the parabola $y = 4 - 16t^2$ at the point $(0.25, 3)$.

4.9 Find the slope of the tangent line to the parabola $y = 4 - 16t^2$ at an arbitrary point $(a, 4 - 16a^2)$. The slope of this tangent corresponds to the velocity of our falling pencil in Section 4.2. Use this slope to determine the velocity of the pencil when $t = 0.0, 0.1, 0.2, 0.3, 0.4$, and 0.5 seconds.

4.10 Find the slope of the tangent line to the parabola $y = 16 - 16t^2$ at an arbitrary point $(a, 16 - 16a^2)$. The slope of this tangent corresponds to the velocity of a pencil dropped from a height of 16 feet. Use this slope to determine the velocity of the pencil when $t = 0.0, 0.25, 0.5, 0.75$, and 1.0 seconds.

4.11 Find the slope of the tangent line to the parabola $y = 4 - 2.65t^2$ at an arbitrary point $(a, 4 - 2.65a^2)$. The slope of this tangent corresponds to the velocity of a pencil dropped from a 4 foot desk on the Moon. Use this slope to determine the velocity of the pencil when $t = 0.0, 1$, and 1.2286 seconds.

4.12 If your home were on Mars, the height function of a pencil dropped from a 4 foot desk would be the parabola $y = 4 - 6.12t^2$.
   a) Set $y = 0$ to find the time t when your pencil strikes the floor.
   b) Find the slope of the tangent line to the height function at an arbitrary time a.
   c) Find the velocity that the pencil strikes the floor.
   d) Compare your answer from part (c) to pencils on the Earth or Moon.

4.4 The Derivative

Recall from algebra class the definition of a function. A *function* is a rule that assigns to each number x in some domain exactly one number y in a range. For example,

$y = x^2$ describes a function by giving a formula. The *domain* of the function is the set of values you can "put in" the function, all real numbers in this case (since every number can be multiplied by itself). The *range* is the set of values that you "get out" of the function, and for the squaring function this is the set $[0, \infty)$. You probably remember many happy hours spent finding the domains and ranges of functions.

In addition to having a domain and range, we know that $y = x^2$ describes a function because for any x we put in we get out exactly one y. If $x = 3$ goes in, $y = 9$ comes out. Put in $x = 0$ and you get out $y = 0$, etc. There is only one way to square a number.

You may remember that functions have graphs that pass the vertical line test. This is the same thing; at any particular x-value, the function should only have a single y-value (which is where it intersects the vertical line).

Not all formulas describe functions. For example, $y = x^{1/2}$ is not a function, because if you put in $x = 9$, the value for y is ambiguous. It could be $y = 3$ or $y = -3$. However, usually when we write $y = \sqrt{x}$ we mean the positive square root; and this *is* a function, because there is only one way to find a *positive* square root.

In Example 4.3 we saw a process that can unambiguously tell us the slope of a tangent line at any point on a curve. That is, there is a formula that can tell us slopes when given x-values. If you name an x-value, I can tell you the slope. Name a different x-value, and I can tell you the new slope.

Since the slope of a tangent line is a rule we can give unambiguously, we know from our experience in algebra that it is a function, and this function has a name. It is called the derivative. Sometimes people refer to the derivative as the "slope generating function."

Definition. The *derivative* of a function $f(x)$ is a new function denoted $f'(x)$ defined by

$$f'(x) = \lim_{h \to 0} \frac{f(x+h) - f(x)}{h}.$$

The process of finding the derivative of a function is called *differentiation*. The notation $f'(x)$ is read "f prime of x."

■ **EXAMPLE 4.4**

Let's find the derivative of $f(x) = x^3$. According to our definition,

$$f'(x) = \lim_{h \to 0} \frac{(x+h)^3 - x^3}{h}.$$

Using the expansion $(x + h)^3 = x^3 + 3x^2h + 3xh^2 + h^3$, we have

$$f'(x) = \lim_{h \to 0} \frac{x^3 + 3x^2h + 3xh^2 + h^3 - x^3}{h}$$

$$= \lim_{h \to 0} \frac{3x^2h + 3xh^2 + h^3}{h}$$

$$= \lim_{h \to 0} \frac{h(3x^2 + 3xh + h^2)}{h}$$

$$= \lim_{h \to 0} (3x^2 + 3xh + h^2)$$

$$= 3x^2.$$

We can use this new function to find the slope of the tangent line at any point on x^3. For example, when $x = 0$, the slope is $f'(0) = 3(0^2) = 0$. Likewise, when $x = 1$, the slope is $f'(1) = 3(1^2) = 3$. This is represented visually in Figure 4.9.

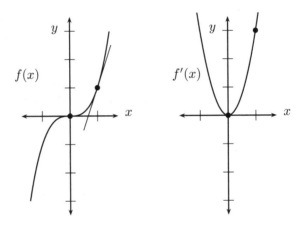

Figure 4.9   Graphs of $f(x)$ and the derivative $f'(x)$.

EXERCISES

4.13   Use the definition to find the derivative of the function $f(x) = x^2 + 3$. Draw a sketch of $f(x)$, or graph it on a calculator, and check that the slopes given by $f'(x)$ look right.

4.14   Use the definition to find the derivative of the function $f(x) = x^2 + 5$. Draw a sketch of $f(x)$, or graph it on a calculator, and check that the slopes given by $f'(x)$ look right.

4.15   Use the definition to find the derivative of the function $f(x) = x^2 + x$. Draw a sketch of $f(x)$, or graph it on a calculator, and check that the slopes given by $f'(x)$ look right.

4.16   Use the definition to find the derivative of the function $f(x) = 3x$. Draw a sketch of $f(x)$, or graph it on a calculator, and check that the slopes given by $f'(x)$ look right.

4.17   Use the definition to find the derivative of the function $f(x) = 3x + 1$. Draw a sketch of $f(x)$, or graph it on a calculator, and check that the slopes given by $f'(x)$ look right.

4.18   Use the definition to find the derivative of the function $f(x) = -2x + 2$. Draw a sketch of $f(x)$, or graph it on a calculator, and check that the slopes given by $f'(x)$ look right.

4.19   Use the definition to find the derivative of the function $f(x) = x^3 + x$. Draw a sketch of $f(x)$, or graph it on a calculator, and check that the slopes given by $f'(x)$ look right.

4.5   Formulas for Derivatives

We saw in Section 4.3 that slopes can be used to compute tangible things such as velocities of objects. In Section 4.4 we created a formal mathematical process for finding slopes, called differentiation. Since derivatives are the key to solving all kinds of mathematical problems, it is natural to want to know the derivative of as many different kinds of functions as possible. However, even the most mathematical of us eventually find it tedious to compute derivatives using the limit definition. Luckily, there are easily recognizable patterns that we can use as shortcuts, or "derivative rules."

The Constant Rule

The simplest derivative rule is that the derivative of a constant function $f(x) = c$ is $f'(x) = 0$. The graph of a constant function $f(x) = c$ is a horizontal line that intersects the y-axis at the value c. Since it is a horizontal line, it has a slope of 0 everywhere. That is, at every x the slope is zero. This is precisely what $f'(x) = 0$ means (it gives 0 for every value of x).

The Line Rule

Non-horizontal lines are just as easy. The derivative of a linear function $f(x) = mx$ is $f'(x) = m$, since f is a line through the origin with slope m. Of course, nothing changes if the y-intercept happens to be different from 0 (i.e., if the line doesn't go through the origin); it still has slope m. We can write this idea as a rule, and we have our first rule of derivatives.

The line rule: If $f(x) = mx + b$, then $f'(x) = m$.

Notice that the line rule contains the constant rule within it. If $f(x) = b$ is a constant function, then the line rule tells of that $f'(x) = 0$. For example, if $f(x) = 7 = 0x + 7$ is a horizontal line (with y-intercept 7), then its derivative is zero.

EXAMPLE 4.5

We can check the line rule carefully, using the definition of derivative. If $f(x) = mx + b$, then the definition of derivative tells us that

$$
\begin{aligned}
f'(x) &= \lim_{h \to 0} \frac{f(x+h) - f(x)}{h} \\
&= \lim_{h \to 0} \frac{\big(m(x+h) + b\big) - \big(mx + b\big)}{h} \\
&= \lim_{h \to 0} \frac{(mx + mh + b) - (mx + b)}{h} \\
&= \lim_{h \to 0} \frac{mx + mh + b - mx - b}{h} \\
&= \lim_{h \to 0} \frac{mh}{h} \\
&= \lim_{h \to 0} m.
\end{aligned}
$$

It feels a little strange to ask what m is getting close to as h becomes closer and closer to 0, since m is just a number, but it's perfectly correct to declare that it gets close to m. You might say that being exactly equal to something is the very best way to get close to it.

So we get $f'(x) = m$. Of course, a slope of m is exactly what our experience tells us to expect for the line $f(x) = mx + b$.

The Sum Rule

It might have occurred to you that the function $f(x) = mx + b$ is actually the sum of two lines, a non-horizontal line $g(x) = mx$ and a horizontal line $h(x) = b$. Using the line rule for each, we get $g'(x) = m$ and $h'(x) = 0$. Notice that if we add these derivatives, $g'(x) + h'(x) = m$, which happens to be the same as $f'(x)$. This is not a coincidence, but is our second rule of derivatives.

The sum rule: The derivative of a sum is the sum of the derivatives:

$$
(f(x) + g(x))' = f'(x) + g'(x).
$$

EXAMPLE 4.6

Since $3x = 2x + x$, we can use the line rule directly to determine that $(3x)' = 3$, but we can also use the sum rule (and the line rule) on the right hand side to get

$(3x)' = (2x)' + (x)' = 2 + 1 = 3$. It doesn't matter which rules we use, we always get the same answer for the derivative.

If your algebra skills are good, you can verify the sum rule using the limit definition:

$$\left(f(x) + g(x)\right)' = \lim_{h \to 0} \frac{\left(f(x+h) + g(x+h)\right) - \left(f(x) + g(x)\right)}{h}$$

$$= \lim_{h \to 0} \frac{\left(f(x+h) - f(x)\right) + \left(g(x+h) - g(x)\right)}{h}$$

$$= \lim_{h \to 0} \left[\frac{f(x+h) - f(x)}{h} + \frac{g(x+h) - g(x)}{h}\right]$$

$$= \lim_{h \to 0} \frac{f(x+h) - f(x)}{h} + \lim_{h \to 0} \frac{g(x+h) - g(x)}{h}$$

$$= f'(x) + g'(x).$$

■ EXAMPLE 4.7

The sum rule works for three (or more) terms as well. For example, take $f(x) = x + x + x$. We can group with parentheses, to make this a sum of just two terms, $f(x) = x + (x + x)$. Then an application of the sum rule tells us

$f'(x) = (x)' + (x + x)'$, and applying the sum rule a second time, we get

$$= (x)' + (x)' + (x)'$$
$$= 1 + 1 + 1$$
$$= 3.$$

The key observation is that this pattern always works, no matter what functions you use. So $\left(f(x) + g(x) + h(x)\right)' = f'(x) + g'(x) + h'(x)$.

The Power Rule

In Example 4.4 we discovered that $(x^3)' = 3x^2$. Before that, in Example 4.3 we determined that $(x^2)' = 2x$. The line rule tells us that $(x^1)' = 1$, since x^1 is another way of writing x. And because $x^0 = 1$ we can see that $(x^0)' = 0$. Perhaps you have already determined that these all adhere to a common pattern.

The power rule: For any integer $n \geq 0$, the derivative of the power function $f(x) = x^n$ is $f'(x) = nx^{n-1}$.

■ EXAMPLE 4.8

Let's find the derivative of $f(x) = x^{10}$. By the power rule, we have

$$f'(x) = 10x^9.$$

EXAMPLE 4.9

Find the derivative of $f(x) = x^3 + x^2 + 3x + 7$.

Solution: By the sum rule, we have

$$f'(x) = \left(x^3\right)' + \left(x^2\right)' + \left(3x\right)' + \left(7\right)'$$
$$= 3x^2 + 2x + \left(3x\right)' + \left(7\right)' \text{ by the power rule}$$
$$= 3x^2 + 2x + 3 \text{ by the line rule and constant rule.}$$

■

The Constant Coefficient Rule

There is a straightforward differentiation rule for multiplying functions by constants.

The constant coefficient rule: If c is a constant, then $\left(cf(x)\right)' = cf'(x)$.

In terms of the graph, multiplying a function by some constant has the effect of stretching (or compressing) it vertically. For example, the function $f(x) = 2\sin x$ is twice as tall as a standard sine curve. What the constant coefficient rule tells us is that the coefficient applies to slopes in the same way. Examine Figure 4.10 to see that this seems plausible.

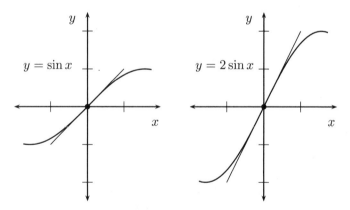

Figure 4.10   Slopes on $2\sin x$ are twice the size of slopes on $\sin x$.

Intuitively, slope is always a calculation of "rise over run." If we multiply a function by the constant 5, then in slope calculations the "rise" changes by the same factor of 5 (because all of the y-values are 5 times as large), but the "run" is the same. The new slope would look something like

$$\text{new slope} = \frac{5\Delta y}{\Delta x} = 5\frac{\Delta y}{\Delta x} = 5 \text{ times old slope.}$$

This agrees with what we already know from the line rule. By the line rule, we know that the derivative of $7x$ is 7. The constant coefficient rule tells us that we can also think of it as 7 times the derivative of x. Of course $7(x)' = 7 \cdot 1 = 7$, so we get the same answer.

In the same way, the constant coefficient rule tells us that $(5x^3)' = 5(x^3)' = 5 \cdot 3x^2 = 15x^2$.

Derivatives of Polynomials

The power rule, sum rule, and constant coefficient rule combine to tell us everything we need to take the derivative of any polynomial. At first, we should probably compute derivatives step by step, but soon we'll be taking derivatives almost effortlessly.

■ EXAMPLE 4.10

Let's find the derivative of $f(x) = 7x^5 + 16x^4 - 32x + 100$. By the sum rule, we know that

$$f'(x) = (7x^5)' + (16x^4)' + (-32x)' + (100)'.$$

The constant rule tells us that $(100)' = 0$, and the constant coefficient rule lets us "pull out" the constant from each of the other terms we are taking the derivative of, so

$$f'(x) = 7(x^5)' + 16(x^4)' - 32(x)' + 0.$$

Finally, applying the power rule,

$$= 7 \cdot 5x^4 + 16 \cdot 4x^3 - 32 \cdot 1x^0$$
$$= 35x^4 + 64x^3 - 32.$$

■ EXAMPLE 4.11

While we are building these differentiation rules, we should not get caught up in the letters we use. Functions do not have to be called f and their variable does not have to be x. The rules apply in the same ways if we use other variable and function names. For example, the derivative of $y(t) = 16 - 4t^2$ is

$$y'(t) = (16)' - (4t^2)' \text{ by the sum rule}$$
$$= 0 - 4(t^2)' \text{ by the constant and constant coefficient rules}$$
$$= -8t \text{ by the power rule.}$$

Here is a summary of some important derivative rules:

Derivative Rules

function	derivative
c (constant)	0
$mx + b$	m
x^n	nx^{n-1}

EXERCISES

4.20   Use derivative rules to verify the derivatives we found for each of the functions from the previous exercise set.

  a) $f(x) = x^2 + 3$
  b) $f(x) = x^2 + x$
  c) $f(x) = 3x$
  d) $f(x) = 3x + 1$
  e) $f(x) = -2x + 2$
  f) $f(x) = x^3 + x$

4.21   Find the derivative of each polynomial.

  a) $f(x) = 9x^5 - x^4 + 8x^3 + 4x + 3$
  b) $x(t) = 2t^3 + 7t^2 + 6t - 1$
  c) $x(t) = 2.5t^3 + 7t^2 + 6.3t - 1$
  d) $x(y) = -y^4 - 3y^3 + 2y^2 + 2y - 2$
  e) $h(t) = 9.9t^4 - 3.7t^3 - 2.2t^2 - 1.1t + 0.2$

4.22   The parabola $y = x(x - 1) = x^2 - x$ has a single point where the slope is 0. Find the value of x where the slope, i.e., the derivative, is 0. Then find the corresponding y. (The point (x, y) identifies the vertex of the parabola.)

4.23   The function $y = x^3 - x$ has two points where the slope is 0. Find both. Sketch the graph of the function by plotting points or using a calculator, and verify that the tangent line looks horizontal at both points.

4.24   Use derivative rules to find an equation of the tangent line to the parabola $y = 3 - x^2$ at the point $(1, 2)$. (Hint: you are given a point, and the derivative tells you the slope.)

4.25   A pencil dropped from a height of 4 feet has a height function that looks like $y(t) = 4 - 16.1t^2$. It only takes about 0.49844 seconds (not a full half second) for the pencil to reach the floor.

  a) Find the derivative of $y(t)$.
  b) Find the slope of the tangent line when $t = 0.49844$.
  c) How fast does the pencil hit the floor?

4.26   A pencil thrown vertically into the air from a height of 4 feet has a height function that looks like $y(t) = 4 + 5t - 16.1t^2$.

   a) Find the derivative of $y(t)$.
   b) Find the slope of the tangent line when $t = 0$.
   c) How fast is the pencil moving when $t = 0$?
   d) At what time t does $y'(t) = 0$?
   e) We know $y'(t) = 0$ means that the slope of the tangent line is zero. What does it physically mean about the motion of our pencil?

4.6   The Product Rule and Quotient Rule

Say we have two functions, $f(x)$ and $g(x)$. If we knew that the derivative of $f(x)$ was 6 and the derivative of $g(x)$ was 5, what would be the derivative of the product $f(x)g(x)$?

If you answered 30, you are probably not alone. But unfortunately you're not right either, and you can easily check for yourself that this is a mistake. You could, for example, take $f(x) = 6x$. That's a function with $f'(x) = 6$. If you also take $g(x) = 5x$, then $g'(x) = 5$. And when we multiply? We get $f(x)g(x) = 6x \cdot 5x = 30x^2$. This means that the derivative of the product $f(x)g(x)$ is then the derivative of $30x^2$, and that is $60x$, and not 30.

It turns out that there is a product rule for derivatives, but it is not the naively simple rule that most people guess.

The product rule: If we can compute the derivatives of $f(x)$ and $g(x)$, the derivative of their product is $\big(f(x)g(x)\big)' = f'(x)g(x) + f(x)g'(x)$.

📖 EXAMPLE 4.12

If we take $f(x) = 6x$ and $g(x) = 5x$, then the product rule tells us that

$$\big(f(x)g(x)\big)' = f'(x)g(x) + f(x)g'(x)$$
$$= 6 \cdot 5x + 6x \cdot 5$$
$$= 30x + 30x$$
$$= 60x.$$

Naturally, this is the same as the answer we get without the product rule (when we do it correctly).

■ **EXAMPLE 4.13**

What is the derivative of

$$f(x) = (5x^5 + 3x^3 + 5x^2 + 3x + 4)(4x^5 + 4x^4 + 5x^3 + x^2 + x)?$$

Solution: In plain English, the product rule tells us, "the derivative of a product is 'the derivative of the first' times 'the second' plus 'the first' times 'the derivative of the second.'" Computing, we have

$$\begin{aligned}
f'(x) &= (5x^5 + 3x^3 + 5x^2 + 3x + 4)'(4x^5 + 4x^4 + 5x^3 + x^2 + x) \\
&\quad + (5x^5 + 3x^3 + 5x^2 + 3x + 4)(4x^5 + 4x^4 + 5x^3 + x^2 + x)' \\
&= (25x^4 + 9x^2 + 10x + 3)(4x^5 + 4x^4 + 5x^3 + x^2 + x) \\
&\quad + (5x^5 + 3x^3 + 5x^2 + 3x + 4)(20x^4 + 16x^3 + 15x^2 + 2x + 1).
\end{aligned}$$

This may seem like an unsatisfyingly complicated answer, but it would be completely adequate if we were in a situation where we didn't need to simplify. Although we devote extensive effort in algebra class to simplification, we don't always need to simplify to solve problems. For example, if all we need to know is the slope of f when $x = 0$, it is straightforward to find that $f'(0) = (3)(0) + (4)(1) = 4$, and the computation is easy even without simplifying. ■

■ **EXAMPLE 4.14**

We can use the product rule when there are more than two factors. We simply apply the product rule more than one time. Consider the function $h(x) = x(x + 1)(x + 2)$. Although this function is easy to handle by multiplying things out, let's use the product rule instead.

To get started, we have to decide which factors of the function will constitute the $f(x)$ part of the product rule and which factors will be the $g(x)$ part. It doesn't really matter, as long as you split the function into two factors that are multiplied together. Let's choose $f(x) = x(x + 1)$ and $g(x) = x + 2$. Then $h(x) = f(x)g(x)$, and we can differentiate:

$$\begin{aligned}
h'(x) &= f'(x)g(x) + f(x)g'(x) \\
&= \big(x(x + 1)\big)'(x + 2) + \big(x(x + 1)\big)(x + 2)' \\
&= \big(x(x + 1)\big)'(x + 2) + \big(x(x + 1)\big) \cdot 1 \\
&= \big(x(x + 1)\big)'(x + 2) + x(x + 1).
\end{aligned}$$

For the derivative of $x(x + 1)$, we use the product rule a second time. Continuing,

$$\begin{aligned}
h'(x) &= \big((x)'(x + 1) + x(x + 1)'\big)(x + 2) + x(x + 1) \\
&= \big(1 \cdot (x + 1) + x \cdot 1\big)(x + 2) + x(x + 1) \\
&= \big((x + 1) + x\big)(x + 2) + x(x + 1) \\
&= (2x + 1)(x + 2) + x(x + 1).
\end{aligned}$$

If we wish to combine terms and simplify a bit, we conclude that $h'(x) = 3x^2 + 6x + 2$. Of course, this is the same answer that we get if we multiply h out first and differentiate directly.

Naturally, we can differentiate functions that are products of four factors (it requires three applications of the product rule), or five factors, or more.

New Derivatives from the Product Rule

We can use the product rule to figure out the derivatives of functions we don't yet know how to differentiate. For example, what is the derivative of $f(x) = 1/x$? We can use the product rule, if we are careful and clever.

If we start with $f(x) = 1/x$, then $xf(x) = 1$, which is a constant. Constant functions are easy to differentiate, so we know immediately that $(xf(x))' = 0$. But we can also find the derivative using the product rule. According to the product rule,

$$(xf(x))' = 1 \cdot f(x) + x \cdot f'(x)$$

Putting these two calculations together (remembering that $f(x) = 1/x$), we get

$$0 = (xf(x))'$$
$$= 1 \cdot \frac{1}{x} + x \cdot f'(x),$$

and it follows that

$$-\frac{1}{x} = xf'(x),$$

and finally that

$$-\frac{1}{x^2} = f'(x).$$

So if $f(x) = 1/x$, the derivative is $f'(x) = -1/x^2$.

The previous argument is an example of a common proof method in mathematics. If we can compute a quantity two different ways, then we know both answers must be equal even if they may not look the same. As in this discussion, often the clever step is figuring out how to arrange things (i.e., knowing to start with $xf(x) = 1$ and then differentiate). Once arranged, the calculations are not necessarily difficult.

▣ EXAMPLE 4.15

Find the derivative of $f(x) = \sqrt{x}$.

Solution: We need to arrange things so we can use the product rule, so let $h(x) = f(x)f(x) = \sqrt{x}\sqrt{x} = x$. We can immediately see that $h'(x) = 1$, but we can also apply the product rule. This means that

$$1 = h'(x)$$
$$= \big(f(x)f(x)\big)'$$
$$= f'(x)f(x) + f(x)f'(x)$$
$$= f'(x)\sqrt{x} + \sqrt{x}f'(x)$$
$$= f'(x)\big(\sqrt{x} + \sqrt{x}\big)$$
$$= f'(x)\big(2\sqrt{x}\big),$$

and dividing, we have

$$f'(x) = \frac{1}{2\sqrt{x}}.$$

■

Just as we can find the derivative of products of functions, there is a rule for taking the derivatives of quotients of functions (that is, when we divide functions). The quotient rule is not easily guessed, but amazingly we can figure out the formula using the product rule.

First we need a quotient. Let $h(x) = \dfrac{f(x)}{g(x)}$. Our goal is to find a formula for $h'(x)$. When we cross-multiply, we get $h(x)g(x) = f(x)$, or reversing the equality we have $f(x) = h(x)g(x)$, and this is a product, so we can differentiate it. On the left, we'll simply write the derivative as $f'(x)$. On the right, we'll use the product rule.

$$f'(x) = \big(h(x)g(x)\big)'$$
$$= h'(x)g(x) + h(x)g'(x)$$

Now, keep the term with $h'(x)$ on the right, and move everything else to the left side.

$$f'(x) - h(x)g'(x) = h'(x)g(x)$$

Reverse the equality and divide by $g(x)$ to get $h'(x)$ alone.

$$h'(x) = \frac{f'(x) - h(x)g'(x)}{g(x)}$$

Next comes a clever part, but we're almost finished. Remember that we started with $h(x) = \dfrac{f(x)}{g(x)}$. Substitute for $h(x)$ to get

$$h'(x) = \frac{f'(x) - \frac{f(x)}{g(x)}g'(x)}{g(x)}$$

$$= \frac{f'(x)}{g(x)} - \frac{f(x)g'(x)}{g(x)g(x)}$$

$$= \frac{f'(x)g(x)}{g(x)g(x)} - \frac{f(x)g'(x)}{g(x)g(x)}$$

$$= \frac{f'(x)g(x) - f(x)g'(x)}{\left(g(x)\right)^2}.$$

This becomes our rule for differentiating quotients.

The quotient rule: If $h(x) = \dfrac{f(x)}{g(x)}$, then $h'(x) = \dfrac{f'(x)g(x) - f(x)g'(x)}{\left(g(x)\right)^2}.$

Because this formula is fairly complex, people usually use one of two ways to remember it. In words we say that the derivative of a fraction is, "Bottom times the derivative of the top, minus top times the derivative of the bottom, all over the bottom squared."

Some people prefer to remember the quotient rule via the (math) poem,

$$\frac{\text{lo } d(\text{hi}) - \text{hi } d(\text{lo})}{\text{lo lo}}.$$

Read aloud, this goes, "Low dee-high, minus high dee-low, over low low (and away we go)!" In the poem, low refers to the bottom function $g(x)$, high refers to the top function $f(x)$, and dee reminds us to take a derivative. So 'low dee-high" is $g(x)f'(x)$. Similarly, "high dee-low" is $f(x)g'(x)$. And we divide by "low low," which is $g(x) \cdot g(x) = \left(g(x)\right)^2$. If you write everything out according to the poem, you get the quotient rule.

▣ EXAMPLE 4.16

Let us apply the quotient rule to a function we already know the derivative of, such as x^2.

We can think of $h(x) = x^2$ as a quotient by writing it as $h(x) = \dfrac{x^2}{1}$. Then, for the purposes of applying the quotient rule, $f(x) = x^2$ and $g(x) = 1$. We calculate

$$h'(x) = \frac{f'(x)g(x) - f(x)g'(x)}{\left(g(x)\right)^2}$$

$$= \frac{(x^2)' \cdot 1 - x^2 \cdot (1)'}{1^2}$$

$$= \frac{2x \cdot 1 - x^2 \cdot 0}{1^2}$$

$$= \frac{2x}{1}$$

$$= 2x,$$

which we know is correct.

◼ EXAMPLE 4.17

Find the derivative of $h(x) = 1/x^k$ where k is a positive integer (1, 2, 3, etc.).

Solution: Take $f(x) = 1$ and $g(x) = x^k$. By the quotient rule,

$$h'(x) = \frac{f'(x)g(x) - f(x)g'(x)}{\left(g(x)\right)^2}$$

$$= \frac{(1)' \cdot x^k - 1 \cdot (x^k)'}{(x^k)^2}.$$

Here we can use the power rule to continue.

$$h'(x) = \frac{0 \cdot x^k - 1 \cdot kx^{k-1}}{(x^k)^2}$$

$$- \frac{-kx^{k-1}}{x^{2k}}$$

To divide powers, we subtract exponents.

$$h'(x) = -kx^{k-1-2k}$$

$$= -kx^{-k-1}$$

◼

This completes the power rule. The computation we just finished verifies that the derivative of x^{-k} is $-kx^{-k-1}$. In other words, the power rule works with negative exponents. We now state the general power rule.

The power rule: If n is any integer (positive, negative, or zero), the derivative of the power function $f(x) = x^n$ is $f'(x) = nx^{n-1}$.

EXAMPLE 4.18

Find the derivative of $h(x) = \dfrac{x^2 + x + 1}{x^3 + 7x^2}$.

Solution: By the quotient rule,

$$h'(x) = \frac{(x^2 + x + 1)'(x^3 + 7x^2) - (x^2 + x + 1)(x^3 + 7x^2)'}{(x^3 + 7x^2)^2}$$

$$= \frac{(2x + 1)(x^3 + 7x^2) - (x^2 + x + 1)(3x^2 + 14x)}{(x^3 + 7x^2)^2}.$$

At this point, you may be inclined to simplify, but you should probably consider whether this gains you much. For example, if you merely need the slope of the function when $x = 1$, it is probably easier to evaluate $h'(1)$ directly without simplifying. You'll be less likely to make an error.

In this example, our goal was simply to find the derivative. We have done that, so we'll stop here. ■

EXERCISES

4.27   Calculate the derivative of each function, once by multiplying out and also using the product rule. Verify that you get the same answer either way.
   a) $f(x) = x^3(x + 4)$
   b) $g(x) = (2x^2 + 7x + 7)(5x^2 + 4x + 5)$
   c) $h(x) = (6x^3 + 7x + 3)(2x^3 + 8x^2 + 4x + 4)$

4.28   For each function in the previous exercise:
   a) Find the slope of the tangent line to the function when $x = 0$.
   b) Find the equation of the tangent line to the function when $x = 0$.
   c) Find the slope of the tangent line to the function when $x = 1$.
   d) Find the equation of the tangent line to the function when $x = 1$.

4.29   Differentiate $h(x) = x(x + 1)(x + 2)$ as we did in Example 4.14, only this time take $f(x) = x$ and $g(x) = (x + 1)(x + 2)$ as your factors for the product rule. Verify that you get the same answer.

4.30   Differentiate $h(x) = x(x + 1)(x + 2)(x + 3)$ with the product rule.

4.31   Use the product rule to find the derivative of $f(x) = 1/x^2$. (Hint: start with $x^2 f(x) = 1$.)

4.32   Use the product rule to find the derivative of $f(x) = \dfrac{1}{x^2 + 3}$.

4.33   Use the product rule to find the derivative of $f(x) = \sqrt{7x + 1}$.

4.34   Find the derivative of each quotient.

a) $\dfrac{1}{x - 1}$

b) $\dfrac{x^2 + x + 1}{(x - 1)(x - 2)}$

c) $\dfrac{1}{x^2 + 3x + 1}$

d) $\dfrac{1}{9x^4 + 3x^3 + 9x^2 + 6x + 6}$

4.35   For each function in the previous exercise:

a) Find the slope of the tangent line to the function when $x = 1$ (if it exists).

b) Find the equation of the tangent line to the function when $x = 1$ (if it exists).

4.36   Find the derivative of each.

a) $-2x + 3x^{-1} + 2x^{-2} - 2x^{-3}$

b) $-2.4x - 3.8x^{-1} - 0.1 - 0.2x^{-2} - 2.8x^{-3}$

c) $\sqrt{3x^2 + 7} - \dfrac{1}{x^2} + \dfrac{7}{x^3}$

d) $\dfrac{\pi}{x^3}$

4.37   Show why the product rule is true. Hint:

$$\frac{f(x + h)g(x + h) - f(x)g(x)}{h}$$
$$= \frac{(f(x + h) - f(x))g(x + h)}{h} + \frac{(g(x + h) - g(x))f(x)}{h}.$$

4.7   The Chain Rule

The sum rule allows us to take the derivative of $f(x) + g(x)$. The product rule tells us how to differentiate $f(x)g(x)$, and the quotient rule lets us take the derivative of $\dfrac{f(x)}{g(x)}$. But there is another way we can combine functions, one that we haven't yet discussed.

Compositions of Functions

Can we take the derivative of a composition of functions? If $h(x) = g(f(x))$, is there a way to say what the derivative will be? For example, if $f(x) = \dfrac{x^2 + 1}{x - 1}$ and

$g(x) = x^2$ then

$$h(x) = \left(\frac{x^2 + 1}{x - 1}\right)^2.$$

Is there a way to compute the derivative of h, a way that lets us use what we know about f and what we know about g? It turns out that there is. It's called the chain rule.

The chain rule: If $h(x) = g(f(x))$, then $h'(x) = g'(f(x))f'(x)$.

■ **EXAMPLE 4.19**

Find the derivative of $h(x) = (x^2)^3$.

Solution: By properties of exponents, $h(x) = x^6$, and the power rule directly tells us that $h'(x) = 6x^5$. We can also use the chain rule to arrive at this. For the purposes of the chain rule, the "inside" function of the composition is $f(x) = x^2$. The "outside" function is $g(x) = x^3$.

By the power rule, $f'(x) = 2x$ and $g'(x) = 3x^2$, and according to the chain rule,

$$h'(x) = g'(f(x))f'(x)$$
$$= 3(f(x))^2 f'(x)$$
$$= 3(x^2)^2 2x$$
$$= 3(x^4)2x$$
$$= 6x^5.$$

■

■ **EXAMPLE 4.20**

Let's apply the chain rule to $h(x) = \left(\frac{x^2 + 1}{x - 1}\right)^2$. Here $f(x) = \frac{x^2 + 1}{x - 1}$ and $g(x) = x^2$. The derivative of g is easy: $g'(x) = 2x$. For the derivative of f, we use the quotient rule:

$$f'(x) = \frac{(x^2 + 1)'(x - 1) - (x^2 + 1)(x - 1)'}{(x - 1)^2}$$
$$= \frac{(2x)(x - 1) - (x^2 + 1)1}{(x - 1)^2}$$
$$= \frac{2x^2 - 2x - x^2 - 1}{(x - 1)^2}$$
$$= \frac{x^2 - 2x - 1}{(x - 1)^2}.$$

According to the chain rule, the derivative of h is

$$
\begin{aligned}
h'(x) &= g'(f(x))f'(x) \\
&= 2(f(x))f'(x) \\
&= 2(f(x))\frac{x^2 - 2x - 1}{(x-1)^2} \\
&= 2\left(\frac{x^2+1}{x-1}\right)\frac{x^2 - 2x - 1}{(x-1)^2}.
\end{aligned}
$$

■ **EXAMPLE 4.21**

In Example 4.15 we showed that the derivative of $g(x) = \sqrt{x}$ is $g'(x) = \dfrac{1}{2\sqrt{x}}$.
Find the derivative of $h(x) = \sqrt{1 + x^2}$.

Solution: In this case, the inside of the composition is $f(x) = 1+x^2$, and the outside
is $g(x) = \sqrt{x}$. By the chain rule,

$$
\begin{aligned}
h'(x) &= g'\left(f(x)\right)f'(x) \\
&= \frac{1}{2\sqrt{f(x)}}f'(x) \\
&= \frac{1}{2\sqrt{1 + x^2}}2x \\
&= \frac{x}{\sqrt{1 + x^2}}.
\end{aligned}
$$

■

Roots and Fractional Exponents

If fractional exponents are a dim memory for you, remember that a square root can
be written as an exponent of $1/2$. This is not crazy. Just as you multiply $\sqrt{x}$ by itself
to get x, when you multiply $x^{1/2} \cdot x^{1/2}$ you get x^1 by adding exponents.
    In Example 4.15 we used the product rule to find the derivative of $\sqrt{x}$, that is,
the derivative of the function $y = x^{1/2}$. Knowing the chain rule, we can find the
derivative of any root.

Let $f(x) = \sqrt[n]{x} = x^{1/n}$, where n is a positive integer $(1, 2, 3, \ldots)$. If we take the nth power of both sides, we learn that $(f(x))^n = x$. Now, x is something we know the derivative of, and its derivative is 1. The left side of the equality is a composition, however, and we can apply the chain rule.

For the purposes of the chain rule, the outside function is $g(x) = x^n$. The inside function is $f(x)$. Taking the derivative, we get

$$1 = g'(f(x))f'(x)$$
$$= n(f(x))^{n-1}f'(x), \text{ and substituting for } f$$
$$= n(\sqrt[n]{x})^{n-1}f'(x).$$

It's really f' that we are interested in, and solving for f', we get

$$f'(x) = \frac{1}{n(\sqrt[n]{x})^{n-1}}.$$

If we use fractional exponents for the root, we can make this a bit simpler:

$$f'(x) = \frac{1}{nx^{\frac{1}{n}(n-1)}}$$
$$= \frac{1}{nx^{1-\frac{1}{n}}}$$
$$= \frac{1}{n}x^{\frac{1}{n}-1}.$$

Thus, the chain rule tells us that the derivative of $f(x) = x^{1/n}$ is $f'(x) = \frac{1}{n}x^{\frac{1}{n}-1}$. This is exactly as we might have guessed from the power rule.

EXAMPLE 4.22

If $f(x) = \sqrt{x} = x^{1/2}$, then $f'(x) = \frac{1}{2}x^{(\frac{1}{2}-1)} = \frac{1}{2}x^{-1/2}$. This agrees with our conclusion in Example 4.15.

EXAMPLE 4.23

Find the derivative of $h(x) = x^{2/3}$.

Solution: We can write $h(x) = \sqrt[3]{x^2}$. This is a composition of two functions. The outside function is $g(x) = x^{1/3}$, and the inside function is $f(x) = x^2$. By the chain rule,

$$h'(x) = g'(f(x))f'(x)$$
$$= \frac{1}{3}(f(x))^{-2/3}f'(x)$$
$$= \frac{1}{3}(f(x))^{-2/3}2x$$
$$= \frac{1}{3}(x^2)^{-2/3}2x, \text{ and multiplying exponents,}$$
$$= \frac{1}{3}x^{-4/3}2x$$
$$= \frac{2}{3}x^{-1/3}.$$

Notice that this agrees with the pattern of the power rule. The derivative of $h(x) = x^{2/3}$ is $h'(x) = \frac{2}{3}x^{\frac{2}{3}-1}$.   ■

It turns out that the power rule works for any fractional exponent. Although we don't prove it here, the power rule works for all exponents, even irrational ones.

The power rule: If $f(x) = x^r$, where r is any real number, then $f'(x) = rx^{r-1}$.

■ **EXAMPLE 4.24**

Find the tangent to $f(x) = x^{\sqrt{2}}$ when $x = 1$.

Solution: To find a tangent line, we need two pieces of information, the point of tangency and the slope of the line. When $x = 1$, we have $y = f(1) = 1^{\sqrt{2}} = 1$, so the point of tangency is $(x, y) = (1, 1)$.

The derivative of the function tells us the slope of the tangent, and the derivative of $f(x) = x^{\sqrt{2}}$ is simply $f'(x) = \sqrt{2}x^{\sqrt{2}-1}$. When $x = 1$, the slope is $f'(1) = \sqrt{2} \cdot 1^{\sqrt{2}-1} = \sqrt{2} \cdot 1 = \sqrt{2}$.

Putting these together using the point-slope form of a line, we obtain the tangent line

$$y - 1 = \sqrt{2}(x - 1).$$

If you prefer the $y = mx + b$ form of a line, you can simplify to get

$$y = \sqrt{2}x + (1 - \sqrt{2})$$
$$\approx 1.4142x - 0.4142.$$

A plot on a calculator or computer, like Figure 4.11, can help us verify that this answer is correct and we have made no mistakes.

■

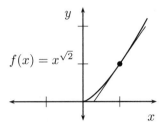

Figure 4.11   Tangent to the power function $y = x^{\sqrt{2}}$.

EXERCISES

4.38   Find the derivative of each function:
  a) $f(x) = \sqrt[5]{x}$
  b) $f(x) = \sqrt[3]{x^2 + 7}$
  c) $y(t) = \sqrt[5]{t^2 - t + 1}$
  d) $h(s) = s^{3/4}$
  e) $y(t) = t^{\pi}$
  f) $y(x) = (3x^2 - 5x + 4)^{2/3}$

4.39   For each function in the previous exercise:
  a) Find the value of function when the independent variable is equal to 1.
  b) Find the slope of the tangent line when the independent variable is 1.
  c) Find the equation of the tangent line to the function at the point where the independent variable is 1. If you have access to a computer or calculator, plot the function and tangent line together to check your answer.

4.8 Slopes and Optimization

In calculus, the term optimization refers to finding the maximum or minimum of some quantity. The maximized quantity can refer to something physical, like the maximum height of a ball thrown into the air. Or it can be non-physical, like the production level of a manufacturing plant that maximizes profit for the company (or minimizes cost).

A Simple Optimization Example

Consider the parabola $f(x) = x(x - 2) = x^2 - 2x$ in Figure 4.12.
    From the graph, it seems obvious that the function has a minimum occurring between $x = 0$ and $x = 2$. You may even intuitively guess that the smallest value happens precisely at $x = 1$. Let's use calculus to verify this.

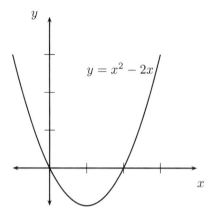

Figure 4.12   Finding the minimum of a parabola.

The key to finding a minimum is to consider slopes, which means we want to use the derivative function. For this parabola, the derivative is $f'(x) = 2x - 2$.

Observation 1: If the derivative is negative, then the slope of the function is "down" and the function gets smaller (or more negative) as we move to the right. A point where the slope is negative is not going to be a minimum because we can move a little bit to the right and find smaller function values.

When is $f'(x) = 2x - 2$ negative? Compute:

$$2x - 2 < 0, \text{ and we can divide both sides by 2 to get}$$
$$x - 1 < 0, \text{ which is the same as}$$
$$x < 1.$$

For any value $x < 1$ we know that the derivative is negative and, consequently, $f(x)$ gets smaller as we move to the right. In mathematical language, we say $f(x)$ is *decreasing*. No x on the left side of 1 can possibly be a minimum of f. Look at the graph, and see that this makes sense.

Observation 2: If the derivative is positive, then the slope of the function is "up" and the function gets bigger (or less negative) as we move to the right. A point where the slope is positive is not going to be a minimum because we can move a little bit to the left and find smaller function values.

When is $f'(x) = 2x - 2$ positive? Compute:

$$2x - 2 > 0, \text{ and we can divide both sides by 2 to get}$$
$$x - 1 > 0, \text{ which is the same as}$$
$$x > 1.$$

For any value $x > 1$ we see that $f(x)$ gets larger as we move to the right, and smaller as we move to the left. In mathematical language, we say $f(x)$ is *increasing*.

No x on the right side of 1 can possibly be a minimum of f. Look at the graph, and see that this makes sense.

Putting these two observations together, we now can be certain that $x = 1$ is the location of a minimum of f, for f is decreasing when $x < 1$ and increasing when $x > 1$.

The value $x = 1$ is the *location* of the minimum. If we were asked to find the minimum *value* of the function, we would want to put that value back into f to find the y-value. For this parabola, the minimum value would be $f(1) = 1^2 - 2 \cdot 1 = -1$. If we were asked to give the lowest *point* on the function, we should give the ordered pair for the point, $(x, y) = (1, -1)$.

Critical Points

Given a function $f(x)$, we are interested in places where f may have a maximum or minimum. In general, functions may have no maximum or minimum, or they may have one or two or possibly many maxima and minima. So far, we have discovered that $f'(x) > 0$ guarantees that x is not the location of a max or min. We also have seen that $f'(x) < 0$ guarantees that x is not the location of a max or min.

Where then should we look for maxima and minima? Author William Priestley suggests we should think about such problems like Sherlock Holmes.[1] There is a famous Holmes quote that reads, "When you have eliminated the impossible, whatever remains, however improbable, must be the truth." We have seen that a max or min is impossible where the derivative is either positive or negative. According to Sherlock Holmes, we eliminate these places and look for maxima and minima at the points that remain, i.e., at the places where the derivative is not positive or negative.

Definition. If $f(x)$ is a function, then the *critical points* of f are the places where $f'(x)$ is not positive or negative. Critical points happen where $f'(x) = 0$ or where $f'(x)$ is undefined.

Critical points are not always maxima and minima, but if we follow the thinking of Sherlock Holmes they are the right places to look for maxima and minima. In mystery parlance, you might call them suspects. They are the places we suspect to find maxima and minima.

■ EXAMPLE 4.25

Let $f(x) = \dfrac{3}{8}x^4 - x^3$. Find any maxima or minima.

[1] W. M. Priestley, "Sherlock Holmes Meets Pierre de Fermat," *Calculus: A Liberal Art*, second edition, Springer-Verlag, New York, 1998.

Solution: To check for critical points, we take the derivative:

$$f'(x) = \frac{3}{2}x^3 - 3x^2$$

$$= x^2 \left(\frac{3}{2}x - 3 \right)$$

$$= 3x^2 \left(\frac{1}{2}x - 1 \right).$$

There are two ways a critical point can occur: the derivative can be zero or it can be undefined. In this example, the derivative is a polynomial, so there is no way for it to be undefined (we'll never divide by zero, take the square root of a negative, etc.).

So we can focus on zeros. This is why we factored the derivative. It makes finding zeros simpler. If we set $f'(x) = 0$, we can see that either $3x^2 = 0$ or $\frac{1}{2}x - 1 = 0$. Consider each, in turn. If $3x^2 = 0$, then $x = 0$. If $\frac{1}{2}x - 1 = 0$, then $x = 2$.

So the function f has critical points when $x = 0$ or when $x = 2$. Accordingly, these are places to look for maxima and minima. Let's consult a graph of f (Figure 4.13).

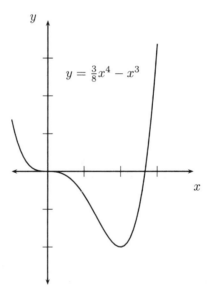

Figure 4.13   Checking the critical points at $x = 0$ and $x = 2$.

It would appear that $x = 2$ is the location of a minimum of f, but $x = 0$ is neither a maximum nor a minimum. When $x = 0$, the slope of the tangent line momentarily becomes zero (that is, horizontal). It looks like the function decreases on the left side of $x = 0$, becomes momentarily horizontal, and continues decreasing to the right until it reaches a minimum at $x = 2$.

We can verify the maxima and minima of a function in practice by checking the function value at a critical point and comparing to function values on both sides. Even without the graph, we could check:

x	$f(x)$
-0.1	.001
0	0
0.1	$-.001$
1.9	-1.972
2	-2
2.1	-1.968

As the graph indicates, the function is a bit positive when $x = -0.1$ and a bit negative when $x = 0.1$, telling us that $(x, y) = (0, 0)$ is a critical point that is not a max or min. However, when checking around $x = 2$, we see that the function is larger (that is, less negative) on both sides, verifying that $x = 2$ is the location of a minimum.

Our conclusion is that the minimum value of f is -2, which occurs at the point $(x, y) = (2, -2)$. There are no maxima.   ∎

EXAMPLE 4.26

Find the maxima and minima of $f(x) = \sqrt[3]{x^2}$.

Solution: To check for critical points, we take the derivative. We can either use the chain rule, thinking of f as a composition of functions, or we can first switch to fractional exponents.

Using fractional exponents, a cube root is the same as an exponent of $1/3$, so $f(x) = \sqrt[3]{x^2} = (x^2)^{1/3} = x^{2/3}$. (Remember that you multiply exponents when taking a power of a power.) So, the derivative is $f'(x) = \dfrac{2}{3}x^{-1/3} = \dfrac{2}{3\sqrt[3]{x}}$.

Critical points occur where the derivative is not positive and not negative. We need to look for values of x with either $f'(x) = 0$ or where $f'(x)$ does not exist.

We can readily see that there are no values that make the derivative zero. The function f' is a fraction, and fractions can only be zero when the numerator (top part) is zero. However, there is one value of x that makes the derivative undefined, $x = 0$, since we can't divide by zero. So $x = 0$ marks a critical point.

We know from the previous example that a critical point may not actually indicate a max or min, so we'll check the values of the function on each side:

x	$f(x)$
-0.1	.215
0	0
0.1	.215

The critical point at $x = 0$ is apparently a minimum of f, because the values of the function are greater on both sides. Let's compare with the graph in Figure 4.14.

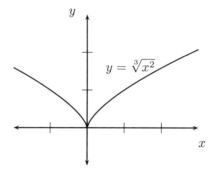

Figure 4.14   Checking the critical point at $x = 0$.

Our graph confirms that $x = 0$ is the location of a minimum of f. The graph also helps us visualize what occurs. At the origin, the tangent line to f becomes vertical, and you may recall that vertical lines have undefined slope (they are the only lines that cannot be written as $y = mx + b$, because there is no m).

The sharp point where the derivative becomes undefined is called a *cusp*, and this cusp forms a minimum for f. The function f has no maximum points.   ∎

In summary, optimization problems are solved by computing the derivative. Points where the derivative of a function is positive or negative will not be maxima or minima. If we eliminate these points, the remaining points are called critical points. Critical points occur where the derivative is zero or undefined and may or may not mark maxima and minima. We have to check each critical point in turn to see what kind of point it is.

EXERCISES

4.40   Find the maxima and minima of each function. Compare with a graph to confirm your answers.

a) $f(x) = x^3 - x$
b) $f(x) = x^2 - 2x + 1$
c) $f(x) = \frac{3}{4}x^4 - 4x^3 + 6x^2$
d) $f(x) = x^{1/3}$
e) $f(x) = \sqrt[3]{(x-1)^2}$
f) $f(x) = 1/x$

4.9 Applying Optimization Methods

I have 200 m of fencing available to create a rectangular pen to hold some sheep. What is the largest area of grass that can be enclosed by my fence? We can use calculus to find the answer.

The key to solving an applied optimization problem is to find a way to model the situation with a function. We know we can use derivatives to optimize functions, that is, to find maxima and minima. To get started with the correct function, it helps to begin with the proper question.

What is it that we are trying to maximize or minimize?

In this case, we want to maximize the area of grass enclosed by the rectangular sheep pen. We need a function that represents the area, and then we can apply calculus.

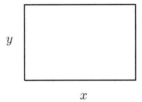

Figure 4.15 A sheep pen.

It is usually a good idea to draw a picture and label it, like Figure 4.15. Taking this as our picture, the area of the pen is simply

$$A = \text{length} \times \text{width} = xy.$$

We would almost be ready to apply calculus, except for one impediment. The functions we have worked with so far have only one variable, but our area is in terms of both x and y. So there is one more step to resolve before we can continue. We need to eliminate one of the variables.

Luckily, we have one more relevant piece of information. We plan to build the pen using 200 m of fence. Nothing requires that we use all of the available fence, but we want the pen to be as large as possible, so it makes sense to use it all. Hence, the perimeter of our rectangle should be 200 m, and we can write this relationship as an equation:

$$200 = 2x + 2y.$$

Solving for y, we get

$$200 - 2x = 2y$$
$$100 - x = y.$$

We can use this to eliminate y. If we substitute for y in our area formula, we get

$$A = xy$$
$$= x(100 - x)$$
$$= 100x - x^2.$$

Now we can apply calculus to this function. Taking the derivative, we get

$$A' = 100 - 2x.$$

Remember, maxima and minima occur at critical points. There are no places where the derivative is undefined (that is, no places where we do something like divide by zero or take the square root of a negative), so we don't have to worry about that. That means the only possible critical points happen when the derivative is zero. Solving:

$$0 = 100 - 2x$$
$$2x = 100$$
$$x = 50.$$

When $x = 50$, the area is $A = x(100 - x) = 50(100 - 50) = 50^2 = 2500 \text{ m}^2$. To verify that this corresponds to the maximum area, compare with the area at values on either side of 50, say 49 and 51.

x	$A(x)$
49	2499 m^2
50	2500 m^2
51	2499 m^2

As we can see, $x = 50$ m corresponds to a maximum. Since $y = 100 - x = 100 - 50 = 50$ m, the largest pen that can be made from 200 m of fence is a 50 m $\times$ 50 m square.

We always need to check the critical points. It is easy to become lazy, since it is sometimes a fair bit of work to get from the initial statement of a problem, to a function, to the derivative, to the critical points. But it would be a shame to accidently *minimize* the value of something we intended to maximize simply because we forgot to check!

Other ways to verify that $x = 50$ m is the location of the maximum pen size would be to realize that $A = 100 - x^2$ is a parabola opening downward (so our critical point marks the vertex of the parabola) or to graph the function on a calculator or computer. The method that we use to check is not so important as remembering to do the checking.

Values at the Boundary

What if you had been challenged to use 200 m of fencing to build a rectangular pen of *minimum* size? The answer is easy. Use 0 m of the fence to build a 0 m $\times$ 0 m pen with a total area of 0 m^2.

What if I insisted, as part of the puzzle, that you use the entire 200 m of fence? Derivatives and critical points can't tell you the answer because we already found that the only critical point happens when $x = 50$ m, and it corresponds to the maximum area.

Looking at a graph, you might conclude that there is no minimum, since the graph of $A(x)$ is a parabola opening downward. But this is silly, because most of the graph has negative function values, and in the real world the area of a rectangular pen can never be negative. In Figure 4.16 we've augmented the graph to emphasize the positive region.

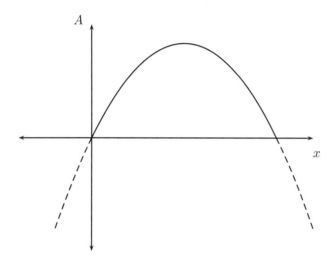

Figure 4.16   The area of a pen cannot be negative.

In mathematical language, this problem is said to be "constrained." It only makes sense on the closed interval $[0, 100]$, and $f(x)$ has two minima on the interval. It has a minimum at each end, at the "boundary" of the interval.

The solutions $x = 0$ m and $x = 100$ m make some practical sense, by the way. When $x = 0$ m, we have $y = 100 - x = 100$ m. Our rectangle has degenerated into two parallel runs of fence slapped together with no space between then. When $x = 100$ m, then $y = 0$ m, and the two strings of fence run the other way. Either solution gives a minimum area for the pen of 0 m^2 and uses all of the fence.

You should convince yourself that this does not contradict Observation 1 on p. 254, that a maximum or minimum cannot occur when the derivative of a function is negative. When we made that observation, we depended on there being points to the left where the function would be higher and points to the right where the function would be lower. Obviously, when we come to the end of the interval there are no more points, and a minimum (as we have observed) or a maximum can occur.

In a similar way, a positive derivative will not prevent a maximum or minimum from occurring at the end of an interval. So Observation 2 is not contradicted either.

Optimization principle: For a function defined on a closed interval, maxima and minima may (only) occur at critical points as well as at the endpoints of the interval. We have to check the function at each of these places.

■ EXAMPLE 4.27

A student has 200 m of fence available to make a garden (Figure 4.17). She wants the shape of the garden to be like the free-throw lane on a basketball court, a rectangle capped by a semicircle. What dimensions make the largest garden (or the smallest)?

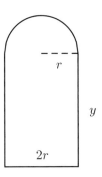

Figure 4.17   A fenced garden.

Solution: Deciding how to label a picture can sometimes be a real challenge. For this figure, it probably makes sense to label the radius of the semicircle, since both the area and perimeter formulas for circles are given in terms of r. This forces one side of the rectangle to be $2r$, and it is left to label the other side, which we have marked with y in this figure.

What are we trying to maximize? The area of the garden is the area of the rectangle plus the area of the semicircle, $A = (2r)y + \frac{1}{2}\pi r^2$. This formula has two variables, so we need to eliminate one before we proceed.

To eliminate a variable, we use the other piece of information that we know, which is that we are to use 200 m of fence. The 200 m perimeter of the garden comes from three sides of the rectangle together with the semicircle:

$$200 = y + 2r + y + \pi r$$
$$= 2y + 2r + \pi r.$$

Solve for either r or y by getting it alone on one side:

$$200 - 2r - \pi r = 2y$$
$$100 - r - \frac{\pi}{2}r = y.$$

If we substitute this into the area formula, we get

$$A = (2r)y + \frac{1}{2}\pi r^2$$

$$= (2r)\left(100 - r - \frac{\pi}{2}r\right) + \frac{1}{2}\pi r^2$$

$$= 200r - 2r^2 - \pi r^2 + \frac{1}{2}\pi r^2$$

$$= 200r - 2r^2 - \frac{1}{2}\pi r^2$$

$$= 200r - \left(2 + \frac{1}{2}\pi\right)r^2.$$

This is a formula that we can optimize. Before we search for critical points, let's first determine if there is an interval that constrains the problem. Clearly, r can't be negative since it represents a real-world distance.

But how large can r become? Since the amount of fence is fixed, as r gets larger it can only be that y becomes smaller. Now, y can't be negative, so the largest r would correspond to $y = 0$. To have a perimeter of 200 m, when $y = 0$ we would need $200 = 2r + \pi r$, and solving, $r = \dfrac{200}{2 + \pi}$.

So our function is constrained to the interval $\left[0, \dfrac{200}{2 + \pi}\right] \approx [0, 38.898]$.

To find critical points, we differentiate:

$$A' = 200 - 2\left(2 + \frac{1}{2}\pi\right)r$$

$$= 200 - (4 + \pi)r.$$

This derivative is never undefined, so critical points can only come from the derivative being zero. Solving:

$$0 = 200 - (4 + \pi)r$$

$$(4 + \pi)r = 200$$

$$r = \frac{200}{4 + \pi}.$$

Our optimization principle tells us that to finish, we need to check the function at this critical point and at the ends of the interval.

r	$A(r)$
0	0
$200/(4 + \pi)$	2800.5
$200/(2 + \pi)$	2376.8

The smallest garden occurs when $r = 0$ and $y = 100$, and it has an area of 0 m². The largest garden comes from making $r = \dfrac{200}{4 + \pi} \approx 28$ m, which (if you check) also makes $y \approx 28$ m, and it has an area of approximately 2800.5 m².   ■

EXERCISES

4.41   For each function find the critical points and classify each as a maximum, minimum, or neither.

   a) $y(x) = 7x + 10$
   b) $y(t) = t(t - 4)$
   c) $y(t) = (t - 1)(t - 4)$
   d) $f(x) = x^3 - x$
   e) $f(x) = \sqrt[3]{(x - 1)^2}$

4.42   For each function and interval, find the points where the function reaches its maximum and minimum.

   a) $y(x) = 7x + 10$ on $[0, 1]$
   b) $y(t) = t(t - 4)$ on $[0, 4]$
   c) $y(t) = (t - 1)(t - 4)$ on $[0, 4]$
   d) $f(x) = 1/x$ on $[1, 2]$
   e) $f(x) = \sqrt[3]{(x - 1)^2}$ on $[2, 4]$

4.43   A rectangular pen runs next to a stream, so one side does not require a fence. Find the dimensions that maximize and minimize the area of the pen assuming 200 m of fence is used.

4.44   A rectangular pen runs along an inside corner of an existing (large rectangular) fence, so two sides do not require a fence. Find the dimensions that maximize and minimize the area of the pen assuming 200 m of fence is used.

4.45   A garden in the shape of the free-throw lane on a basketball court is built with one side against an existing wall, so that side needs no fence as in Figure 4.18. What are the dimensions that maximize the area of the garden?

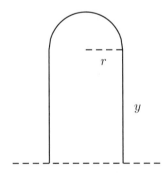

Figure 4.18   A fenced garden against a wall.

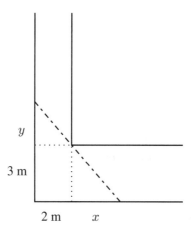

Figure 4.19   A ladder goes around a corner.

4.46   Imagine carrying a ladder down a hallway when you come to a right-angle corner, as in Figure 4.19. Assume that the ladder is arriving from a hallway that measures 2 m across and entering a hallway that measures 3 m across. If the ladder is too long, it may not make the turn.

    a) Taking the ladder to be the hypotenuse of a right triangle measuring $y + 3$ along one leg and $x + 2$ along the other, give a formula for the length of the ladder in terms of x and y.

    b) Use similar triangles to relate x and y.

    c) Substitute for y in the answer you got from part (a) to write the length of the hypotenuse as a function of x alone.

    d) We can find the length of the longest ladder that fits around the corner by minimizing the function in part (c). This makes sense, because the ladder has to fit around the tightest part of the corner. But minimizing this function is inconvenient, because the function has a square root in it. It is not too hard to minimize the *square* of the function, however. Find the value of x that minimizes the square of the function from part (c).

    e) How long is the longest ladder that can make the turn?

4.47   A wire 90 cm long is divided into three straight pieces of wire. Two pieces are the same length, x, which leaves the remaining piece of length $90 - 2x$. Use the wire to form an isosceles triangle.

    a) What is the height of the isosceles triangle as a function of x, taking the "odd" side as the base?

    b) What is the area of the isosceles triangle, as a function of x?

    c) What values of x maximize or minimize the area of the triangle?

4.48   A string consisting of n 9-volt batteries is connected in series to a 100 ohm circuit. Assume the current supplied (in amps) depends on the number of batteries, n, according to the formula $I(n) = \dfrac{9n}{100 + 0.15n^2}$.

a) How much current does a single 9-volt battery supply?
b) Find the derivative of $I(n)$.
c) Find n where $I'(n) = 0$.
d) How many batteries would we use to get the maximum amount of current, and how much current will they provide?
e) How many batteries would we use to get the minimum amount of current (you may assume we use fewer than 100 batteries)?

4.10   Differential Notation and Estimates

We have grown accustomed to prime notation for derivatives; if $f(x)$ is a function then $f'(x)$ is its derivative. This was the notation of Joseph-Louis Lagrange who lived 1736–1813, and it is one of the most popular derivative notations.

Leonhard Euler (1707–1783) used the capital letter D to indicate derivatives. So the derivative of the function $f(x)$ is Df, or $D_x f$ when we want to be explicit about the dependent variable being x.

Isaac Newton, who lived 1642–1727 and was one of the inventors of calculus, used dot notation. If we recall our example of dropping a pen, we had the position formula

$$y = 4 - 16t^2.$$

To indicate a derivative, Newton placed a dot over the dependent variable, y. In his notation, the derivative (which we now know indicates the velocity) looks like

$$\dot{y} = -32t.$$

Gottfried Wilhelm Leibniz, who lived 1646–1716 and is credited along with Newton for the invention of calculus, had yet another notation: differential notation. If $y = f(x)$, then Leibniz's notation indicates the derivative by the symbol dy/dx.

To understand what Leibniz's symbol is trying to represent, remember how we came to the idea of derivative. The derivative tells us the slope of a tangent line, and we determine the slope of a tangent line from the slopes of secant lines using a limit.

In normal conversation, we say

$$\text{slope} = \frac{\text{rise}}{\text{run}}.$$

But if we take $y = f(x)$, the "rise" is simply the change in y, which we might write as Δy. Similarly, the "run" is the change in x, or Δx. In this context, the slope of a secant line is

$$\text{slope of secant} = \frac{\Delta y}{\Delta x}.$$

The slope of the tangent is defined to be the limit of the slopes of the secants over shorter and shorter intervals, that is, the slope as Δx gets closer and closer to 0:

$$\text{slope of tangent} = \lim_{\Delta x \to 0} \frac{\Delta y}{\Delta x}.$$

The Leibniz notation is meant to reflect this process, so

$$\frac{dy}{dx} = \lim_{\Delta x \to 0} \frac{\Delta y}{\Delta x}.$$

Where we see Δx and Δy we should think of secant lines. A small change in y is divided by a small change in x. Where we see dx and dy, we should infer tangent lines, i.e., the result of taking a limit.

It is common for people to think of dy/dx as an infinitesimal change in y divided by an infinitesimal change in x. Although this is (formally) a lie, since the real number line is usually not considered to contain infinitesimal values other than 0, it is often a useful and tangible way to think of dy/dx.

Derivatives for Computing Estimates

Derivatives can be useful for estimating small changes in a function. Consider, for example, measuring a square with a ruler. Say we measure the side length to be 5 cm, as in Figure 4.20. Then we know the area to be 25 cm^2.

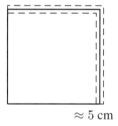

$$\approx 5 \text{ cm}$$

Figure 4.20   Measuring a 5 cm $\times$ 5 cm square.

Rulers, however, are not perfect. We never measure a length to be exactly 5 cm. There is always a margin of error. Perhaps we actually know the length to be 5 cm $\pm$ 0.1 cm.

Of course, a change in the side length means a corresponding change in the area of the square. We can compute this directly: the largest possible area is 5.1 cm $\times$ 5.1 cm $= 26.01$ cm^2, and the smallest is 4.9 cm $\times$ 4.9 cm $= 24.01$ cm^2.

We can also estimate this in a calculus context by letting x represent the length of the side we are measuring so that $y = f(x) = x^2$ is the area. Our goal is to estimate the change in $f(x)$ that comes from changing (or mis-measuring) x by a small amount.

We've emphasized that the derivative is a limit of slopes of secant lines. Another way of saying this is that when Δx is small, the slope of the secant is approximately the same as the slope of the tangent. When Δx is small,

$$\frac{\Delta y}{\Delta x} \approx \frac{dy}{dx}.$$

Multiplying by Δx, we get a formula that estimates the change in the function.

The derivative approximation rule: $\Delta y \approx \dfrac{dy}{dx}\Delta x.$

In our case, since $y = f(x) = x^2$, we have $dy/dx = 2x$. We measured $x = 5$ cm, but we may have a measurement error as large as $\Delta x = \pm 0.1$ cm. To estimate the error in the area, we compute:

$$\begin{aligned}
\Delta y &\approx \frac{dy}{dx}\Delta x \\
&= 2x\Delta x \\
&= 2 \cdot 5 \text{ cm} \cdot (\pm 0.1 \text{ cm}) \\
&= \pm 1 \text{ cm}^2.
\end{aligned}$$

So, we estimate the area of the square to be $25 \text{ cm}^2 \pm 1 \text{ cm}^2$; it may be as low as 24 cm^2 or as high as 26 cm^2. This estimate matches almost exactly what we obtained by direct computation.

To understand how this estimate works, it may help to look carefully at the graph of $f(x) = x^2$ near the point $x = 5$.

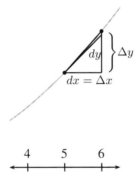

Figure 4.21   The function $f(x) = x^2$ near $x = 5$.

In this picture, Δy indicates the true change in the value of a function, the true difference that we could compute by evaluating the function at two points and subtracting. For example, if this is the graph of $y = x^2$ with $\Delta x = 1$, then $\Delta y = 36 - 25 = 11$.

The value dy is the corresponding change that happens in the tangent line. We know from the derivative that the slope at 5 is $f'(5) = 2 \cdot 5 = 10$, so the equation of the tangent line (in point-slope form) is $y - 25 = 10(x - 5)$, or in the slope-intercept form, $y = 10x - 25$. Evaluating at $x = 6$ and $x = 5$, we subtract to get $dy = 35 - 25 = 10$. This is a bit smaller than the true Δy of the function, and we can see this on the graph.

We took $\Delta x = 1$ cm, a pretty large change. Imagine how much closer the estimate would be if we took $\Delta x = 0.1$ cm. The difference between Δy and dy would be very small. This is the key to approximating with derivatives. For small changes in x, the tangent line is a close approximation to the function, so (vertical) changes measured on the tangent line do a good job of estimating (vertical) changes in the function.

The symbol dy is called the *differential* of the function. This is why the symbol dy/dx is often referred to as *differential notation* and why the method of approximating with derivatives is called *differential approximation*.

■ **EXAMPLE 4.28**

Estimate $\sqrt{8.95}$ using only the operations available on a 4-function calculator.

Solution: If we let $f(x) = \sqrt{x}$, then our task is to estimate the value of $f(8.95)$. Fortunately, the value $f(9) = 3$ is easy to calculate. We can apply a differential approximation using $\Delta x = -0.05$ to see how much f changes as we move to the nearby x-value:

$$\Delta y \approx \frac{dy}{dx} \Delta x$$

$$= \frac{1}{2} x^{-1/2} \Delta x$$

$$= \frac{1}{2x^{1/2}} \Delta x$$

$$= \frac{1}{2\sqrt{x}} \Delta x,$$

and using $x = 9$ and $\Delta x = -0.05$, we get

$$= \frac{1}{2\sqrt{9}} - 0.05$$

$$= \frac{1}{6} - 0.05$$

$$\approx -0.0083.$$

Having estimated the change in function value, we can now calculate that $\sqrt{8.85} \approx \sqrt{9} - 0.0083 = 2.9917$. We used only addition, subtraction, multiplication, and division (the operations available on a 4-function calculator). ■

It may have never occurred to you, but even powerful computer chips often know how to do only simple operations. More complicated calculations, such as roots and the values of trigonometric functions, are typically produced by approximation methods programmed in software.

EXERCISES

4.49   The side length of a cube is measured to be $x = 15$ cm with a margin of error of ± 0.5 cm. Estimate the change in the volume that results from measurement error.

4.50   The side length of a cube is measured to be $x = 15$ cm with a margin of error of ± 0.5 cm. Estimate the change in the surface area that results from measurement error.

4.51   The radius of a circle is measured to be $r = 30$ cm with a margin of error of ± 1 cm. Estimate the change in area that results from measurement error.

4.52   The radius of a sphere is measured to be $r = 30$ cm with a margin of error of ± 0.1 cm. Estimate the change in volume that results from measurement error.

4.53   The Earth is approximately spherical, with a radius of 6378.1 km. A 'belt' of 40074.8 km would wrap around the "waist" of the Earth. If 1 m of slack were added to the belt, how high would the belt rise above the Earth? (Hint: you are given a circumference C with change ΔC, and asked to estimate Δr.)

4.54   The Moon has a 'waist size' of 10916.4 km. If 1 m of slack were added to its 'belt,' how high would the belt rise above the surface of the Moon?

4.55   Use differential approximation to estimate each value:
   a) $\sqrt{4.1}$
   b) $\sqrt[3]{8.1}$
   c) $\sqrt[3]{26}$
   d) $\sqrt{2}$ (Hint: $100/49$ is a perfect square.)
   e) $\sqrt{5}$ (Hint: find a ratio of perfect squares near the value 5.)

4.11   Marginal Revenue, Cost, and Profit

If we are a company that manufactures some item, then the *marginal cost* of the item is said to be the cost of increasing our production level by one item. For example, if we can produce 10 hand-held radios by spending $33.67 for parts and labor, and 11 radios would cost us $35.77, then we can compute the marginal cost to be $35.77 − $33.67 = $2.10. At a production level of 10 radios, the marginal cost for making one more radio is $2.10.

   Perhaps for our particular production line, the cost of making x radios is modeled by the function

$$C(x) = \frac{2x^2 + 19x + 14}{x + 2}.$$

We can make a couple of easy observations. One observation is that the cost of making zero radios is $C(0) = 7$. In most manufacturing situations there are some expenditures even when nothing is produced. These are called *fixed costs*. We can also verify the marginal cost for producing the eleventh radio by computing $C(11) - C(10) \approx 35.77 - 33.67 = \$2.10$.

■ **EXAMPLE 4.29**

Marginal cost generally depends on the production level, and it is common that the marginal cost decreases as we make more and more items. In fact, you've probably heard the phrase "efficiencies of scale." Verify that the twenty-first item costs less to produce than the eleventh.

Solution: We'll use the cost function. The cost of the twenty-first item is the difference between the cost for twenty-one items and the cost for twenty items:

$$C(21) - C(20) \approx 56.30 - 54.27$$
$$= 2.03.$$

The cost for the twenty-first radio is $\$2.03$, which is less than the $\$2.10$ that the eleventh radio costs to produce. ■

Definition. If $C(x)$ is the cost of producing x items, then the *marginal cost* function is the derivative of the cost function. That is, $MC(x) = C'(x)$.

■ **EXAMPLE 4.30**

Since the cost function for radios is

$$C(x) = \frac{2x^2 + 19x + 14}{x + 2},$$

the marginal cost function is

$$
\begin{aligned}
MC(x) &= C'(x) \\
&= \frac{(x+2)(2x^2 + 19x + 14)' - (2x^2 + 19x + 14)(x+2)'}{(x+2)^2} \\
&= \frac{(x+2)(4x + 19) - (2x^2 + 19x + 14)1}{(x+2)^2} \\
&= \frac{(4x^2 + 27x + 38) - (2x^2 + 19x + 14)}{(x+2)^2} \\
&= \frac{2x^2 + 8x + 24}{(x+2)^2}.
\end{aligned}
$$

It follows that the marginal cost when producing 10 radios is $C'(10) \approx \$2.11$, and the marginal cost when producing 20 radios is $C'(20) \approx \$2.03$.

Notice that we don't get precisely the same answer when using the derivative as we did by direct computation, but the results are very close. This is because our derivative definition of marginal cost is really a differential approximation of the cost for one more item.

Recall how differential approximation works. For small values of Δx, we know that

$$\frac{\Delta C}{\Delta x} \approx \frac{dC}{dx}$$

or equivalently,

$$\Delta C \approx \frac{dC}{dx} \Delta x.$$

If we take $\Delta x = 1$ (we are interested in the change in cost that comes from producing just one more item), we get $\Delta C \approx dC/dx$. The cost of one more item is approximately the derivative of the cost function. The intuitive definition of marginal cost and the calculus definition are approximations.

Revenue and profit work the same way.

Definition. If $R(x)$ denotes the revenue from producing x items, then the *marginal revenue* is $MR(x) = dR/dx$. If $P(x)$ denotes the profit from producing x items, then the *marginal profit* is $MP(x) = dP/dx$.

The marginal revenue is approximately the change in revenue that comes from producing one more item, and the marginal profit is approximately the change in profit from one more item.

The marginal profit function is the difference of the marginal revenue and marginal cost functions. That is, $MP(x) = MR(x) - MC(x)$.

This agrees with our intuition. Since profit is revenue minus cost, the profit from one more item should be the revenue for the item after its costs are subtracted.

Proof. The rules of derivatives make this easy to verify:

$$\text{Profit} = \text{Revenue} - \text{Cost}$$
$$P(x) = R(x) - C(x)$$

and taking derivatives of both sides,

$$P'(x) = R'(x) - C'(x)$$
$$MP(x) = MR(x) - MC(x).$$

Maximizing Profit

In a business setting, raising revenues is good. Cutting costs is good. But maximizing profit is best, because that is what puts money in our pockets. If our business is modeled by a cost function and a revenue function, can we determine the production level that maximizes our profit?

Recall from the optimization principle (p. 261) that maxima of a function occur at critical points. We will want to check places where the derivative of $P(x)$ is zero or undefined.

In general we need to worry about derivatives being undefined, but this is not a great worry for the profit function $P(x)$. Remember, the derivative tells us the marginal profit, which is approximately the profit for making one more item. It would be an uncommon scenario where the profit for the next item couldn't be determined. Consequently, we can assume that critical points of the profit function occur because the derivative is zero.

Theorem. Profit maxima (and minima) occur at production levels for which the marginal cost and marginal revenue coincide, that is, at values of x for which $MC(x) = MR(x)$.

Proof. Maxima (and minima) occur at critical points, i.e., where the derivative is zero or undefined. Since there is not a worry that marginal profit is undefined, maxima (and minima) will occur where the derivative is zero. Calculating:

$$0 = P'(x)$$
$$= MP(x)$$
$$= MR(x) - MC(x) \text{ by the } MP = MR - MC \text{ theorem,}$$

and, adding $MC(x)$ to both sides, it follows that

$$MC(x) = MR(x).$$

■

We can verify this result intuitively with the following thought experiment. Assume we are manufacturing radios and our current production level is x radios.

If the marginal cost, $MC(x)$, is less than the marginal revenue, $MR(x)$, then we should increase our production level by at least one more radio. It will increase our profit (by increasing our revenue more than our costs). On the other hand, if the marginal cost is more than the revenue, i.e., $MC(x) > MR(x)$, we should produce fewer radios. The cost savings will more than make up for the loss in revenue.

So, the maximum profit can only occur where the marginal cost neither lags nor exceeds the marginal revenue.

A similar argument works for minimum profit. The minimum profit will also occur at a production level where the marginal cost and marginal revenue coincide. If we invoke the precept of Sherlock Holmes yet again, we have to remember that places where the marginal cost and marginal revenue coincide are only "suspects" for the maximum profit. We still have to check each one.

◨ **EXAMPLE 4.31**

Yo-yos sell for \$2 each. The cost of producing x yo-yos is $C(x) = 50 + 0.01x^2$. How many yo-yos should be produced to maximize profit?

Solution: Note that we haven't been given a revenue function, but we have been given enough information to figure it out for ourselves. Since yo-yos sell for \$2, if we sell x of them, our revenue function is $R(x) = 2x$.

To maximize profit, we want to consider when $MC(x) = MR(x)$, so we compute the derivatives, $MC(x) = 0.02x$ and $MR(x) = 2$. Setting them equal, we learn that

$$MC(x) = MR(x)$$
$$0.02x = 2$$
$$x = 100.$$

The profit for 100 yo-yos is $R(100) - C(100) = 200 - (50 + 100) = \$50$. This could be a maximum, but it could also be a minimum (or even just a lucky point where the marginal cost and revenue happened to coincide by accident). We can verify that it is a maximum in several ways, but probably the two most obvious checks are:

We could calculate the profit for 99 yo-yos and for 101 yo-yos and discover that in each case the profit is less than \$50 (i.e., less than the profit for 100 yo-yos). It is in fact \$49.99 at both $x = 99$ and $x = 101$.

We could use a computer or calculator to graph the profit function $P(x) = 2x - (50 + 0.01x^2)$, on an interval around 100 and see that $x = 100$ is the location of the highest point. ◼

◨ **EXAMPLE 4.32**

A plane holds 450 seats. Tickets cost \$400 each, and the cost of operating the plane with x passengers is $C(x) = -0.001x^3 + 0.9x^2 + 130x + 3000$. How full should the plane be to maximize profit?

Solution: As in the previous example, we have to realize that a \$400 ticket price implies that the revenue function is $R(x) = 400x$. To find critical points of the profit function, we take derivatives and set marginal cost and marginal revenue equal.

$$MC(x) = MR(x)$$
$$-0.003x^2 + 1.8x + 130 = 400$$

Subtracting 400 from both sides gives

$$-0.003x^2 + 1.8x - 270 = 0.$$

Since this is a quadratic equation we can solve with the quadratic formula to get

$$x = \frac{-1.8 \pm \sqrt{1.8^2 - 4(-0.003)(-270)}}{-0.006}$$

$$= \frac{-1.8 \pm \sqrt{0}}{-0.006}$$

$$= 300.$$

We still need to verify that selling 300 tickets maximizes profit, so we check values on each side:

x	$P(x)$
299	23999.999
300	24000.000
301	24000.001

It looks as though $x = 300$ is not a maximum. (It is not a minimum either.) Admittedly, the values are very close, so close that we should suspect or at least be cautious about rounding error. Let's check again, with values a little further from 300:

x	$P(x)$
290	23999.00
300	24000.00
310	24001.00

Indeed, it looks like $x = 300$ is not a maximum or minimum for profit. That leaves our question unanswered: How many tickets should we sell to maximize profit? The answer is "all of them."

Here's how we see that: Since the critical point did not provide an answer, we must look at the boundary, that is, the smallest and largest possible values of x. The fewest number of tickets we can sell is 0, and the most is 450, so this entire problem takes place on the interval $[0, 450]$.

Check the profit at the endpoints of the interval:

x	$P(x)$
0	-3000.00
450	27375.00

The profit for selling 450 seats is $27,375.00, which is greater than the value at $x = 0$ (which generates a loss) and the value at $x = 300$ where the critical point occurred. A full plane generates the greatest revenue. ∎

EXERCISES

4.56   Graph the revenue function $R(x)$ and the cost function $C(x)$ for Example 4.31 together on one graph.

    a) Sketch the tangent to each curve at the location of the maximum, $x = 100$, to demonstrate that the slopes of both curves agree there.

    b) Draw a vertical line between the two curves at $x = 100$. The length of this line represents the profit. You should notice that the maximum vertical distance occurs where you've drawn the line.

4.57   Graph the revenue function $R(x)$ and the cost function $C(x)$ for Example 4.32 together on one graph.

    a) Sketch the tangent to each curve at the location of the maximum, $x = 300$, to demonstrate that the slopes of both curves agree there.

    b) Draw a vertical line between the two curves at $x = 300$, and another at $x = 450$. The lengths of these lines represent the profit at each capacity. You should notice that the larger vertical distance occurs at $x = 450$.

4.58   At a production level of $x = 200$ items, $MC(x) = 7$ while $MR(x) = 10$.

    a) To maximize profit, how many items should we produce: fewer than 200, exactly 200, or more than 200 items?

    b) Which has greater slope at $x = 200$: the cost function $C(x)$ or the revenue function $R(x)$?

4.59   I intend to produce computers that sell for \$400 each. My cost for producing x computers is $C(x) = 1000 + x^2/80$.

    a) How many computers should I produce to maximize my profit?

    b) How much profit will I make?

4.60   Assume that the price I can charge for a product drops if I flood the market with items, so that the price I can charge is $p(x) = 2000/x$.

    a) What is the revenue generated by x items, $R(x)$?

    b) What is the marginal revenue function, $MR(x)$?

    c) Assume that the cost function is an increasing function. Does that mean that the marginal cost function, $MC(x)$, is less than 0, exactly 0, or greater than 0?

    d) Assuming that I produce at least 1 item, prove that the maximum profit occurs when $x = 1$.

4.12   Exponential Growth

Since derivatives tell us rates of change, it should be little surprise that calculus is important for the study of different kinds of growth. For example, linear growth (which describes something whose size looks like a line when you graph it over time) is very easy to describe with calculus. Its derivative is a constant (the slope of the line).

One of the most important kinds of growth is exponential growth. It describes many biology situations, such as the growth of a bacteria colony or the spread of a disease. It has a role in computing physical properties such as the decay of radioactive elements, or changes in temperature. It also has applications to finance.

Simple and Compound Interest

If you invest a sum, say $1000, and you receive simple interest at a rate of 5%, then each year you are paid 5% of your $1000 as interest. Over time, your investment will grow as in the following table:

years	balance
0	1000
1	$1000 + 50$
2	$1000 + 50 + 50$
3	$1000 + 50 \cdot 3$
4	$1000 + 50 \cdot 4$
$\vdots$	$\vdots$
t	$1000 + 50t$

No matter how many years you maintain your investment, the principal will always be $1000, and the interest payment will be 5% of that $1000. This kind of arrangement is typical for a bond that makes regular payments to you.

If we write r for the fractional interest rate (i.e., $r = 0.05$ denotes a 5% rate), and P for the principal invested, then over time the investment is worth

$$A(t) = P + (Pr)t$$
$$= P(1 + rt).$$

With an investment that pays compound interest, each interest payment is added to the principal. If you invest $1000, earning 5% annually, your principal becomes $1050 after the first year. Consequently the second year's interest payment will be larger, because it will be calculated based on both the original $1000 investment and the previously awarded interest—5% of the entire $1050. Over time, your compounded investment will grow like:

years	balance
0	1000
1	$1000 + 50$
2	$1050 + 52.50$
3	$1102.50 + 55.13$
4	$1157.63 + 57.88$
⋮	⋮
t	?

Determining a formula for your investment after t years is a little harder than for simple interest, but not too bad if you write things down the right way. After 1 year, your balance is

$$\$1000 + \$1000 \cdot 0.05 = \$1000 \cdot 1 + \$1000 \cdot 0.05$$
$$= \$1000(1 + 0.05)$$
$$= \$1000(1.05).$$

In the second year another interest payment is added, 5% of the new amount, increasing your balance to

$$\$1000(1.05) + \big(\$1000(1.05)\big) \cdot 0.05 = \big(\$1000(1.05)\big) \cdot 1 + \big(\$1000(1.05)\big) \cdot 0.05$$
$$= \$1000(1.05)(1 + 0.05), \text{ by factoring}$$
$$= \$1000(1.05)(1.05)$$
$$= \$1000(1.05)^2.$$

Each year, the new balance is derived from the previous balance by multiplying by 1.05, and the amount after t years is

$$A(t) = \$1000(1.05)^t.$$

Again, let r denote the fractional interest rate. If P is the original principal invested and t is years, then the amount the investment is worth after t years (compounded every year) is

$$A(t) = P(1 + r)^t.$$

A Review of Exponentials and Logarithms

Since compound interest is a kind of exponential growth, we take a moment and review some of what we learned from algebra class about exponential functions and their inverse functions, logarithms.

Definition. Let a be any positive real number, and define $f(x) = a^x$. Then f is said to be an *exponential function* with *base a*. If $a > 1$, then f is an *exponential growth* function, and if $0 < a < 1$, then f is an *exponential decay* function.

■ **EXAMPLE 4.33**

The function $f(t) = 1.05^t$ is an exponential growth function. The function $g(t) = 0.95^t$ is an exponential decay function. Some sample values of each are computed in the following tables.

t	1.05^t		t	0.95^t
-1	0.952		-1	1.053
0	1		0	1
1	1.05		1	0.95
2	1.103		2	0.903
3	1.158		3	0.857

The exponent of an exponential can be any real number, though if it is not an integer we will usually use a calculator or a computer to aid with the calculation. On scientific calculators, the button for computing general exponentials is usually marked with a caret ^ or the expression y^x, and it can compute values such as $1.05^{3.3} = 1.17469$ and $0.95^{10.4} = 0.58658$ (these are approximate values).

You may also recall from algebra class that we have identity laws for exponentials. We summarize some here:

$$a^x b^x = (ab)^x \qquad \text{multiplying with same exponent}$$

$$a^x a^y = a^{x+y} \qquad \text{multiplying with same base}$$

$$\frac{a^x}{b^x} = \left(\frac{a}{b}\right)^x \qquad \text{dividing with same exponent}$$

$$\frac{a^x}{a^y} = a^{x-y} \qquad \text{dividing with same base}$$

$$\frac{1}{a^y} = a^{-y} \qquad \text{reciprocals}$$

$$\left(a^x\right)^y = a^{xy} \qquad \text{power of a power}$$

$$\sqrt[x]{a} = a^{1/x} \qquad \text{roots are fractional exponents}$$

We can readily verify these with a calculator. For example, if $a = 1.3$, $b = 5.5$, and $x = 3$, then

$$a^x b^x = 1.3^3 5.5^3 = 2.197 \cdot 166.375 = 365.526,$$

while

$$(ab)^x = (1.3 \cdot 5.5)^3 = 7.15^3 = 365.526.$$

We get the same answer either way.

There is one special base that is so important that we reserve a letter for it, and that is the number $e = 2.71828182845904523536\ldots$. Like π, e has a nonterminating, nonrepeating decimal expansion. Scientific calculators have a key that is usually marked either `exp` or with the expression e^x which calculates exponentials with base e. With your calculator, you can verify a few values of this function.

t	e^t
-1	0.3679
0	1
1	2.7183
1.5	4.4817
2	7.3891

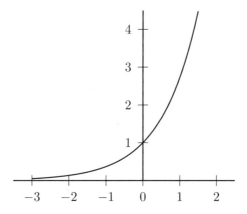

Figure 4.22   The natural exponential function $f(x) = e^x$.

If you inspect the graph of the exponential function in Figure 4.22, you will see that it passes a horizontal line test (horizontal lines intersect the graph in a single point). You may remember from algebra class that functions that pass the horizontal line test have *inverse functions*. The inverse of the exponential with base a is called the logarithm with base a, and it is sometimes written as $g(x) = \log_a x$.

Inverse functions "undo" each other, just as $f(x) = \sqrt{x}$ and $g(x) = x^2$ undo each other. For example, start with $x = 9$. Then $f(9) = \sqrt{9} = 3$, and the reverse process that takes us from 3 back to 9 is $g(3) = 3^2 = 9$.

For the calculus student, the most important logarithm function is the logarithm with base e, which is called the *natural logarithm*, since it is the inverse to the natural exponential function. The notation for the natural logarithm is $g(x) = \ln x$. Scientific calculators have a key labeled `ln` that computes natural logarithms, not to be confused with the `log` key which computes common (base 10) logarithms.

Using a calculator, we can verify the inverse relationship of the natural exponential and natural logarithm functions. First, compute $e^{1.5} = 4.4817$. Then check that $\ln(4.4817) = 1.5$. It works the other way too, that is, computing the logarithm first and then doing the exponential. For example, compute that $\ln(10) = 2.3026$ and verify that $e^{2.3026} = 10.0$.

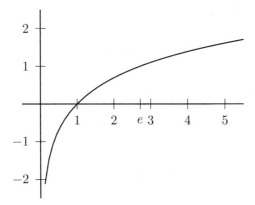

Figure 4.23   The natural logarithm function $f(x) = \ln x$.

Some values of the natural logarithm are given in the table below. You may wish to check that you can produce them with your calculator. Notice that they agree with the graph of the natural logarithm in Figure 4.23.

t	$\ln(t)$
0.5	−0.6931
1	0
1.5	0.4055
2.0	0.6931
e	1
5.0	1.6094

Some computer programs use $\log(x)$ to denote the natural logarithm. If you suspect this, you can check by computing $\log(10)$. If you get an answer of 2.302585, then it is the natural logarithm. The common logarithm would return 1.0.

As with exponentials, there are identity laws for logarithms that help us with computations. Versions of these work with logarithms of any base, but we focus on the natural logarithm.

$$\ln(e^t) = t \qquad\qquad \text{property of inverses}$$
$$e^{\ln t} = t \qquad\qquad \text{property of inverses}$$
$$\ln(ab) = \ln(a) + \ln(b) \qquad\qquad \text{log of a product}$$
$$\ln\left(\frac{a}{b}\right) = \ln(a) - \ln(b) \qquad\qquad \text{log of a quotient}$$
$$\ln(a^t) = t\ln(a) \qquad\qquad \text{log of a power}$$
$$\log_a(t) = \frac{\ln(t)}{\ln(a)} \qquad\qquad \text{change of base rule}$$

◧ **EXAMPLE 4.34**

If I invest $1000 and receive 5% interest compounded annually, how long will it be before I have $5000?

Solution: We computed previously that the amount of money after t years is

$$A(t) = \$1000(1.05)^t.$$

We need to find the value of t when $A(t) = \$5000$. Solving,

$$\$5000 = \$1000(1.05)^t$$
$$5 = (1.05)^t.$$

Here, we can take the logarithm of both sides.

$$\ln 5 = \ln\left(1.05^t\right)$$

Apply the identity for log of a power to the right-hand side.

$$\ln 5 = t \ln 1.05$$

Now get t alone on one side.

$$t = \frac{\ln 5}{\ln 1.05}$$
$$\approx \frac{1.6094}{0.0488}$$
$$\approx 33.0$$

It takes about 33 years for this $1000 to grow to $5000. ■

Derivatives of Exponentials and Logarithms

We saw that $1000 invested at a *simple* interest rate of 5% has a value over time of $A(t) = P(1 + rt) = 1000(1 + .05t) - 1000 + 50t$. What happens if we take the derivative of this function?

The answer is $A'(t) = 50$. Does this represent anything useful? It does indeed, and recalling the differential notation can make it a bit more apparent. We could just as correctly have written,

$$\frac{dA}{dt} = 50.$$

The derivative tells us the (instantaneous) rate of change of the amount over the time. Consider the units that apply here. A change in A, which we write as ΔA is measured in dollars. A change in t is measured in years. The fraction dA/dt is the (instantaneous) slope of the function, measured in dollars/year. It tells us how fast our money is growing at any moment, at a rate of 50 dollars/year.

We don't yet have a formula for the derivative of an exponential function, such as $A(t) = 1000(1.05)^t$. It is tempting to try to apply the power rule, but that is a mistake. A power, like x^2, is not an exponential, like 2^x. A different rule applies.

It turns out that the slope of an exponential function is always proportional to the height of the function at the point of tangency (i.e., there is some constant number that you multiply the value of the function by to get the slope). We get a sense of this if we remember that

$$\frac{dA}{dt} \approx \frac{\Delta A}{\Delta t}.$$

If we take $\Delta t = 1$, then ΔA is the amount an investment will grow over the period of 1 year, and that amount is the interest rate times the current amount. If $A(t) = 2000$ for some t and $r = 0.05$, then we know that

$$\frac{dA}{dt} \approx \frac{\Delta A}{\Delta t} = \frac{0.05 \cdot 2000}{1}.$$

This is an approximation of the slope, but for the derivative we want the exact slope. To discover this, we'll take a detour and look at the natural exponential function, e^x.

You may wonder what is so 'natural' about the natural exponential function. It is natural because it has a simple differentiation formula.

The natural exponential rule: The derivative of $f(x) = e^x$ is $f'(x) = e^x$. The natural exponential function is its own derivative.

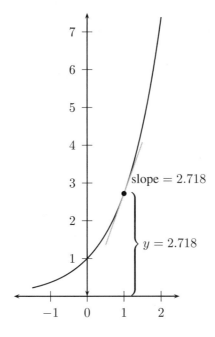

Figure 4.24   The natural exponential function is its own derivative.

At any point on the graph of $y = e^x$, you'll find the slope of the tangent line is the same as the height of the graph. Check, for example, by sketching a tangent at the point $(0, 1)$, and notice that your line has slope 1 (it extends up at a 45 degree angle). In Figure 4.24 the tangent at the point $(1, e)$ has slope e, as we expect.

The natural exponential is the only exponential equal to its derivative. The derivative of any other exponential function contains an extra constant.

The general exponential rule: If $a > 0$ is any real number, then the derivative of $f(x) = a^x$ is $f'(x) = a^x \ln a$.

■ EXAMPLE 4.35

Find the derivative of $A(t) = 1000(1.05)^t$.

Solution: By the general exponential rule,

$$A'(t) = 1000\big((1.05)^t\big)'$$
$$= 1000(1.05)^t \ln 1.05$$
$$\approx 1000(1.05)^t 0.04879.$$

As promised, the derivative is proportional to the function $A(t)$. That is,

$$A'(t) = A(t) \ln 1.05,$$

and the constant of proportionality is $\ln 1.05 \approx 0.04879$. Also note that 0.04879 is approximately the same as the interest rate of 0.05. Of course, 0.05 appears in the derivative estimate we would get from using $\Delta t = 1$, i.e., if we computed $A'(t) \approx A(t+1) - A(t) = 1000(1.05)^{t+1} - 1000(1.05)^t = 1000(1.05)^t(0.05)$.   ■

■ EXAMPLE 4.36

What is the slope of the tangent to $A(t) = 2000(1.07)^t$, when $t = 0$? What about when $t = 3$?

Solution: The derivative of A is $A'(t) = 1000(1.07)^t \ln 1.07$, so the slope when $t = 0$ is $A'(0) = 1000(1.07)^0 \ln 1.07 = 1000 \ln 1.07 \approx 67.66$. The slope when $t = 3$ is $A'(3) = 1000(1.07)^3 \ln 1.07 \approx 82.88$.   ■

Like the natural exponential function, the natural logarithm function has a nice derivative.

The natural logarithm rule: The derivative of $f(x) = \ln x$ is $f'(x) = 1/x$.

Figure 4.25 shows a graph of $y = \ln x$ together with a tangent line at $x = 2$. Notice that the slope appears to be $1/2$, as the natural logarithm rule states. Ask yourself, what should the slope be when $x = 1$? Does the graph look right to you at that point?

The natural logarithm rule "fills in" a hole that we had in our differentiation rules. For each power, $f(x) = x^n$ there is a function that we can take the derivative of to get

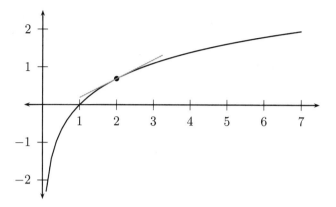

Figure 4.25   The derivative of $\ln x$ is $1/x$.

f, except in the case where $n = -1$. For example, $\dfrac{d}{dx}\dfrac{x^2}{2} = x^1$, and $\dfrac{d}{dx}\dfrac{x^3}{3} = x^2$.

In general, $\dfrac{d}{dx}\dfrac{x^{n+1}}{n+1} = x^n$, as long as $n \neq -1$. So, until we learned the derivative of the logarithm, we did not have a function whose derivative was $x^{-1} = 1/x$.

■ EXAMPLE 4.37

The number of computer clock cycles a particular computer program needs to sort n items of data is given by the formula $T(n) = 7000 + 1440n \ln(n)$. For example, sorting $n = 2000$ data items requires about $T(2000) \approx 21.9$ million clock cycles. At what rate is the sort time changing when $n = 2000$?

Solution: Since we are asked for a rate, we know that the answer is a derivative. It is reasonable to ask "what is the variable?" and "what is the function?" In this case, the variable is n, the number of items we wish to sort. The function is T, the time required to sort n items.

The rate we want is then the change in T with respect to n, or $\dfrac{dT}{dn}$. Using the sum and constant coefficient rules we have

$$\frac{dT}{dn} = \frac{d}{dn}\left(7000 + 1440n \ln n\right)$$

$$= 0 + 1440\frac{d}{dn}(n \ln n),$$

and applying the product rule and natural logarithm rule yields

$$= 1440\left(1 \cdot \ln n + n \cdot \frac{1}{n}\right)$$

$$= 1440(\ln n + 1).$$

If we evaluate at $n = 2000$ we get $1440(8.6) \approx 12{,}385$ cycles/item. The computer program requires approximately 12,385 cycles to accommodate another sort item.

To check, let's compute directly the difference between sorting $n = 2001$ items and $n = 2000$ items (without calculus).

$$T(2001) - T(2000) \approx 21909985 - 21897599$$
$$\approx 12386$$

This agrees well with our derivative calculation. ■

EXERCISES

4.61   If I invest $1500 and receive a return of 6% simple interest for two years, how much money will I have?

4.62   If I invest $5000 and receive a return of 8% simple interest,
  a) How much interest will I receive each year?
  b) How many years will be required to receive $45,000 of interest?

4.63   If I invest $5000 and receive a return of 8% compounded annually, how many years will it take to grow to $50,000?

4.64   The "rule of 72" says that if you receive a compound annual interest rate of $p\%$, then the number of years it takes for your money to double is $72/p$ years. So, a 6% return will double your money in approximately $72/6 = 12$ years.
  a) Compute the number of years it takes $100 to double to $200 at an annual compound rate of 6% and compare it to the rule of 72.
  b) Compute the number of years it takes $100 to double to $200 at an annual compound rate of 3% and compare it to the rule of 72.
  c) Compute the number of years it takes $100 to double to $200 at an annual compound rate of 12% and compare it to the rule of 72.

4.65   Every exponential function $f(t) = A \cdot b^t$ can be rewritten using the natural exponential function in the form $f(t) = Pe^{rt}$. Find suitable P and r so that $1000(1.05)^t = Pe^{rt}$ by applying these steps:
  a) Let $t = 0$ to find P.
  b) Using P from part (a), let $t = 1$ and solve for r.
  c) Write $f(t)$ using the natural exponential.

4.66   If we invest $1000 at a 6% interest rate compounded monthly, then the balance after t years is $A(t) = 1000 \left(1 + \dfrac{0.06}{12}\right)^{12t} = 1000(1.005)^{12t}$. Find suitable P and r so that $1000(1.005)^{12t} = Pe^{rt}$.

4.67   Compute the derivative of each function.
  a) $f(t) = te^t$
  b) $A(t) = 100e^{-t}$.
  c) $A(t) = e^{-1.8t}$
  d) $g(x) = (\ln x)e^x$
  e) $y(t) = t(\ln t - 1)$
  f) $h(x) = \dfrac{\ln x}{x}$
  g) $f(x) = \ln(3x)$
  h) $f(x) = 2\ln(3x + 4)$

4.68   Find the equation for the tangent line to $y = e^x$ at the value $x = 1$, as illustrated in Figure 4.24.

4.69   Find the equation for the tangent line to $y = \ln x$ at the value $x = 2$, as in Figure 4.25.

4.70   During the Cold War, nuclear detonators were made from radioactive Polonium-210. Because it is highly radioactive, Polonium-210 quickly decays into other elements. If a detonator initially contains 11 mg of Polonium-210, then the amount (in mg) remaining after t days is $A(t) = 11e^{-0.00502t}$.
  a) How much Polonium-210 remains after 138 days? What can you conclude about the half-life of Polonium-210?
  b) At what rate (in mg/day) is the Polonium-210 decaying into other elements when the detonator is new?
  c) At what rate (in mg/day) is the Polonium-210 decaying into other elements when the detonator is 180 days old?

4.13   Periodic Functions and Trigonometry

Think about riding a Ferris wheel. What are the aspects of the ride that make it enjoyable? As you turn about the wheel, you initially rise up the back side, and the lift of the wheel presses you down into the seat. There's the moment where you "hang" at the top, and then you fall down the front side.

The sensations of a Ferris wheel come predominantly from the interaction of gravity with the centrepital (due to rotation) force you experience, but you might wonder related questions like "how fast am I falling (vertically) when I go around the wheel?" When is the vertical acceleration the greatest? When is it the least?

The Circular Functions

You may recall from trigonometry that, on a circle of radius r, a point (x, y) is completely determined by the angle θ of a ray from the origin, as in Figure 4.26.

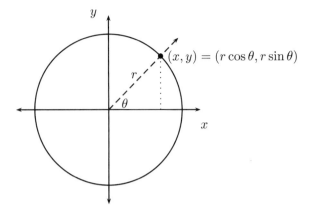

Figure 4.26   The circular functions $y = r\sin\theta$ and $x = r\cos\theta$.

The ratio of y to r is given by the sine function,

$$\sin\theta = \frac{y}{r}.$$

Another way to say this is that if you know θ and r, then the y coordinate must be $y = r\sin\theta$. Similarly, the cosine function relates the x coordinate to the radius:

$$\cos\theta = \frac{x}{r},$$

Equivalently, $x = r\cos\theta$.

Derivatives of Sine and Cosine

If we are to use sine and cosine with our calculus, we're going to need to know the derivative of each. Although these differentiation formulas can be proven, here we will guess the formulas from graphs.

Note, in calculus we almost always take angles to be measured in radians (that is, a complete circle is 2π radians rather than $360°$). Just as you measure temperature in Celsius (or when things really matter, in Kelvin) when doing chemistry, if you want your mathematics to turn out right in calculus class, your best bet is to work in radians.

Look carefully at the sine function in the top graph of Figure 4.27. A small piece of tangent line has been added at each of the multiples of $\frac{\pi}{2}$. Notice at $\frac{-\pi}{2}$, for example, that the slope of the tangent is 0. At the origin, the tangent line appears to have slope 1. Now compare with the points added to the bottom (cosine) graph. The slope of each tangent to sine has been plotted on the cosine graph.

When the slope of sine is zero, the cosine graph has value zero. When the slope of sine is 1, the cosine has value 1. In each case, the slope of the sine curve is precisely

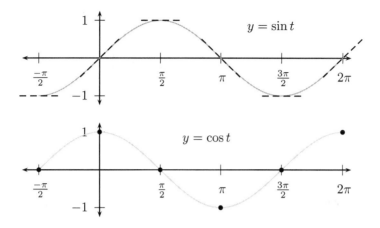

Figure 4.27 Graphs of $y = \sin t$ and $y = \cos t$.

the value on the cosine curve. Thus, the cosine function tells the slope of the sine function. That means that the derivative of $y = \sin t$ is the function $y = \cos t$.

The sine rule: The derivative of $f(x) = \sin x$ is $f'(x) = \cos x$.

EXAMPLE 4.38

What is the tangent line to $y = \sin x$ when $x = \pi/6$?

Solution: For a tangent line, we need a point and a slope. The y coordinate of the point of tangency is $y = \sin(\pi/6) = 1/2$. The slope is given by the derivative, $m = \cos(\pi/6) = \sqrt{3}/2$. Using the point-slope form of a line, the tangent is

$$y - y_0 = m(x - x_0)$$

$$y - \frac{1}{2} = \frac{\sqrt{3}}{2}(x - \pi/6).$$

In slope-intercept form, this is

$$y = \frac{\sqrt{3}}{2}x + \left(\frac{1}{2} - \frac{\pi\sqrt{3}}{12}\right).$$

There are several ways to determine the derivative of cosine, and perhaps the most obvious is to do what we did for sine, draw a graph and guess. But we can also derive the rule for cosine from the existing sine rule, because sine and cosine are related by a cofunction identity. For any angle θ, the complementary angle is $90° - \theta$, and $\cos\theta = \sin(90° - \theta)$. Figure 4.28 illustrates this relationship on a triangle.

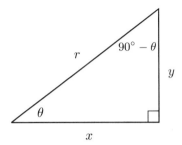

Figure 4.28   Complementary angles: $\cos\theta = \sin(90° - \theta) = x/r$.

If we know the complementary angle formula, we can derive a derivative formula for cosine. First we switch to radians, and then we differentiate using the chain rule:

$$\frac{d}{dx}\cos x = \frac{d}{dx}\sin(\pi/2 - x).$$

On the right hand side, the outside function is $g(x) = \sin x$ and the inside function is $f(x) = \pi/2 - x$. By the chain rule,

$$\begin{aligned} &= g'(f(x))f'(x) \\ &= \cos(\pi/2 - x)(-1) \\ &= -\cos(\pi/2 - x). \end{aligned}$$

There is one last trick, and that is to use a cofunction identity again to replace $\cos(\pi/2 - x)$ with $\sin x$,

$$= -\sin x.$$

The cosine rule: The derivative of $f(x) = \cos x$ is $f'(x) = -\sin x$.

■ EXAMPLE 4.39

What is the derivative of $y = x^2\cos(3x + 1)$?

Solution: All of our usual rules (product rule, chain rule, etc.) apply to trig functions. So, beginning with the product rule,

$$\frac{d}{dx}\left(x^2\cos(3x + 1)\right) = 2x\cos(3x + 1) + x^2\frac{d}{dx}\cos(3x + 1),$$

and continuing with the chain rule,

$$\begin{aligned} &= 2x\cos(3x + 1) + x^2\left(-\sin(3x + 1)\cdot 3\right) \\ &= 2x\cos(3x + 1) - 3x^2\sin(3x + 1). \end{aligned}$$

■

We now have the tools we need to answer questions about Ferris wheels. The Beijing Great Wheel, planned to be the tallest Ferris wheel in the world if completed, will have a wheel of diameter 99 m and reach a total height of 208 m. One revolution of the wheel requires 20 min (1200 s).

A function that models the height of a point on this wheel, with t given in seconds, is

$$y(t) = 109 + 99 \sin\left(\frac{2\pi}{1200}(t - 300)\right).$$

To verify this, check that $t = 0$ gives a height of 10 m (the lowest the wheel achieves):

$$109 + 99 \sin\left(\frac{2\pi}{1200}(0 - 300)\right) = 109 + 99 \sin\left(-\frac{\pi}{2}\right)$$
$$= 109 + 99(-1)$$
$$= 10.$$

After 10 min (600 s), the function reaches its maximum height,

$$109 + 99 \sin\left(\frac{2\pi}{1200}(600 - 300)\right) = 109 + 99 \sin\left(\frac{\pi}{2}\right)$$
$$= 109 + 99(1)$$
$$= 208.$$

◼ EXAMPLE 4.40

What is the vertical velocity of a point moving around the Great Wheel at time $t = 0, 150, 300$, or 600 s?

Solution: We need to compute the derivative.

$$y'(t) = 99\left(\cos\left(\frac{2\pi}{1200}(t - 300)\right) \cdot \frac{2\pi}{1200}\right)$$
$$= 99\frac{2\pi}{1200}\cos\left(\frac{2\pi}{1200}(t - 300)\right)$$
$$= \frac{33\pi}{200}\cos\left(\frac{\pi}{600}(t - 300)\right)$$

This yields vertical velocity values (in units of m/s) of

t	$y'(t)$
0	0
150	0.3665
300	0.5134
600	0

■

EXERCISES

4.71   Find the derivative of each function.

a) $f(x) = \dfrac{x}{\sin x}$

b) $f(x) = \sin \dfrac{1}{x}$

c) $x(t) = 30\cos(2\pi t)$

d) $g(t) = \sqrt{\sin(t/7)}$

e) $h(t) = \cos(\ln x)$

4.72   In Figure 4.29 the top function is $y = \cos t$, and the bottom function is $y = \sin t$. Where each piece of tangent line occurs on the top graph, draw a corresponding point on the bottom graph representing the slope. Use these values to guess the derivative of cosine.

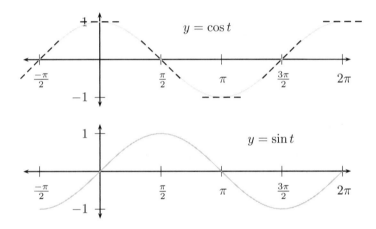

Figure 4.29   Graphs of $y = \cos t$ and $y = \sin t$.

4.73   Write $\tan x = \dfrac{\sin x}{\cos x}$ and use the quotient rule to find a derivative formula for the tangent function.

4.74   Write $\sec x = \dfrac{1}{\cos x}$ and differentiate to find a derivative formula for the secant function.

4.75   Write $\csc x = \dfrac{1}{\sin x}$ and differentiate to find a derivative formula for the cosecant function.

4.76   Write $\cot x = \dfrac{\cos x}{\sin x}$ and differentiate to find a derivative formula for the cotangent function.

4.77 From trigonometry, we know the identity $\sin 2t = 2 \sin t \cos t$.
  a) Find the derivative of $\sin 2t$.
  b) Find the derivative of $2 \sin t \cos t$.
  c) What must be true of your answers from parts (a) and (b)?
  d) Plot your answers from a and b on the same graph.

4.78 The horizontal position of a point on the Beijing Great Wheel is modeled by the function

$$x(t) = -99 \cos \left(\frac{2\pi}{1200}(t - 300) \right).$$

  a) Find the horizontal position at times $t = 0$, 150, 300, and 600 s.
  b) Find the horizontal velocity at times $t = 0$, 150, 300, and 600 s.

4.14 The Fundamental Theorem of Calculus

Although computing areas may seem like an abstract geometrical task, the (calculus) techniques for finding general areas are also used to solve physical problems such as finding the height of a rock thrown into the air or even a rocket accelerating into space. The same techniques can be used to compute physical work, such as the work required to lift an elevator and its cabling to the top of a shaft or the work required to pump the water out of a swimming pool.

If you consider the kinds of shapes you already know how to find the area of, you may find that the list is pretty short. Most of us know that the area of a parallelogram (and consequently a rectangle and a square) is base $\times$ height, and the area of a triangle is $\frac{1}{2} \times$ base $\times$ height (because two triangles together make a parallelogram). We can compute the area of any shape made from straight sides by cutting it into triangles.

Can you compute the area of any curved shapes? Many people know that the area of a circle with radius r is $A = \pi r^2$. Unless you've been a student of calculus, that's probably the extent of your knowledge of curved areas.

We would like to be able to find areas of many curved regions. For instance, what is the area under the function $y = x^2$ as x goes from 0 to 1? This is the shaded area in Figure 4.30.

We can see right away that whatever the area under the curve is, it must be less than 1, since the shaded region fits entirely inside a 1×1 square. In fact, it fits entirely inside a triangle with vertices at $(0, 0)$, $(1, 0)$, and $(1, 1)$, so the area must be less than $1/2$.

Can we say what the area must be bigger than? A rectangle with vertices $(0.5, 0)$, $(0.5, 0.25)$, $(1, 0.25)$, and $(1, 0)$ fits entirely within the shaded region, and the rectangle has area $1/8$, so we know the area must be larger than that.

There is a standard (clever) way to estimate unknown areas using shapes that are easy to understand, and it works much like the exploring we are doing here. You find easy-to-compute areas that contain your region and you find easy-to-compute areas contained entirely inside your region. The answer you desire must be somewhere in between. We always make our easy-to-compute areas from rectangles.

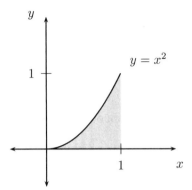

Figure 4.30   The area under x^2.

For example, divide the interval $[0, 1]$ into five congruent subintervals, at the values $x_0 = 0$, $x_1 = 0.2$, $x_2 = 0.4$, $x_3 = 0.6$, $x_4 = 0.8$, and $x_5 = 1.0$. In each little interval, find the largest value of the function $f(x) = x^2$. For example, on the interval $[0, 0.2]$ the highest value of the function is $f(0.2) = 0.2^2 = 0.04$. Use that highest value to create a rectangle, as in Figure 4.31.

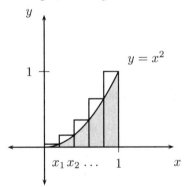

Figure 4.31   Rectangles that fit over the function $y = x^2$.

We can tabulate an (over-)estimate for the shaded area:

width	height	area
0.20	0.04	0.008
0.20	0.16	0.032
0.20	0.36	0.072
0.20	0.64	0.128
0.20	1.00	0.200
	total area	0.440

A collection of rectangular areas, summed to estimate the area of a curved region, is called a Riemann sum, after the German mathematician Bernhard Riemann (1826–

1866). Riemann is famous, not only for his work on calculus, but for work in number theory (where the Riemann hypothesis is well known) and for foundational work in the mathematical branch now known as Riemannian geometry.

Notice that this Riemann sum comes to a bit less than 0.5, which was our first over-estimate for the shaded area. You can probably imagine what would happen if we were to use more rectangles. With 10 rectangles, we get an estimate for the area of 0.385 (check for yourself; the calculation is not difficult). As we use more and more "upper" rectangles, the estimates will continue getting lower and lower, becoming closer and closer to the area we are interested in.

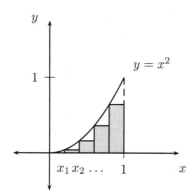

Figure 4.32   Rectangles that fit under the function $y = x^2$.

We can do something similar with rectangles lying under the curve, as in Figure 4.32. For example, in the subinterval $[0, 0.2]$, the smallest value of the function $y = x^2$ is 0. The only rectangle that could fit "under" the curve is a flat line, with no area at all. On $[0.2, 0.4]$ the smallest value of the function is $0.2^2 = 0.04$. As before, we can tabulate an estimate of these "lower" rectangles:

width	height	area
0.20	0.00	0.000
0.20	0.04	0.008
0.20	0.16	0.032
0.20	0.36	0.072
0.20	0.64	0.128
	total area	0.240

As we did with upper rectangles, we can use more and more lower rectangles to get better (under-)estimates for the area. For example, 10 rectangles yields an area estimate of 0.285 (check for yourself).

Although we can get as close as we like, no finite number of upper or lower rectangles will exactly measure the area we want. Fortunately, there is a deep connection between derivatives and areas.

Fundamental Theorem of Calculus. Let $f(x)$ be a continuous function over the closed interval $[a, b]$. If $F(x)$ is any function with $F'(x) = f(x)$, then the area under f on the interval $[a, b]$ is $F(b) - F(a)$.

■ EXAMPLE 4.41

What is the area below the function $f(x) = x^2$ on the interval $[0, 1]$?

Solution: The Fundamental Theorem of Calculus tells us that to compute this area exactly, we only need to find some function $F(x)$ whose derivative is x^2. The function $F(x) = \frac{x^3}{3}$ works. Once we find F, we evaluate it at the ends of the interval and subtract, so the area is $F(1) - F(0) = \frac{1^3}{3} - \frac{0^3}{3} = 1/3$.

Notice that $1/3$ is between the lower and upper estimates that we computed earlier using lower and upper rectangles. ■

■ EXAMPLE 4.42

What is the area below the function $f(x) = \dfrac{x^3}{2} - 3x^2 + 4x$ on the interval $[0, 2]$? See Figure 4.33.

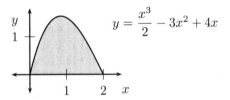

Figure 4.33   Area under $\dfrac{x^3}{2} - 3x^2 + 4x$ from 0 to 2.

Solution: We need a function F with $F'(x) = \dfrac{x^3}{2} - 3x^2 + 4x$. If we take $F(x) = \dfrac{x^4}{8} - x^3 + 2x^2$, that works exactly. By the Fundamental Theorem, the area is $F(2) - F(0) = 2 - 0 = 2$. ■

EXERCISES

4.79 Estimate the area under the function $f(x) = x^2$ on the interval $[0, 1]$ using ten equal subintervals and computing the Riemann sum using upper rectangles.

4.80   Estimate the area under the function $f(x) = x^2$ on the interval $[0, 1]$ using ten equal subintervals and computing the Riemann sum using lower rectangles.

4.81   The function $F(x) = \frac{x^3}{3} + 7$ has the property that $F'(x) = x^2$. Use this fact to find the area under the function $f(x) = x^2$ for $0 \leq x \leq 1$.

4.82   Let $f(x) = x - 2$.
   a) Find a function $F(x)$ whose derivative is $f(x)$.
   b) Find the area under the curve for $2 \leq x \leq 5$.
   c) Draw a graph of the function and find the area directly. (Hint: the region is a triangle.)

4.83   For each function $f(x)$, find a function $F(x)$ with $F'(x) = f(x)$.
   a) $f(x) = x^2 + x + 1$
   b) $f(x) = 4x^3$
   c) $f(x) = 4(x + 1)^3$
   d) $f(x) = 4(2x + 1)^3$
   e) $f(x) = e^{3x+2}$
   f) $f(x) = \dfrac{1}{7x}$

4.15   The Riemann Integral

Area is an important mathematical concept, and like other mathematical notions it has its own notation and careful definition. Informally, you can take the symbols

$$\int_a^b f(x)\, dx$$

as representing the area under the function $y = f(x)$ between the values of a and b. In words we read this aloud as "the integral from a to b of f of x dee-ecks."

We call such an integral a *definite integral*, since it is defined on an actual (definite) interval $[a, b]$. The numbers a and b are called the *lower limit* and *upper limit* of the integral, respectively. Computing this area is "finding" or "taking the integral of f." The process of finding an integral is called *integration*.

In the last section, we took for granted that we could find the area under a curve such as $y = x^2$ or $y = \sin x$ and then started looking at rectangles. But mathematicians are typically too careful to assume an answer will exist because it looks good in a picture. Mathematical objects can be surprisingly subtle. (For example, there is an object that has a finite volume and an infinite surface area. If you are curious, you may want to research Gabriel's Horn or Torricelli's trumpet.)

So, while we started with the intuitive idea of area and explored it by looking at rectangles, it turns out that mathematically that it should work the other way around. Mathematically, there is no question that any interval $[a, b]$ can be divided into subintervals. Any function defined on an interval $[a, b]$ can be evaluated at points of our

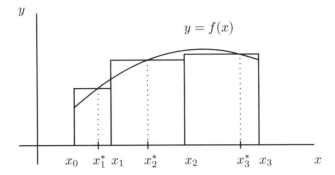

Figure 4.34 An area estimated by general rectangles.

choosing (including in each subinterval). So we can always form rectangles that match the height of a function in subintervals of $[a, b]$.

While we can't guarantee that our intuitive idea of area is correct, the process of taking rectangle estimates is completely rigorous. So this is actually how we define integrals (and we intuitively understand them to be areas). The definition of the *Riemann integral* is a limit of a sum of rectangles.

We have a lot of freedom about how we choose our rectangles. For example, it is convenient for all the rectangles to have the same width, but that is not necessary. In Figure 4.34 we can see an estimate made with three non-uniform rectangles.

The points that set the heights of the rectangles can be the left endpoint of each subinterval, or the right endpoint, or the midpoint. In fact, when you use more and more rectangles (when you take a limit), it doesn't matter what point in each subinterval you use to set the height of the rectangle. We don't even have to choose the same way in every interval, we just need *some* point in each interval. It has become convention to use "star" notation to indicate this. The value x_1^* can be anywhere in $[x_0, x_1]$. In general, x_i^* can be anywhere in the ith interval, $[x_{i-1}, x_i]$.

In Figure 4.34 the height of the function at x_1^* would be written $f(x_1^*)$, so the area of the first rectangle is height $\times$ width $= f(x_1^*)(x_1 - x_0)$. The areas of the other rectangles are computed similarly, so the entire area estimate with all three rectangles would be

$$f(x_1^*)(x_1 - x_0) + f(x_2^*)(x_2 - x_1) + f(x_3^*)(x_3 - x_2).$$

As we use more and more rectangles, the sums become tedious to write, so we usually use some abbreviation. We write Δx_i for the quantity $x_i - x_{i-1}$, which is the width of the ith subinterval. This makes our sum simpler:

$$f(x_1^*)\Delta x_1 + f(x_2^*)\Delta x_2 + f(x_3^*)\Delta x_3.$$

In general, an area estimate with n rectangles looks like

$$f(x_1^*)\Delta x_1 + f(x_2^*)\Delta x_2 + f(x_3^*)\Delta x_3 + \cdots + f(x_n^*)\Delta x_n.$$

Finally, as we use more (progressively narrow) rectangles, the area estimates often approach a single value. Another way to say this is that the sums have a limit as n (the number of rectangles) approaches infinity. Mathematically, we use this limit to define the integral:

$$\int_a^b f(x)\, dx = \lim_{n \to \infty} \left[f(x_1^*)\Delta x_1 + f(x_2^*)\Delta x_2 + \cdots + f(x_n^*)\Delta x_n \right].$$

In plain English, a Riemann integral is defined to be the limit of sums of (very narrow) rectangles. If the sums converge to some answer, then we define that to be the integral and refer to it as the area under the curve.

When you understand that the integral is defined in terms of sums, the notation for an integral makes more sense. It is an elongated 'S', standing for "sum." Gottfried Wilhelm Leibniz (1646–1716) introduced the concept of the integral, and the notation, but it was Bernhard Riemann (1826–1866) who defined the integral formally and rigorously.

With this new notation, we can restate the major theorem we use to compute integrals, the Fundamental Theorem of Calculus, expounded by Barrow, Leibniz, and Newton:

Fundamental Theorem of Calculus. Let $f(x)$ be a continuous function over the closed interval $[a, b]$. If $F(x)$ is any function with $F'(x) = f(x)$, then

$$\int_a^b f(x)\, dx = F(b) - F(a).$$

The function $F(x)$ in the Fundamental Theorem is called an *antiderivative* of $f(x)$. We say "an" antiderivative instead of "the" antiderivative, because antiderivatives are not unique. For example, $F(x) = x^2$ and $F(x) = x^2 + 14$ are both antiderivatives of $f(x) = 2x$.

The more antiderivatives we are able to compute, the more integrals we can find. That makes antiderivatives and antiderivative formulas important to us. It also means that we'll want a notation for antiderivatives.

Definition. For a function $f(x)$, the symbol $\int f(x)\, dx$ represents an antiderivative of f. It is also referred to as the *indefinite integral* of f.

Even though we say "the" indefinite integral, we must remember that a function has many antiderivatives. Sometimes we refer to the *general antiderivative* by giving a formula for all of these antiderivatives at once. For example, the general antiderivative of $f(x) = 2x$ is $F(x) = x^2 + C$, where we understand C to be any constant.

Integral Rules

Many of the derivative rules that we studied in Sections 4.5–4.7 have corresponding antiderivative (integral) rules.

The power rule: If $n \neq -1$, then

$$\int x^n \, dx = \frac{x^{n+1}}{n+1} + C.$$

For example, $\int x^6 \, dx = x^7/7 + C$.

There are also an integral sum rule and a constant coefficient rule.

The sum rule: The integral of a sum is the sum of the integrals:

$$\int f(x) + g(x) \, dx = \int f(x) \, dx + \int g(x) \, dx.$$

The constant coefficient rule: If c is a constant, then

$$\int cf(x) \, dx = c \int f(x) \, dx.$$

EXAMPLE 4.43

Find the general antiderivative of the polynomial $f(x) = 12x^3 - 5x^2 + 1$.

Solution: If we carefully apply the preceding rules, we get

$$\int 12x^3 - 5x^2 + 1 \, dx = \int 12x^3 \, dx + \int -5x^2 \, dx + \int 1 \, dx$$

$$= 12 \int x^3 \, dx + -5 \int x^2 \, dx + (x + C_1)$$

$$= 12 \left(\frac{x^4}{4} + C_2 \right) - 5 \left(\frac{x^3}{3} + C_3 \right) + (x + C_1)$$

$$= 3x^4 - \frac{5}{3}x^3 + x + C_1 + 12C_2 - 5C_3$$

$$= 3x^4 - \frac{5}{3}x^3 + x + C.$$

In the last line of our calculation, we recognize that C_1, C_2, and C_3 are all constants, and a sum of different constants (even when scaled by multiplying or dividing by some fixed number) is merely another constant.

We can easily check that we are correct. If we take the derivative of our answer, we get $f(x)$ back. ∎

EXERCISES

4.84 Find each indefinite integral (general antiderivative).

a) $\displaystyle\int 5x^2 + 4x + 4\,dx$

b) $\displaystyle\int 7\cos x - \frac{1}{x}\,dx$

c) $\displaystyle\int \frac{3}{t^2} + \frac{1}{5t} - \sqrt{t}\,dt$

d) $\displaystyle\int \sin(x+3) + \frac{1}{x-2} + \sqrt{x+1}\,dx$

e) $\displaystyle\int 2xe^{x^2}\,dx$

f) $\displaystyle\int \cos(7x+1) - 2x\sqrt{x^2+1}\,dx$

4.85 Find each area

a) $\displaystyle\int_0^6 x^2 + x + 1\,dx$

b) $\displaystyle\int_0^1 4x^3\,dx$

c) $\displaystyle\int_{-1}^1 4(x+1)^3\,dx$

d) $\displaystyle\int_0^1 4(2x+1)^3\,dx$

e) $\displaystyle\int_1^6 \frac{1}{7x}\,dx$

4.16 Signed Areas and Other Integrals

You might wonder what an integral represents for a function that takes both positive and negative values. For example, we can use the Fundamental Theorem of Calculus to compute

$$\int_0^4 \frac{x^3}{2} - 3x^2 + 4x\,dx = \left[\frac{x^4}{8} - x^3 + 2x^2\right]_{x=0}^{x=4}$$

$$= \left(\frac{4^4}{8} - 4^3 + 2\cdot 4^2\right) - \left(\frac{0^4}{8} - 0^3 + 2\cdot 0^2\right)$$

$$= 0 - 0$$

$$= 0.$$

Figure 4.35 illustrates this situation.

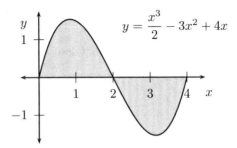

Figure 4.35   Signed area under $\dfrac{x^3}{2} - 3x^2 + 4x$ from 0 to 4.

The integral is zero because the area under the curve over the interval $[0, 2]$ is counted as *positive* area, and the area above the function on the interval $[2, 4]$ counts as *negative* area. There are precisely matching amounts of positive and negative area, and they cancel each other out.

We can be sure this is correct if we think about the Riemann definition of integral. Consider dividing the interval $[0, 4]$ into small subintervals, and creating a rectangle approximation. The area of the ith rectangle is

$$f(x_i^*) \Delta x_i.$$

If x_i^* lands in the left half of the interval, then $f(x_i^*)$ is positive. In the right half of the interval, $f(x_i^*)$ is negative. The width of a rectangle, Δx_i, is always positive.

Since we get a sum of "positive rectangles" from the left half of the interval and "negative rectangles" from the right half of the rectangle, the definition verifies that (when we take the limit) the integral will be the sum of the "signed areas." In this case, we computed this to be zero, and we can see this is zero in the figure. There is an important principal here.

If a quantity is represented by a limit of Riemann sums, then it is computed by taking an integral.

In geometry, we don't speak of positive and negative areas. Areas are always positive. But for functions that take positive and negative values, the Riemann sum essentially counts positive areas above the x-axis and negative areas below. So the integral does the same.

Although it is intuitive to think of integrals as (signed) areas, anything that can be represented as the limit of a Riemann sum will be an integral, even if you can't imagine it as an area.

EXAMPLE 4.44

Imagine that you are in a car driving away from town. To keep a record of your speed, you've modified your car with a scrolling paper roll (like a seismograph) attached to the speedometer. When you go faster, the "seismograph" pen moves up, and when you go slower, the pen moves down. Perhaps it creates a graph like Figure 4.36.

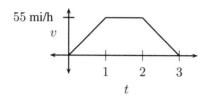

Figure 4.36   Speed of a car recorded over time.

On this trip, the car starts from rest. We can see that, because the curve starts at $(0, 0)$. The pen makes a mark on the graph for $t = 0$ indicating that the speed is $y = 0$ mi/h. Over the first hour of the trip, the car accelerates evenly until it is traveling 55 mi/h, and it maintains that speed for the next hour before slowing to a stop over the next hour.

Consider dividing the interval into subintervals, and creating the rectangles of a Riemann sum, as in Figure 4.37.

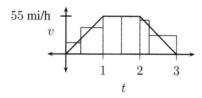

Figure 4.37   Speed of a car recorded over time.

Each rectangle is formed the same way, by taking some reading of the speedometer (measured in mi/h) for the height of the rectangle and a time interval (measured in hours) for the width of the rectangle.

What does the area of a rectangle represent? Looking at units, we can see that the height of the rectangle is mi/h, so the height of the rectangle is a rate. The width of the rectangle is in hours, so it represents an elapsed time. Since distance = rate × time, the area of the rectangle appears to represent a distance.

When the car moves at a constant rate, as it does between $t = 1$ and $t = 2$, then the area of a rectangle is exactly how far the car moves in an interval. For example, if the car travels 55 mi/h for 0.5 h, then it covers a distance of 27.5 mi.

When the car is changing speed (either speeding up or slowing down), then we can't expect things to work out so exactly. But imagine using more and more rectangles, so that the time intervals are very short (and the speedometer doesn't have time to change much in any interval). If the speedometer shows approximately 40 mi/h for 10 seconds, which is $1/360$ of an hour, then a good estimate for the distance traveled would be

$$40\frac{\text{mi}}{\text{h}} \cdot \frac{1}{360}\,\text{h} = \frac{1}{9}\,\text{mi}.$$

If we consider cutting the trip into many 10-second time intervals, we can estimate the total distance traveled by the car as the sum of these rectangular areas. The Riemann sum approximates the length of the trip, and the estimate is better when the intervals are shorter. It seems clear that the true length of the trip is found in the limit: the area under the curve measures the distance the car has moved.

In Section 4.2 we saw that the derivative of position is velocity. In this example we saw that the integral of velocity is (change in) position. This is clear if we think about what the Fundamental Theorem of Calculus (FToC) says.

The FToC tells us that the area over an interval and under a function can be computed by finding an antiderivative and subtracting the values at each end of the interval. Since the velocity of a car is the derivative of the position of the same car, the FToC guarantees that the integral of the velocity is the change in position (the distance the car has moved). In symbols:

$$\int_a^b v(t)\,dt = \text{dist}(b) - \text{dist}(a).$$

These ideas apply to any rate. If $f(t)$ is the cash flow of a company (in dollars/day), then $\int_0^{30} f(t)\,dt$ is the total change in cash during a 30-day period, i.e., the total net money earned or lost. If $E(t)$ is the energy consumption of the United States (in quadrillion BTU/year), then $\int_0^{10} E(t)\,dt$ is the total energy consumed in a decade (in quadrillion BTU).

Theorem. If $f(t)$ denotes a rate of change at time t in some quantity F, then

$$\int_a^b f(t)\,dt$$

denotes the total change in F between time a and time b.

■ EXAMPLE 4.45

A lead ball is dropped from a tower. How fast is the ball moving after 1.5 seconds?

Solution: To calculate this, we need to know one empirical fact. The acceleration of gravity is (essentially) a constant: -9.8 m/s^2. Acceleration is the rate of change in velocity, so integrating the acceleration function tells us the total change in velocity of the ball:

$$\int_0^{1.5} -9.8 \, dt = [-9.8t]_0^{1.5}$$
$$= (-9.8)(1.5) - (-9.8)(0)$$
$$= -14.7 \text{ m/s}.$$

Since the ball was dropped, it started at 0 m/s. The velocity changed by -14.7 m/s as it fell, so the final velocity after 1.5 s is -14.7 m/s. The negative answer indicates that the ball is falling (its height is decreasing). ■

■ **EXAMPLE 4.46**

A lead ball is thrown upward at 5 m/s from the top of a tower and then falls. How fast is the ball moving after 1.5 seconds?

Solution: In this example the acceleration of gravity is still -9.8 m/s^2, and the total change in velocity is still the same integral with the same result, -14.7 m/s. However, since the ball started with a velocity of 5 m/s, the final velocity is $5 \text{ m/s} - 14.7 \text{ m/s} = -9.7 \text{ m/s}$. ■

EXERCISES

4.86 A math professor is pushed from the top of a 30 m building.
a) How fast is the professor falling after 0 seconds?
b) How fast is the professor falling after 0.25 seconds?
c) How fast is the professor falling after t seconds?

4.87 A math professor launched from a cannon has a velocity function of $v(t) = 9 - 9.8t$ m/s, where t is measured in seconds from the time the cannon fires.
a) Is the cannon pointed up or down? How do you know?
b) At what time t does the professor have a velocity of 0 m/s? Why does the professor stop moving?
c) How high is the professor after 0.25 seconds?
d) How high is the professor after 1 second?
e) How high is the professor after t seconds?
f) When does the professor hit the ground?

4.88 The rate of fossil fuel consumption in the United States (in quadrillions of British thermal units per year) is approximately $E(t) = 12e^{0.02t}$, where t is years since 1900. How much fuel was used between January 1, 1900 and January 1, 1910?

4.89   A yo-yo company has a profit rate of $20,000 per day when a new advertising campaign begins. Profits rise $1,000 per day, each day, for the next six weeks.

      a) What is the profit rate in dollars per day after the six weeks have transpired?

      b) What was the profit rate t days after the advertisements began running?

      c) What was the total amount of money made during the six week campaign?

4.17 Application: Rocket Science

If we know some simple facts, we can estimate some pretty cool things using only basic calculus. For example, we can figure out roughly how high the Space Shuttle flies, knowing only a few facts about the Shuttle. Calculus really is rocket science.

Although the space shuttle no longer flies, when it did, the first part of the ascent was accelerated by the three main engines as well as two solid rocket boosters (which then detached and fell in the ocean).

In the previous section, we learned that integrating an acceleration function will give us the total change in velocity of an object. In the same way, integrating a velocity function gives us the total change in position of an object. So, if we can determine the acceleration function of the Space Shuttle, then determining the height should be a matter of integrating two times, once to get the velocity and again to get the height.

The facts we need to know about the Shuttle are:

- On the launch pad, the mass of the Shuttle is two million kilograms, 2.0×10^6 kg.

- This mass includes 1.18 million kilograms of fuel (1.18×10^6 kg) that will burn up during the first two minutes of flight. (This includes fuel from both the main engines and the solid rocket boosters.)

- The solid rocket boosters produce a total of 23.6 million Newtons of force, or 2.36×10^7 kg m/s^2.

- The main engines produce 5.0 million Newtons of force, or 5.0×10^6 kg m/s^2.

Newton's Second Law

Newton's second law of motion tells us that force is mass times acceleration, or symbolically,

$$F = ma.$$

We are interested in the acceleration generated by the rocket engines, so we solve and substitute the facts that we know:

$$a = F/m$$
$$= \frac{2.36 \times 10^7 \text{ kg m/s}^2 + 5.0 \times 10^6 \text{ kg m/s}^2}{m}$$
$$= \frac{2.86 \times 10^7 \text{ kg/s}^2}{m}.$$

If we knew the mass of the Shuttle, then this calculation would be complete. You may be thinking that we already know that the mass of the Shuttle is 2.0×10^6 kg, but that is not quite correct. The mass certainly starts at two million kilograms, but only two minutes later more than half of the mass has been expended in the form of 1.18 million kilograms of burned fuel. That average rate of fuel use amounts to

$$\frac{1.18 \times 10^6 \text{ kg of fuel}}{120 \text{ s}} \approx 9833 \text{ kg/s.}$$

So the mass of the Shuttle is a function of time. We don't know that the Shuttle uses fuel at a constant rate, but that's probably not a bad guess. That means that after t seconds, the mass in kilograms is $m(t) = 2.0 \times 10^6 - 9833t$ kg. Using this in the acceleration equation, the acceleration due to the engines is

$$a_e(t) = \frac{2.86 \times 10^7 \text{ kg m/s}^2}{m(t)}$$
$$= \frac{2.86 \times 10^7 \text{ kg m/s}^2}{2.0 - 9833t \text{ kg/s}}$$
$$= \frac{2.86 \times 10^7}{2.0 - 9833t} \text{ m/s}^2.$$

Of course, the engines are not the only acceleration that a rocket experiences. Gravity is also pulling it toward Earth. Fortunately, even for rockets enduring a launch, the acceleration of gravity is easy to compute. It is effectively constant, with the well-known value -9.8 m/s^2. Combining this with the acceleration due to the engines, we get the total acceleration on the rocket (as a function of time):

$$a(t) = \frac{2.86 \times 10^7}{2.0 \times 10^6 - 9833t} - 9.8 \text{ m/s}^2.$$

We should be able to integrate this to get the velocity and height of the rocket.

Velocity of a Rocket

We know that the velocity of the rocket, $v(t)$, will be an antiderivative of $a(t)$, so we integrate:

$$v(t) = \int a(t)\, dt$$

$$= \int \frac{2.86 \times 10^7}{2.0 \times 10^6 - 9833t} - 9.8\, dt,$$

which by the rules of integrals is

$$= 2.86 \times 10^7 \int \frac{1}{2.0 \times 10^6 - 9833t}\, dt - \int 9.8\, dt.$$

An antiderivative for the second integral is easy to find, since it is the integral of a constant, so let's focus on the first integral. It is a bit trickier, but not extremely hard. Let $F(t) = \frac{-1}{9833} \ln(2.0 \times 10^6 - 9833t)$. Then by the chain rule,

$$F'(t) = \frac{-1}{9833} \cdot \frac{1}{2.0 \times 10^6 - 9833t}(-9833).$$

So F is an antiderivative that we can use to compute the integral. Thus

$$v(t) = 2.86 \times 10^7 \int \frac{1}{2.0 \times 10^6 - 9833t}\, dt - \int 9.8\, dt$$

$$= 2.86 \times 10^7 \cdot \frac{-1}{9833} \ln(2.0 \times 10^6 - 9833t) - 9.8t + C$$

$$\approx -2909 \ln(2.0 \times 10^6 - 9833t) - 9.8t + C.$$

To find the value of C, remember that $v(0) = 0$ since the rocket starts resting on the launch pad. That means

$$0 = v(0) = -2909 \ln(2.0 \times 10^6) + C,$$

and solving, we get $C = 2909 \ln(2.0 \times 10^6) \approx 42206$. So, our velocity function must be

$$v(t) = -2909 \ln(2.0 \times 10^6 - 9833t) - 9.8t + 42206.$$

To find the velocity after 2 minutes, for example, we compute

$$v(120) = -2909 \ln(2.0 \times 10^6 - 9833 \cdot 120) - 9.8 \cdot 120 + 42206$$

$$\approx 1418 \text{ m/s}.$$

In miles per hour, this is

$$1418 \text{ m/s} \times \frac{60 \text{ s}}{1 \text{ min}}\, \frac{60 \text{ min}}{1 \text{ h}}\, \frac{100 \text{ cm}}{1 \text{ m}}\, \frac{1 \text{ in}}{2.54 \text{ cm}}\, \frac{1 \text{ ft}}{12 \text{ in}}\, \frac{1 \text{ mi}}{5280 \text{ ft}} = 3172 \text{ mi/h}.$$

Height of a Rocket

We're only one step away from deriving the height of the rocket. In the last step, we computed that the velocity function of the rocket is

$$v(t) = -2909 \ln(2.0 \times 10^6 - 9833t) - 9.8t + 42206 \text{ m/s.}$$

Since velocity is the derivative of position (the height) we need only integrate one more time. You may quiver at the thought of finding an antiderivative for $v(t)$, but at least some parts of it are easy. The last term is simple:

$$\int 42206 \, dt = 42206t + C.$$

And the linear term is too:

$$\int -9.8t \, dt = -4.9t^2 + C.$$

The remaining term is going to take a little more work, but the rules of integrals at least let us factor out the constant in front:

$$\int -2909 \ln(2.0 \times 10^6 - 9833t) \, dt = -2909 \int \ln(2.0 \times 10^6 - 9833t) \, dt.$$

At its heart, the expression we need to integrate is a logarithm. Alas, we haven't learned an antiderivative for $\ln t$. But this doesn't mean that no one has, and if you were to consult a table of integrals or use a computer program, you would find that $t(\ln t - 1)$ is an antiderivative for $\ln t$.

We know how to take derivatives, so nothing stops us from checking this for ourselves. Let $F(t) = t(\ln t - 1)$. Then by the product rule,

$$F'(t) = 1(\ln t - 1) + t\left(\frac{1}{t} - 0\right)$$
$$= \ln t - 1 + 1$$
$$= \ln t.$$

So, we know an antiderivative for $\ln t$. Can we build that up to an antiderivative for the expression we need to integrate, namely, $\ln(2.0 \times 10^6 - 9833t)$? If we start from $F(t)$, our antiderivative for $\ln t$, we might think to try $F(2.0 \times 10^6 - 9833t)$, and that's a good guess (though not quite right).

To see that it doesn't quite work, use the chain rule:

$$\frac{d}{dt}F(2.0 \times 10^6 - 9833t) = F'(2.0 \times 10^6 - 9833t) \cdot \frac{d}{dt}(2.0 \times 10^6 - 9833t)$$
$$= \ln(2.0 \times 10^6 - 9833t) \cdot \frac{d}{dt}(2.0 \times 10^6 - 9833t)$$
$$= -9833 \ln(2.0 \times 10^6 - 9833t).$$

This would be exactly right, except for the extra constant factor of -9833 in front. Hopefully, you see that we can make one last adjustment, dividing by the constant -9833 to get an antiderivative that works perfectly. The antiderivative we need is

$$\frac{1}{-9833} F(2.0 \times 10^6 - 9833t) = \frac{2.0 \times 10^6 - 9833t}{-9833} \left(\ln(2.0 \times 10^6 - 9833t) - 1 \right).$$

Wow, that was work! But it was work that brings us to our final goal. We can now integrate the velocity to find the position of the rocket at any time t.

$$y(t) = \int v(t)\, dt$$

$$= \int -2909 \ln(2.0 \times 10^6 - 9833t) - 9.8t + 42206\, dt$$

$$= -2909 \frac{2.0 \times 10^6 - 9833t}{-9833} \left(\ln(2.0 \times 10^6 - 9833t) - 1 \right)$$
$$\quad - 4.9t^2 + 42206t + C$$

To discover the value of C, simply observe that the rocket begins on the ground. So when $t = 0$, the height of the rocket is $y(0) = 0$. It might be best to use a calculator for this part:

$$0 = -2909 \frac{2.0 \times 10^6}{-9833} \left(\ln(2.0 \times 10^6) - 1 \right) + C,$$

and, solving,

$$C \approx -7.993 \times 10^6.$$

That makes the height function of the rocket

$$y(t) = -2909 \frac{2.0 \times 10^6 - 9833t}{-9833} \left(\ln(2.0 \times 10^6 - 9833t) - 1 \right) - 7.993 \times 10^6 \text{ m.}$$

To calculate the height of the rocket after two minutes (which is 120 seconds), evaluate $y(120)$ to get approximately 62,000 meters. Is this a reasonable answer? Let's convert it to miles:

$$62000 \text{ m} \frac{100 \text{ cm}}{1 \text{ m}} \frac{1 \text{ in}}{2.54 \text{ cm}} \frac{1 \text{ ft}}{12 \text{ in}} \frac{1 \text{ mi}}{5280 \text{ ft}} = 38.5 \text{ mi.}$$

If we check our numbers against NASA results, NASA says that after about 2 minutes of flight, the space shuttle has finished its first stage. It is at a height of 28 miles, and it is traveling 3000 mph. We computed a height of 38.5 miles and (in the previous section) a velocity of 3172 mph. It's not a perfect match, but we are pretty close.

If you feel bad that our conclusion was off by about 10 miles, it may console you to know that at 3000 mph, the space shuttle covers 10 miles in approximately 12 seconds. So don't think of our calculation as being off by 10 miles. Think of it as being off by 12 seconds.

EXERCISES

4.90   Assume that the space shuttle has a mass of $m = 2.0 \times 10^6$ kg, and the engines produce $F = 2.86 \times 10^7$ kg m/s^2 of force. *For this exercise, imagine that the shuttle stays the same mass, rather than getting lighter as it expends fuel.*

   a) Use Newton's second law, $F = ma$, to compute the acceleration the shuttle feels from the engines.

   b) Subtracting gravity, what is the total acceleration the shuttle feels?

   c) Knowing that $v(t) = \int a(t)\, dt$, what is the velocity of the shuttle when $t = 120$ s?

4.91   Compute the derivative of each function.

   a) $\ln t$

   b) $\ln(2 - t)$

   c) $\ln(2 - 9t)$

   d) $\ln(2.0 \times 10^6 - 9833t)$

4.92   Compute the derivative of each function.

   a) $t \ln t$

   b) $-t + t \ln t$

   c) $-\dfrac{1}{9}(2 - 9t)\big(-1 + \ln(2 - 9t)\big)$

   d) $-\dfrac{1}{9833}(2.0 \times 10^6 - 9833t)\big(-1 + \ln(2.0 \times 10^6 - 9833t)\big)$

4.93   Consider the height function we derived for the rocket:

$$y(t) = -2909\frac{2.0 \times 10^6 - 9833t}{-9833}\big(\ln(2.0 \times 10^6 - 9833t) - 1\big) - 7.993 \times 10^6.$$

   a) Plot the rocket's height at times $t = 0, 10, 20, 30, \ldots, 120$ s.

   b) Find the velocity function of the rocket by computing $v(t) = y'(t)$.

4.18   Infinite Sums

Here's a mathematical riddle. "I am a real number. I am not negative, but I am less than any positive number. Who am I?"

   The answer, of course, is zero. But if we change the setting slightly, it can feel less obvious. For example, perhaps you are acquainted with the infinite repeating decimal number $1.999999\ldots = 1.\bar{9}$. Do you know what number this is? As a riddle, we might have said, "I am a real number. I am not greater than 2, but I am greater than any number less than 2. Who am I?" Naturally, the answer is 2. Yet for many people, it feels strange to say "One point nine-repeating is another way of writing the number 2."

What do we mean when we write a number like $1.999999\ldots$? Remember that in our decimal number system, each place in a number has a corresponding value. In this number, the 1 digit is in the ones place. Then a 9 follows in the tenths place, another 9 in the hundredths place and so on. Just as the number 9.81 means $9 + 8\frac{1}{10} + 1\frac{1}{100}$, the expression $1.999999\ldots$ means

$$1 + 9\frac{1}{10} + 9\frac{1}{100} + 9\frac{1}{1000} + \cdots.$$

If you think there might be something shifty about adding up infinitely many values, consider that in another context this probably doesn't upset you at all. You are probably comfortable with the fact that $1/3 = 0.333333\ldots = 0.\overline{3}$, and $1/7 = 0.\overline{142857}$. If you think about what those decimal expansions mean, each *must* refer to an infinite sum.

Geometric Series

An infinite sum is called a *series*. Our goal is to find the sum of different kinds of series, including the series for $1.\overline{9}$ and $0.\overline{3}$. As a step toward that goal, consider this version of Zeno's famous dichotomy paradox: Imagine yourself standing on the number line. You are at the number 0 and facing towards the 1. Now step half the distance to 1. This puts you on $1/2$. Step half the distance again, to $3/4$. Step half the distance again, and again, and so on, forever. Where do you end up?

Hopefully, you see that, in the language of our riddle, you end up at a real number no larger than 1 but beyond any of the numbers less than 1. That is, you are at the number 1. As a series, this conclusion could be written as

$$\frac{1}{2} + \frac{1}{4} + \frac{1}{8} + \frac{1}{16} + \cdots = 1.$$

To be very careful about this, however, consider where you are after each step. After your first step (of size $1/2$), your distance from 1 was also $1/2$. After your second step, (of size $1/4 = 1/2^2$), you were at $3/4$ and your distance from 1 was only $1/4 = 1/2^2$. Each step is half as large as the previous, so the nth step is of size $1/2^n$ and it leaves you short of 1 by a distance of only $1/2^n$. As n becomes larger and larger, you get closer and closer to 1 (though you never reach it in any finite number of steps). In the language of calculus, we solve the riddle by taking a limit, and here the limit is 1.

Mathematically, the sum of a series is always defined this way. We add up a finite (but ever growing) number of terms, then see if the finite sums approach some limit. If they do, the series is said to be *convergent*, and the limit is called the *sum* of the series.

The series $\frac{1}{2} + \frac{1}{2^2} + \cdots$, $0.\overline{3}$, and $1.\overline{9}$ are all members of an important family of (convergent) series, the geometric series. A *geometric series* is a series of the form

$$a + ar + ar^2 + ar^3 + ar^4 + \cdots.$$

The Zeno series corresponds to $a = 1/2$ and $r = 1/2$. The value $0.\overline{3}$ corresponds to $a = 3/10$ and $r = 1/10$. The value $1.\overline{9}$ is not a geometric series, but the decimal part, $0.\overline{9}$, is a geometric series with $a = 9/10$ and $r = 1/10$.

Since the sum of a series comes from taking a limit of finite sums, it is important to be able to compute finite sums. For the geometric series, the finite sums are given by a well known formula.

Geometric Sum Formula. For real numbers a, r where $r \neq 1$,

$$a + ar + ar^2 + \cdots + ar^{n-1} = a\frac{1 - r^n}{1 - r}.$$

There are two typical ways to derive this fact. If you are comfortable with polynomial long division, then computing $\dfrac{1 - r^n}{1 - r}$ is a matter of dividing $r^n - 1$ by $r - 1$. Otherwise, this formula can be checked by computing the product

$$\begin{aligned}
&(a + ar + ar^2 + \cdots + ar^{n-1})(1 - r) \\
&= (a + ar + ar^2 + \cdots + ar^{n-1}) - (a + ar + ar^2 + \cdots + ar^{n-1})r \\
&= (a + ar + ar^2 + \cdots + ar^{n-1}) - (ar + ar^2 + ar^3 + \cdots + ar^n) \\
&= a - ar^n \\
&= a(1 - r^n).
\end{aligned}$$

Since $(a + ar + ar^2 + \cdots + ar^{n-1})(1 - r) = a(1 - r^n)$, we can divide both sides by $(1 - r)$ to get the finite sum formula.

Now that we know how to add finite geometric sums, we can ask what happens when n approaches infinity. The most interesting case is when $|r| < 1$, since in that case r^n gets closer and closer to 0. To see this, consider any number of size (absolute value) less than 1. If you square your number and cube it, and so on, it becomes smaller and smaller.

Since $r^n \to 0$, we now know the limit of a geometric series.

Geometric Series Formula. For real numbers a, r, with $|r| < 1$,

$$a + ar + ar^2 + \cdots + ar^n + \cdots = a\frac{1}{1 - r}.$$

■ **EXAMPLE 4.47**

The Zeno series is a geometric series with $a = 1/2$ and $r = 1/2$. The sum of the series, therefore, is

$$a\frac{1}{1 - r} = \frac{1}{2}\frac{1}{1 - \frac{1}{2}} = \frac{1}{2}\frac{1}{\frac{1}{2}} = 1.$$

▣ EXAMPLE 4.48

The value $0.\overline{3}$ represents a geometric series with $a = 3/10$ and $r = 1/10$. The sum is therefore

$$a\frac{1}{1-r} = \frac{3}{10}\frac{1}{1-\frac{1}{10}} = \frac{3}{10}\frac{1}{\frac{9}{10}} = \frac{3}{10}\frac{10}{9} = \frac{1}{3}.$$

▣ EXAMPLE 4.49

The value $1.\overline{9}$ is not a geometric series, but $0.\overline{9}$ is. Here $a = 9/10$ and $r = 1/10$. The sum of the series part is

$$a\frac{1}{1-r} = \frac{9}{10}\frac{1}{1-\frac{1}{10}} = \frac{9}{10}\frac{1}{\frac{9}{10}} = 1.$$

And finally, the value is $1.\overline{9} = 1 + 0.\overline{9} = 1 + 1 = 2$.

▣ EXAMPLE 4.50

What value does $x = 0.\overline{13} = 0.13131313\ldots$ represent?

Solution: This question is slightly harder, but one way to write this as a series is

$$0.13131313\ldots = \frac{13}{100} + \frac{13}{10000} + \frac{13}{10^6} + \cdots$$
$$= \frac{13}{100} + \frac{13}{100}\frac{1}{100} + \frac{13}{100}\frac{1}{100^2} + \cdots .$$

This is a geometric series with $a = \frac{13}{100}$ and $r = \frac{1}{100}$. The sum is

$$a\frac{1}{1-r} = \frac{13}{100}\frac{1}{1-\frac{1}{100}} = \frac{13}{100}\frac{1}{\frac{99}{100}} = \frac{13}{99}.$$

Here's another approach, using algebra. Let $x = 0.\overline{13}$. Our job is to find x. Notice that $100x = 13.131313\ldots = 13.\overline{13}$. Now compute:

$$100x = 13.\overline{13}$$
$$x = 0.\overline{13},$$

and subtracting,

$$99x = 13.$$

Solving, we see again that $x = 13/99$.   ■

The Harmonic Series

Are there series that don't converge? Of course. The simple series $1 + 1 + 1 + \cdots$ fails to converge. Rather than getting close to any fixed number, it becomes larger and larger as you take more terms. Such a series is called *divergent*. In this example, we say that the series "diverges to infinity."

It is fairly clear that if the terms of a series don't get smaller and smaller, the series will be divergent. But what if the terms do get smaller? Is that enough to make the series converge? As it turns out, there are divergent series whose individual terms get smaller and smaller. The most famous is the harmonic series.

Definition. The *harmonic series* is the sum

$$1 + \frac{1}{2} + \frac{1}{3} + \frac{1}{4} + \frac{1}{5} + \cdots .$$

As always, the key to understanding the limit of any series is to look at n terms, and then see what happens as n gets larger and larger. If we take the first n terms of the harmonic series, we get $1 + \frac{1}{2} + \cdots + \frac{1}{n}$.

To see how big this sum is, we'll use a geometric argument. Draw rectangles having area $1, \frac{1}{2}, \ldots, \frac{1}{n}$, and stack them next to each other on the number line, starting at the value $x = 1$, as indicated by the five rectangles in Figure 4.38. The jth rectangle should have width 1 and height $\frac{1}{j}$, so that each area is the jth term of the harmonic series.

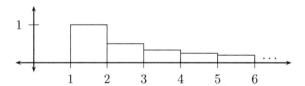

Figure 4.38   The harmonic series as areas of rectangles.

Now that we are thinking of this sum as an area, our knowledge of calculus and integrals can be applied. Notice what happens if we insert the curve $y = 1/x$ into the picture, as in Figure 4.39.

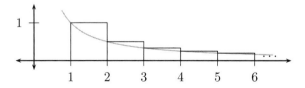

Figure 4.39   Harmonic series compared with $y = 1/x$.

Look first at what happens on the interval $[1, 2]$. On this interval, we know that $y = 1/x$ will never be larger that 1. So, the area of the rectangle is larger than the area under the curve. In the language of integrals, we can write

$$1 > \int_1^2 \frac{1}{x}\,dx.$$

The same thing happens in the next interval. In $[2,3]$, we know that $y = 1/x$ will never be larger than $\frac{1}{2}$, which is precisely the height of the second rectangle. So the area under the curve is smaller than the area of the rectangle. If we combine this with what we know about the first rectangle, we find that the sum of the first two rectangles is more than the area under the curve over the interval $[1,3]$:

$$1 + \frac{1}{2} > \int_1^3 \frac{1}{x}\,dx.$$

We can continue in the same way. The area of five rectangles is greater than the area under the curve over the interval $[1,6]$. In general, the area of n rectangles will be greater than the area under $y = 1/x$ over the interval $[1, n+1]$. Written out,

$$1 + \frac{1}{2} + \cdots + \frac{1}{n} > \int_1^{n+1} \frac{1}{x}\,dx.$$

Here's the point. We know how to compute the area under the curve, because we know how to do integrals. All we need is an antiderivative for $f(x) = 1/x$, and $F(x) = \ln x$ works fine. Applying the Fundamental Theorem of Calculus,

$$\int_1^{n+1} \frac{1}{x}\,dx = \ln(n+1) - \ln 1 = \ln(n+1) - 0 = \ln(n+1).$$

What does this tell us about the harmonic series? If we take n terms and add them together, it is true that we don't know exactly what the result will be. But we do know that whatever that sum is, it will be more than $\ln(n+1)$. Since $\ln(n+1)$ goes to infinity as n gets larger, the harmonic series must also go to infinity as n grows.

Theorem. The harmonic series diverges. That is,

$$1 + \frac{1}{2} + \frac{1}{3} + \frac{1}{4} + \frac{1}{5} + \cdots = \infty.$$

EXERCISES

4.94   Let $x = 0.24\overline{9} = 0.24999\ldots$.
   a) Write x as a fraction.
   b) What other decimal expression represents x?

4.95   Write $2.\overline{2} = 2.222\ldots$ as a fraction.

4.96 Write $0.\overline{123} = 0.123123123\ldots$ as a fraction in lowest terms.

4.97 Interpret this math joke.

> An infinite number of mathematicians walk into a bar. The first mathematician says to the bartender, "I'd like 1 glass of beer." The second mathematician says, "I'd like 1/2 glass of beer." The third mathematician says, "I'd like 1/4 glass of beer." At that point, the bartender gets annoyed, slams a couple of mugs on the bar, and says, "Look. Here's two glasses of beer. Now you all get out of here!"

   a) Why does this joke work?
   b) When one of the authors told this joke to a group of advanced students, a student remarked, "It's a good thing the third mathematician didn't ask for 1/3 of a glass!" Why did that get a laugh from the other students?

4.19 Exponential Growth and Doubling Times

If you've studied science, you may know that radioactive substances have a "half life," a period during which half of the substance will decay into another element. For example, Polonium-210, which is used as a detonator for nuclear bombs, has a half-life of 138 days. After 138 days, a 1 gram sample of Polonium-210 decays until half of it is lead. After another 138 days, half of the remaining Polonium-210 will have decayed into lead (meaning that there is only 1/4 gram of Polonium-210 left in the sample). An expensive aspect of maintaining a large nuclear arsenal is continuously replacing decayed detonators.

If you've studied finance (or worked Exercise 4.64), you may know that compound interest endows investments with a doubling time. You might have learned the "rule of 72," which estimates how long it takes for your money to double. According to the rule, if you divide 72 by your compound interest rate, the result is approximately the doubling time. For example, if you earn 6% on your money, it should double in $72/6 = 12$ years.

Although the rule of 72 is often a good estimate, as math students, we can compute doubling times precisely. Let P be the amount of principal invested, and assume it grows at 6%, compounded annually. Then the amount of money after t years, $A(t)$, is given by the formula

$$A(t) = P(1.06)^t.$$

If we want to know when the investment will double, we simply set $A(t) = 2P$ and find t. Calculating,

$$2P = P(1.06)^t$$
$$2 = (1.06)^t$$

and to solve this, we take a logarithm of both sides,

$$\ln 2 = \ln\left(1.06^t\right) = t \ln 1.06.$$

Finally, we divide by $\ln 1.06$ to get t:

$$t = \frac{\ln 2}{\ln 1.06} = 11.90.$$

A similar calculation works with any investment that pays compound interest. That's because compound interest is an example of *exponential growth*, and anything that grows exponentially has a doubling time. To see this, let $A(t) = Pa^t$ be a generic exponential function with base $a > 1$ (it doesn't grow if $a < 1$). Then we can compute the time it takes for doubling:

$$2P = Pa^t$$
$$2 = a^t$$

and, using logarithms, as before,

$$t = \frac{\ln 2}{\ln a}.$$

You can verify that the time required to go from $2P$ to $4P$ is of the same duration, as is the time to go from $3P$ to $6P$, etc.

Other than compound interest, can you think of anything else that grows exponentially? Anything that "grows by a percentage" is exponential growth. For example, over the long term inflation is expected to grow around 3% per year. That means if an item costs \$10 this year, it costs 3% more money a year from now, or $\$10 + \$10(0.03) = \$10(1.03) = \$10.30$. In two years, the cost is $\$10(1.03)^2$, and after t years the cost is $\$10(1.03)^t$.

Population is another quantity that often grows exponentially. Some percentage of the population tends to be older (or die randomly from accidents). Another percentage is of childbearing age and temperament to have offspring. The difference is the growth rate of the population. It fluctuates, but over the long term it has tended to have an exponential shape.

Figure 4.40 shows government estimates for the amount of fossil fuel used in the United States since 1900. The curve through the data points has the formula $f(t) = e^{2.48+0.0221t} = 11.9e^{0.0221t}$, which we recognize as an exponential curve.

If fossil fuel use has grown (approximately) exponentially, it should have a doubling time, and we can calculate it. First, $f(0) = 11.9$, since fossil fuel use in 1900 was approximately 11.9 quadrillion British thermal units. To find the doubling period, we solve to find the t for which fuel use had doubled to 23.8 quadrillion BTUs:

$$23.8 = f(t)$$
$$23.8 = 11.9e^{0.0221t}$$
$$2 = e^{0.0221t}$$

and taking logs of both sides,

$$\ln 2 = 0.0221t$$
$$t = \frac{\ln 2}{0.0221} \approx 31.4.$$

In the United States over the last century, fossil fuel use doubled approximately every 31.4 years. Is that an alarming rate? Perhaps not: the growth rate might be

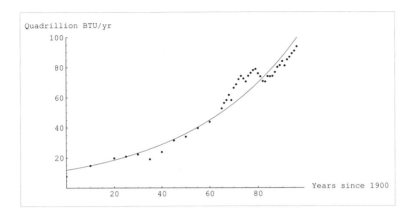

Figure 4.40 US fossil fuel use since 1900.

thought leisurely at only a bit over 2% per year. Yet there are some reasons to think it is alarming, and our knowledge of series can give us some perspective on the question.

Let x be the total amount of fossil fuel used in one doubling period (in this case 31.4 years). How much was used during the period before? If you answered $\frac{x}{2}$, you are correct. How much was used during the period before that? Yes, $\frac{x}{4}$. Looking backward, counting by doubling times, we get a progression that looks like a geometric series, as in Figure 4.41.

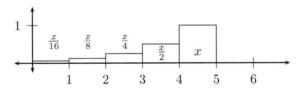

Figure 4.41 Fossil fuel use divided into doubling periods.

Compare the amount of fuel used in the final doubling period to the periods that came before. What happens if we start adding the previous amounts? We get

$$\frac{x}{2} + \frac{x}{4} + \frac{x}{8} + \frac{x}{16}.$$

This is a finite part of a geometric series, with $a = \frac{x}{2}$ and $r = \frac{1}{2}$. The sum of these four terms is less than the sum of the entire series, which we can calculate:

$$\frac{x}{2} + \frac{x}{4} + \frac{x}{8} + \frac{x}{16} + \cdots = \frac{x}{2}\frac{1}{1 - \frac{1}{2}} = x.$$

What did we learn? Not only is the amount of fossil fuel used in the final doubling period equal to twice as much as the preceding period, *it is more than all of previous history added together.* If this kind of growth continues, what will happen during the next doubling period? In Figure 4.42 the final doubling period again exceeds all of previous history, and that includes the period we just finished!

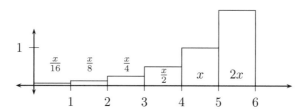

Figure 4.42   Fossil fuel use after one more doubling period.

Exponential growth negates many arguments for complacency about energy policy. For example, what if there are twice as many resources as we estimate still in the ground, coal and oil and gas still undiscovered and untapped? That buys us more time, but not hundreds of years, merely one more 31.4 year doubling period. It is clear that repeatedly doubling consumption is not a pattern that can continue indefinitely.

◼ **EXAMPLE 4.51**

The United Nations has estimated that the world human population was approximately 6 billion in the year 2000 and was increasing at the rate of approximately 1 percent per year at that time. Based on this information, find the function representing world human population as a function of the number of years past 2000. When will population reach 10 billion, according to this model?

Solution: Let $A(t)$ be the world human population at time t, where t is number of years after 2000. Since population grows by a percentage, we recognize it as exponential growth, and we can write $A(t) = Pe^{rt}$ for some value of P and some value of r. We are told that the population in 2000 is 6 billion, thus $6 \times 10^9 = A(0) = Pe^{r \cdot 0} = P$, and it follows that $A(t) = 6 \times 10^9 e^{rt}$.

We are also told that the rate of change of population is 1 percent per year. Since derivatives tell us rates of change, this means that

$$\frac{dA(t)}{dt} = (0.01)A(t),$$

and substituting for $A(t)$, we have

$$\frac{d}{dt}\left(6 \times 10^9 e^{rt}\right) = (0.01)(6 \times 10^9)e^{rt}$$

$$(6 \times 10^9)re^{rt} = (0.01)(6 \times 10^9)e^{rt}.$$

Divide both sides by $6 \times 10^9 e^{rt}$ to get

$$r = 0.01.$$

Therefore, the model for global population growth is

$$A(t) = 6 \times 10^9 e^{(0.01)t},$$

and to find the time when $A(t) = 10^{10}$ we solve the equation

$$10^{10} = 6 \times 10^9 e^{(0.01)t},$$

obtaining $t \approx 51$. The world's human population will reach 10 billion in the year 2051, according to this model.   ■

EXERCISES

4.98   Suppose that you have a bank account earning 5 percent interest per year, compounded annually. If you have $2000 today, when will you have $4000?

4.99   Assume the inflation rate remains 3.4% per year.
  a) How many years will it take for prices to double?
  b) How many years will it take for prices to quadruple?
  c) How many years will it take for prices to grow 10-fold?

4.20   Beyond Calculus

In this chapter we have seen that calculus is the mathematics we use to understand change. When quantities change, derivatives tell us the rate. If we know a rate, integrals allow us to transform that knowledge into a total change. Our study has allowed us to maximize and minimize functions, to analyze the history of fossil fuel use, and even to compute (roughly) the height of a rocket.

Yet as much as we have learned, our view of calculus has been very one-dimensional, literally. We worked solely with functions of a single variable. The world is filled with beautiful complicated things that can't be expressed as functions of one variable. For example, when a rocket launches into orbit, it doesn't fly straight up as we assumed in Section 4.17. It follows some arc into space that results in an elliptical shaped orbit. Truly modeling the height of a rocket requires a three-dimensional view and functions with variables x, y, z, and t, along with corresponding multi-dimensional calculus principles. There are whole courses of study that take the ideas of calculus and put them in a multi-variable context.

Many of the physical models we use for the universe around us are based on relationships defined in terms of derivatives. For example, Newton's Law of Cooling says that the rate of change in the temperature of an object is proportional to the difference in temperature between the object and its surroundings. Informally, it says that very hot objects cool at a faster rate than slightly hot objects (but it still takes hot objects longer to completely cool since they have further to go). Formally, Newton's Law of Cooling is a *differential equation,* an equation containing derivatives of a function,

$$\frac{dy}{dt} = k(A - y).$$

To solve a differential equation is to find a function y that makes the equation true. For example, in the equation above, $y = y(t) = A + Be^{-kt}$ works. Take a derivative for yourself and check. After a first course in calculus, many people go on to one or more courses in differential equations.

One can also study calculus where the functions are defined on the complex numbers (i.e., numbers that have both real and imaginary parts). The complex numbers are an especially beautiful domain in which to do calculus, and many elegant results exist in this context. One elegant fact is that any function that has a (single) derivative defined at every complex number is guaranteed to have a power series, which is a way of saying that it can be written as an infinite polynomial. For example, the exponential function has a power series,

$$e^x = 1 + x + \frac{x^2}{1 \cdot 2} + \frac{x^3}{1 \cdot 2 \cdot 3} + \frac{x^4}{1 \cdot 2 \cdot 3 \cdot 4} + \cdots.$$

Calculus has helped us gain understanding in nearly every domain of mathematics, including number theory, geometry, game theory, topology, numerical analysis, and probability. It is an essential tool for physics, chemistry, biology, engineering, economics, and statistics. It is one of humankind's most powerful and awe-inspiring creations.

EXERCISES

4.100 Let $y(t) = 70 + 20e^{-2t}$, which might describe the temperature of an object that cools from $90°$F to room temperature at $70°$F.

   a) Find $2\big(70 - y(t)\big)$.

   b) Find $\dfrac{dy}{dt}$ and verify that you get the same answer as in part (a).

4.101 Let $y(t) = A + Be^{-kt}$, where A, B, and k are constants.

   a) Find $k(A - y(t))$.

   b) Find $\dfrac{dy}{dt}$ and verify that you get the same answer as in part (a).

4.102 Let $T(x) = 1 + x + \frac{x^2}{1 \cdot 2} + \frac{x^3}{1 \cdot 2 \cdot 3} + \frac{x^4}{1 \cdot 2 \cdot 3 \cdot 4} + \frac{x^5}{1 \cdot 2 \cdot 3 \cdot 4 \cdot 5}$, the sum of the first six terms in the power series for e^x.

   a) Compute $T(0)$, and verify that you get $e^0 = 1$.

   b) Compute $T(1)$, and verify that the answer you get is close to $e = e^1 \approx 2.718281828459045$.

   c) Compute $T(2)$, and compare to the value of e^2 from a calculator or computer.

CHAPTER 5

NUMBER THEORY

God made integers, all else is the work of man.

L<small>EOPOLD</small> K<small>RONECKER</small> (1823–1891)

Now we turn to our second major topic in detail, the theory of numbers. While calculus studies continuously changing processes, number theory deals primarily with a discrete set, the integers. The integers are both familiar and mysterious. People have been fascinated by the integers since the beginning of history. We will investigate properties of the integers, and prove some of these properties.

5.1 What Is Number Theory?

What are numbers? In different times and places, numbers have meant different things. Here are some examples.

- Counting numbers: 1, 2, 3, ...

- Whole numbers: 0, 1, 2, ...

- Integers: $0, \pm 1, \pm 2, \ldots$

- Rational numbers: numbers of the form $\frac{a}{b}$, where a and b are integers and $b \neq 0$

- Real numbers: the numbers on a *number line*, e.g., the x-axis

- Complex numbers: numbers of the form $a + bi$, where a and b are real numbers, and $i^2 = -1$

Although people study all these numbers, the term *number theory* has come to mean the study of the integers. So that will be the focus of this chapter.

5.2 Divisibility

Of the four common operations on integers—addition, subtraction, multiplication, and division—division presents special difficulties. The reason is simple. When we add, subtract, or multiply integers, the result is an integer; but when we divide one integer by another, the result is not guaranteed to be an integer. Because of this, much of number theory is devoted to problems of division.

Definitions and Examples

Definition. For integers a and b, with $a \neq 0$, we say that a *divides* b if there is some integer c such that $b = ac$. If a divides b, we write $a \mid b$. If a does not divide b, we write $a \nmid b$.

In other words, a divides b if b is a multiple of a.

◼ EXAMPLE 5.1

(a) $7 \mid 343$ because $343 = 7 \cdot 49$.

(b) $5 \nmid 24$ because there is no *integer* c such that $24 = 5c$. Of course, $24 = 5 \times 4.8$, but 4.8 is not an integer.

(c) $-3 \mid (-33)$, because $-33 = -3 \cdot 11$.

Simple Properties of Divisibility

In this section we will work out a few simple rules of divisibility, simple but useful. Some of them may seem obvious to you, but mathematicians like to prove everything, so we will proceed accordingly.

Let a, b, and c be integers, and suppose that $a \mid b$. Then $a \mid bc$.

Proof. Since $a \mid b$, there is some integer d such that $b = ad$. Then

$$bc = (ad)c = a(dc).$$

By the definition, since there is an integer dc such that $bc = a(dc)$, we have $a \mid bc$. ∎

■ **EXAMPLE 5.2**

Suppose that x is some integer. Show that $7 \mid 343x$.

Solution: This is a straightforward application of the last theorem. We saw above that $7 \mid 343$. So applying the theorem, with $a = 7$, $b = 343$, and $c = x$, we conclude that $7 \mid 343x$. ■

i. If $a \mid b$ and $a \mid c$, then $a \mid (b + c)$ and $a \mid (b - c)$.

ii. If $a \mid b$ and $c \mid d$, then $ac \mid bd$.

iii. If $a \mid b$ and $b \mid c$, then $a \mid c$.

Important Note: Does this seem dry to you? Too abstract and formal? Do statements like "if $a \mid b$ and $c \mid d$, then $ac \mid bd$" make your eyes glaze over? There is a cure. Pick numbers for the letters. Find numbers a and b so that a divides b. Find numbers c and d such that c divides d. Convince yourself that ac divides bd. If that isn't enough, find some more examples. We are dealing with familiar numbers. If you do enough examples, this will make sense to you.

Proof. (i) Let $a \mid b$ and $a \mid c$. Since $a \mid b$, there is some integer e such that $b = ae$. Since $a \mid c$, there is some integer f such that $c = af$. Thus

$$b + c = ae + af = a(e + f)$$

so $a \mid (b + c)$. Similarly

$$b - c = ae - af = a(e - f)$$

so $a \mid (b - c)$.

(ii) Let $a \mid b$ and $c \mid d$. Since $a \mid b$, there is some integer e such that $b = ae$. Since $c \mid d$, there is some integer f such that $d = cf$. Using these last two facts, we have

$$bd = (ae)(cf) = (ac)(ef).$$

So $ac \mid bd$.

(iii) Let $a \mid b$ and $b \mid c$. Since $a \mid b$, there is some integer e such that $b = ae$. Since $b \mid c$, there is some integer f such that $c = bf$. Thus

$$c = bf = (ae)f = a(ef).$$

So $c = a(ef)$, which implies that $a \mid c$. ■

Definition. Let n be an integer. Then n is *even* if $2 \mid n$, and n is *odd* if $2 \nmid n$.

EXAMPLE 5.3

The sum of any two even numbers is even. Proof: Suppose that b and c are even numbers. We apply the result (i) above, with $a = 2$. We have $2 \mid b$ and $2 \mid c$, so $2 \mid (b + c)$, i.e., $b + c$ is even.

EXAMPLE 5.4

Let x be even, and y be divisible by 3. Then xy is divisible by 6.

Solution: We have $2 \mid x$ and $3 \mid y$. Applying result (ii) above, with $a = 2$, $b = x$, $c = 3$, and $d = y$, we find that $(2 \cdot 3) \mid xy$; in other words, 6 divides xy. ■

EXAMPLE 5.5

Any integer divisible by 10 is divisible by 5.

Solution: Suppose that x is divisible by 10. Applying (iii) above, with $a = 5$, $b = 10$, and $c = x$, we conclude that $5 \mid x$. ■

EXAMPLE 5.6

An integer is divisible by 5 if and only if its last digit is 0 or 5.

Solution: Let's assume that n is a positive integer. If this is true for positive integers, it is true for all integers, since n and $-n$ have the same divisors.

Let a be the last digit of n. We will see that 5 divides n precisely when 5 divides a. First note that $n - a$ ends in 0, so is divisible by 10. By the last example, 5 also divides $n - a$, say $n - a = 5b$. So we have $n = 5b + a$. Using (i) above:

If 5 divides a, then 5 divides $n = 5b + a$.

If 5 divides n, then 5 divides $a = n - 5b$.

So $5 \mid n$ if, and only if, $5 \mid a$. Since a is a digit, i.e., between 0 and 9, it is divisible by 5 precisely when it is 0 or 5. ■

Division and Remainders

In the last section, we introduced the notion of divisibility: For two integers a and b, either $b \mid a$ or $b \nmid a$. But this is not the end of the story.

The Division Theorem. Let a and b be integers, with b positive. Then there exist unique integers q and r such that

i. $a = bq + r$, and

ii. $0 \le r < b$.

Does this look familiar? Think q for quotient and r for remainder. This theorem is the basis for long division. Of course, b divides a if and only if $r = 0$.

⬛ EXAMPLE 5.7

Apply the Division Theorem to $a = 237$ and $b = 13$.

Solution: If you remember your long division, you can divide 237 by 13, and get quotient 18, remainder 3. So let $q = 18$ and $r = 3$. We can verify the two statements of the theorem:

i. $237 = 13 \cdot 18 + 3$

ii. $0 \le 3 < 13$.

But the theorem says more than this; it says that q and r are *unique.* There is only one choice of q and r that satisfy the two conditions above. (Unlike in our everyday, imprecise language, in mathematics unique doesn't mean "unusual" or "very." It means *one.* It is from a French word, meaning "single.")

There are other choices of q and r that satisfy condition (i) above. Here are two:

Let $q = 17$ and $r = 16$. You can check that $237 = 13 \cdot 17 + 16$.

Let $q = 19$ and $r = -10$. Check that $237 = 13 \cdot 19 + (-10)$.

You can probably see that there are many other choices for q and r, but none of them also satisfies the condition $0 \le r < 13$. That is why we can talk about *the* quotient and *the* remainder. This uniqueness will be important to us in future sections. ⬛

Many people find long division hard. Even mathematicians get lazy. Can't we do this on a calculator? Simple calculators do not find quotients and remainders for us, but they will divide numbers. We can use this, with an alternate version of part (i) of the theorem:

$$\frac{a}{b} = q + \frac{r}{b}.$$

(Make sure you understand how this follows from (i).)

Looking at our last example, use your calculator to divide 237 by 13. Your result depends on how many digits your calculator displays, but will be something like 18.230769. This gives the quotient as 18. Now

$$18.230769 \approx \frac{237}{13} = 18 + \frac{r}{13}.$$

Hence $r/13$ must be (approximately) .230769. So if we multiply .230769 by 13, we should get r. The result may be something like $r = 2.999999$. Since we know r to be an integer, we round to $r = 3$.

EXAMPLE 5.8

Show that the product of two odd numbers is odd.

Solution: The Division Theorem gives us a useful way to think about odd numbers. It tells us that any number n can be written as $n = 2x$ or $n = 2x + 1$, depending on its remainder. Numbers of the form $2x$ are even, and numbers of the form $2x + 1$ are odd. Suppose then that we have two odd numbers, say, $2x+1$ and $2y+1$. Then their product is $(2x + 1)(2y + 1) = 4xy + 2x + 2y + 1 = 2(2xy + x + y) + 1$. Hence, the product is of the form $2z + 1$, with $z = 2xy + x + y$, so is odd.   ■

Finally, if you look at the statement of the Division Theorem, you will see that a can be any integer, even a negative one. How does this work? Let us consider an example: $a = -14$, $b = 3$. Since 14 has remainder 2 upon division by 3, you may think that -14 has remainder -2. But the remainder has to be 0, 1 or 2, so this can't work. In fact, we have $-14 = 3 \cdot (-5) + 1$ and $0 \leq 1 < 3$, so the uniqueness part of the theorem tells us that the quotient is -5 and the remainder is 1. In general, we have to be careful with negative integers here. We will later see how to make them easier to deal with.

EXERCISES

5.1   True or false: the sum of any two odd numbers is odd.

5.2   Does 7 divide 6998? (Hint: add 2.)

5.3   In one of our examples, we stated a *divisibility rule* for 5, i.e., a rule that allows us to quickly check whether a number is divisible by 5. Find divisibility rules for each of the following.
   a) 2
   b) 25 (Hint: look at the last two digits.)
   c) 4

5.4 Using the notation of the Division Theorem, find q and r for each of the following.

    a) $a = 82, b = 7$
    b) $a = 34526, b = 23$
    c) $a = -17, b = 5$

5.5 You have two numbers. Dividing the larger by the smaller gives you a quotient of 3 and a remainder of 8. If the sum of the numbers is 80, what are the numbers?

5.6 The remaining problems give assertions about arbitrary integers a, b, c, and d. For each, decide whether it is always true. If it can be false, give a *counterexample*, a selection of integers for which the statement is false.

    a) If $a \mid b$ and $a \mid (b + c)$, then $a \mid c$.
    b) If $a \mid b$ and $a \nmid (b + c)$, then $a \nmid c$.
    c) If $a \mid c$ and $b \mid c$, then $ab \mid c$.
    d) If $a \mid bc$ and $a \nmid b$, then $a \mid c$.
    e) If $a \mid b$ and $b \mid a$, then $a = b$.
    f) If $a \mid b$ and $ac \mid bd$, then $c \mid d$.

5.3 Irrational Numbers

The Problem

Mathematicians have long struggled with the question of what a number is. Early civilizations mostly accepted positive integers and fractions; zero and negative numbers took much longer to gain acceptance as full-fledged numbers.

One place where numbers can be found is in measuring lengths. Which numbers appear as lengths? It is not hard to show that every positive integer and fraction can be represented as a length. To many people, it seems self-evident that these are the only lengths. Roughly 2500 years ago it was discovered that this is *not* the case.

The length that led to this discovery is not obscure: it is the length of the diagonal of a square of side 1 (Figure 5.1). Using the Pythagorean Theorem, this length is $\sqrt{2}$.

Figure 5.1 $\sqrt{2}$ as a length.

The Theorem

There are no integers a and b such that $\sqrt{2} = a/b$.

Proof. We will give a proof by contradiction. We will assume that the theorem is false, and show that this leads to a logical contradiction.

So let us assume that integers a and b exist such that $\sqrt{2} = a/b$. Recall that we can reduce any fraction to one in *lowest terms* by dividing numerator and denominator by any common factors. Let us assume that we have done that, so that a and b don't have any common factors.

Now square the above equation to get $2 = (a/b)^2$. Multiplying both sides of this equation by b^2 gives

$$2b^2 = a^2.$$

This means that a^2 is even—divisible by 2. We saw in the last section that the product of two odd numbers is always odd, so a cannot be odd, i.e., a is divisible by 2. Hence $a = 2c$ for some integer c. Substituting this into the above equation, we get $2b^2 = (2c)^2 = 4c^2$. Hence $b^2 = 2c^2$. Therefore, b^2 is even, so b is also divisible by 2. We have arrived at our contradiction: both a and b are divisible by 2, contradicting the assumption that a and b have no common divisors. Therefore, there do not exist integers a and b such that $\sqrt{2} = a/b$. ∎

Numbers that can be expressed as ratios of two integers are called *rational numbers*. Ones that cannot be so expressed are *irrational*. So the theorem tells us that $\sqrt{2}$ is irrational. Much later, in the 19th century, it was shown that, in some sense, most lengths are irrational. But when the first irrational number was found, it no doubt caused much consternation. The discovery was made by the Pythagoreans, in the 5th or 6th century BCE. Like many of the greatest discoveries in science, it was negative; something was shown to be impossible.

This theorem also implies a result in geometry concerning commensurable line segments. Two line segments are *commensurable* if there is a third line segment that can fit into each of the two a whole number of times. For example, line segments of length $1/2$ and $1/3$ are commensurable, because a line segment of length $1/6$ fits into the one of length $1/2$ exactly three times and into the one of length $1/3$ exactly twice. Early on, many mathematicians thought that all pairs of line segments were commensurable. This idea was often used in geometric proofs. But it turned out to be wrong.

Corollary. The side of a unit square and its diagonal are not commensurable.

A *corollary* is a theorem that follows easily from another theorem. So let us see how this follows from the last theorem.

Again, we proceed by contradiction. Suppose that the side and diagonal of a unit square are commensurable, that we have a line segment of length x that can fit into each of them an integral number of times. Suppose that it fits into the diagonal a times, and into the side b times. Then we have $ax = \sqrt{2}$ and $bx = 1$. Dividing the two equations yields $a/b = \sqrt{2}$, which we know to be impossible for integers a and b. So our two line segments must be incommensurable.

EXERCISES

5.7   Which of the following do you think are rational numbers?
   a) $\sqrt{3}$
   b) $\sqrt{4}$
   c) $117/3459753$
   d) π (Hint: the Internet is a great resource.)

5.8   True of false? A number is rational if, and only if, its decimal expansion terminates. (Recall that the decimal expansion of $1/4$, for example, is .25.)

5.9   Prove that $10\sqrt{2}$ is irrational.

5.10   Prove that $\sqrt{2} + 10$ is irrational.

5.11   Prove that if r and s are rational numbers, then $r + s$, $r - s$, and rs are rational numbers, and if $s \neq 0$, then r/s is a rational number.

5.12   One of the authors had an idea when he was very young. He was measuring a length with a ruler. The ruler had marks every inch, half-inch, and quarter-inch. The length didn't quite match one of the marks, so he had to mentally create another mark at an eighth-inch place. This worked, and gave him his idea. If it didn't work, then he could go to sixteenths of an inch, thirty-seconds of an inch, then sixty-fourths, etc., until he got an *exact* match.
   a) What is wrong with this idea?
   b) What numbers can be exactly measured in this way?

5.13   Look up the origin of the word *corollary*. How does it relate to a little theorem presented after a big theorem?

5.4   Greater Common Divisors

Definitions and Examples

Definition. Let a and b be integers, not both 0. The *greatest common divisor (gcd)* of a and b is an integer d that satisfies the following two conditions.

   i. $d \mid a$ and $d \mid b$, i.e., d is a common divisor of a and b.

   ii. If $c \mid a$ and $c \mid b$, then $c \leq d$, i.e., of all common divisors, d is the largest.

We will write the gcd of a and b as $\gcd(a, b)$.

▉ EXAMPLE 5.9

Find the greatest common divisors.

(a) $\gcd(15, 12)$

(b) $\gcd(37, 64)$

(c) $\gcd(-14, 49)$

(d) $\gcd(100, 1000)$

(e) $\gcd(0, 1203)$

Solution: For (a), let us list all the divisors of 12 and 15. The set of divisors of 12 is $\{\pm1, \pm2, \pm3, \pm4, \pm6, \pm12\}$, and the set of divisors of 15 is $\{\pm1, \pm3, \pm5, \pm15\}$. From this, we see that the greatest common divisor is 3.

(b) Check that 37 cannot be factored into smaller numbers, and 64 is a power of 2. Since they have no common factors, the highest common divisor must be 1; therefore $\gcd(37, 64) = 1$.

(c) This one is easy. Spend a moment to convince yourself that $\gcd(-14, 49) = 7$.

(d) Since $100 \,|\, 1000$, we have $\gcd(100, 1000) = 100$.

(e) Since $1203 \,|\, 1203$ and $1203 \,|\, 0$, the greatest common divisor of 1203 and 0 is 1203.

Note, for future use, that the last argument generalizes: $\gcd(0, a) = a$ for *any* positive integer a. This is important in the following discussion. ▉

Computing GCDs: The Euclidean Algorithm

How do we find gcd's? You may have learned a method that involves finding the prime factorizations of a and b. We will take a look at that method later, but now we examine another method, one that goes all the way back to Euclid. In fact, this method is called the Euclidean Algorithm. It is based on the next theorem.

Let a and b be integers, not both 0. If $a = bq + r$, then $\gcd(a, b) = \gcd(b, r)$.

Proof. We will show that the set of common divisors of a and b is identical to the set of common divisors of b and r. This implies that the greatest common divisor of a and b is equal to the greatest common divisor of b and r.

First, suppose that x is a divisor of both a and b. We want to show that $x \,|\, b$ and $x \,|\, r$. We already know that $x \,|\, b$, so we need only show that $x \,|\, r$. Since $x \,|\, a$ and $x \,|\, b$, there are integers c and d such that $xc = a$ and $xd = b$. Using these and $a = bq + r$, we have

$$r = a - bq = (xc) - (xd)q = x(c - dq).$$

So $r = x(c - dq)$; hence $x \,|\, r$. Therefore x is a common divisor of b and r.

Finally, suppose that y is a common divisor of b and r. We need only show that y is a common divisor of a and r, i.e., that $y \mid a$. We can use the properties of divisibility from Section 5.2. Since y divides b, then y divides every multiple of b. So, in our case, y divides bq. From Section 5.2 we also know that, if y divides two numbers, then y divides their sum. In our case, $y \mid bq$ and $y \mid r$, so $y \mid (bq + r)$, i.e., $y \mid a$. This completes our proof.   ■

◨ **EXAMPLE 5.10**

Find $\gcd(116, 34)$.

Solution: We will use the last theorem to find the gcd, with $a = 116$ and $b = 34$. What are q and r? By the Division Theorem of Section 5.2, the quotient and remainder work. (That is of course why we call them q and r.) So let us divide 116 by 34:

$$116 = 34 \cdot 3 + 14.$$

Hence the last theorem tells us that $\gcd(116, 34) = \gcd(34, 14)$. We can repeat this argument:

$$34 = 14 \cdot 2 + 6, \text{ so } \gcd(34, 14) = \gcd(14, 6);$$

$$14 = 6 \cdot 2 + 2, \text{ so } \gcd(14, 6) = \gcd(6, 2);$$

$$6 = 2 \cdot 3 + 0, \text{ so } \gcd(6, 2) = \gcd(2, 0).$$

Finally $\gcd(2, 0) = 2$. Putting this all together, $\gcd(116, 34) = 2$.   ■

This method is called the Euclidean Algorithm.

The Euclidean Algorithm. Let a and b be positive integers. Define integers $q_1, \ldots, q_t$ and $r_1, \ldots, r_t$ by successive divisions:

$$
\begin{aligned}
a &= bq_1 + r_1, & 0 \le r_1 < b \\
b &= r_1 q_2 + r_2, & 0 \le r_2 < r_1 \\
r_1 &= r_2 q_3 + r_3, & 0 \le r_3 < r_2 \\
r_2 &= r_3 q_4 + r_4, & 0 \le r_4 < r_3 \\
&\ \ \vdots & \\
r_{t-2} &= r_{t-1} q_t + r_t, & 0 \le r_t < r_{t-1} \\
r_{t-1} &= r_t q_t + 0.
\end{aligned}
$$

Then $\gcd(a, b) = r_t$. In other words, keep on dividing until the remainder is 0. The gcd is the last nonzero remainder.

■ **EXAMPLE 5.11**

Find $\gcd(93, 37)$.

Solution:

$$93 = 37 \cdot 2 + 19$$
$$37 = 19 \cdot 1 + 18$$
$$19 = 18 \cdot 1 + 1$$
$$18 = 1 \cdot 18 + 0$$

So $\gcd(93, 37) = 1$.   ■

Why does the Euclidean Algorithm work? It is simply the gcd relation applied over and over again. In the notation of the algorithm,

$$\gcd(a, b) = \gcd(b, r_1) = \gcd(r_1, r_2) = \cdots = \gcd(r_{t-1}, r_t) = \gcd(r_t, 0) = r_t.$$

Why is the Euclidean Algorithm important? Because it is fast. The other popular method of finding gcd's is based on factoring numbers, which is slow. But division is fast. The difference in speed becomes evident when computers try to find gcd's of very large numbers. The computers available to Euclid were of course cruder than ours, but the problem of finding fast ways to do computations is very old.

EXERCISES

5.14   Find the gcd's.
   a) $\gcd(11, 22)$
   b) $\gcd(12, 32)$
   c) $\gcd(314, 159)$
   d) $\gcd(4144, 7696)$
   e) $\gcd(480097, 48)$
   f) $\gcd(480099, 48)$

5.15   Let $a = bx + 1$. Show that $\gcd(a, b) = 1$. (Hint: run the Euclidean Algorithm.)

5.16   In the definition of greatest common divisor, we insist that a and b cannot both be 0. Why do you think this condition is there?

The rest of the exercises concern least common multiples. The *least common multiple* of integers a and b, written $\text{lcm}(a, b)$, is the smallest positive integer d such that both a and b divide d. It is denoted by $\text{lcm}(a, b)$.

5.17   Fill in the table below.

a	b	$\gcd(a,b)$	$\mathrm{lcm}(a,b)$
6	8	2	24
12	15		
13	8		
17	187		

5.18 Least common multiples arise in the handling of different calendars. This is why the Chinese and the Mayans were masters of the lcm.

    a) Suppose we have two calendars: a solar calendar with 365 days, and a lunar calendar of 30 days. Suppose that today is June 5 and day 23 of the lunar calendar. How long would it take before we again have the next day 23 of the lunar calendar on June 5? (Ignore leap days.)

    b) The Mayans used two different calendars, the 260-day almanac and a 365-day year. The time it took to cycle both calendars, to come around to the same date in both, was called the "calendar round." How long was the calendar round?

5.19 There is a formula relating least common multiples and greatest common divisors. See if you can find it. (Hint: look for a pattern in the table above.)

5.20 Suppose that we have a pool table divided into $24 = 4 \times 6$ squares (Figure 5.2). We start a ball from the lower left, aiming it to cross each little square on its diagonal. The ball continues, perhaps bouncing off some sides, until it hits a corner, as in the figure. This problem asks you to explore similar tables, to find some patterns. In the process, you will fill in the table below.

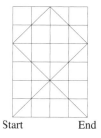

Start   End

Figure 5.2 A number-theoretic pool table.

    a) The second column of the table is for the number of little squares the ball crosses before ending in a corner. You can check that this number is 12 for the 4×6 table. Fill in this column, for every row but the last. Try to find the pattern, then use it to fill in the value for the last row.

    b) The third column of the table is for the corner the ball ends up in. See if you can fill this column. [Hint: consider first when it ends up left or right (using the number of little squares crossed), then whether it ends up top or bottom.]

c) The fourth column of the table is for the number of sides the ball hits, *including* the starting and ending corners. See if you can fill this column. (Hint: first consider how many times it crosses between left and right, then top and bottom.)

Table Dimensions	# of Little Squares Crossed	End Corner	Sides Hit
4×6	12	lower right	5
2×3			
2×4			
3×5			
6×8			
45×120			

5.5 Primes

Definitions and Examples

In studying the divisibility of integers, one soon encounters special integers that cannot be divided into smaller ones. You may be familiar with this notion.

Definition. Let n be an integer greater than 1. Then n is *prime* if its only positive divisors are 1 and n. If n is not prime, it is called *composite*.

▉ EXAMPLE 5.12

The first three primes are 2, 3, and 5. The numbers 4 and 6 are composite, because they are divisible by 2.

Notes:

i. The number 1 is not considered prime or composite.

ii. There is only one even prime, 2. Any other even number is divisible by 2, so cannot be prime.

iii. Similarly, the only prime divisible by 3 is 3 itself, and the only prime divisible by 5 is 5 itself, etc.

Computing Primes

The notes (ii) and (iii) above are the basis for a very old (c. 240 BCE) method of computing all the primes up to any integer N, called the *Sieve of Eratosthenes*. We will demonstrate the sieve to determine all the primes up to 50.

Write all numbers 1, 2, ..., 50.

$$
\begin{array}{cccccccccc}
1 & 2 & 3 & 4 & 5 & 6 & 7 & 8 & 9 & 10 \\
11 & 12 & 13 & 14 & 15 & 16 & 17 & 18 & 19 & 20 \\
21 & 22 & 23 & 24 & 25 & 26 & 27 & 28 & 29 & 30 \\
31 & 32 & 33 & 34 & 35 & 36 & 37 & 38 & 39 & 40 \\
41 & 42 & 43 & 44 & 45 & 46 & 47 & 48 & 49 & 50
\end{array}
$$

We will proceed to systematically cross out all the composite numbers, being left at the end with only the primes. We can cross out all the even numbers except 2, since, as noted above, they are all composite. We will also cross out 1, since it is not a prime, and circle 2, because it is a prime.

$$
\begin{array}{cccccccccc}
\cancel{1} & ② & 3 & \cancel{4} & 5 & \cancel{6} & 7 & \cancel{8} & 9 & \cancel{10} \\
11 & \cancel{12} & 13 & \cancel{14} & 15 & \cancel{16} & 17 & \cancel{18} & 19 & \cancel{20} \\
21 & \cancel{22} & 23 & \cancel{24} & 25 & \cancel{26} & 27 & \cancel{28} & 29 & \cancel{30} \\
31 & \cancel{32} & 33 & \cancel{34} & 35 & \cancel{36} & 37 & \cancel{38} & 39 & \cancel{40} \\
41 & \cancel{42} & 43 & \cancel{44} & 45 & \cancel{46} & 47 & \cancel{48} & 49 & \cancel{50}
\end{array}
$$

Now 3 is the smallest number not circled or crossed out, so it must be a prime. Circle it, and cross out all the multiples of 3 that are not already crossed out.

$$
\begin{array}{cccccccccc}
\cancel{1} & ② & ③ & \cancel{4} & 5 & \cancel{6} & 7 & \cancel{8} & \cancel{9} & \cancel{10} \\
11 & \cancel{12} & 13 & \cancel{14} & \cancel{15} & \cancel{16} & 17 & \cancel{18} & 19 & \cancel{20} \\
\cancel{21} & \cancel{22} & 23 & \cancel{24} & 25 & \cancel{26} & \cancel{27} & \cancel{28} & 29 & \cancel{30} \\
31 & \cancel{32} & \cancel{33} & \cancel{34} & 35 & \cancel{36} & 37 & \cancel{38} & \cancel{39} & \cancel{40} \\
41 & \cancel{42} & 43 & \cancel{44} & \cancel{45} & \cancel{46} & 47 & \cancel{48} & 49 & \cancel{50}
\end{array}
$$

Now 5 is the smallest number not circled or crossed out, so it must be a prime. Circle 5, and cross out all its multiples that are not already crossed out.

$$
\begin{array}{cccccccccc}
\cancel{1} & ② & ③ & \cancel{4} & ⑤ & \cancel{6} & 7 & \cancel{8} & \cancel{9} & \cancel{10} \\
11 & \cancel{12} & 13 & \cancel{14} & \cancel{15} & \cancel{16} & 17 & \cancel{18} & 19 & \cancel{20} \\
\cancel{21} & \cancel{22} & 23 & \cancel{24} & \cancel{25} & \cancel{26} & \cancel{27} & \cancel{28} & 29 & \cancel{30} \\
31 & \cancel{32} & \cancel{33} & \cancel{34} & \cancel{35} & \cancel{36} & 37 & \cancel{38} & \cancel{39} & \cancel{40} \\
41 & \cancel{42} & 43 & \cancel{44} & \cancel{45} & \cancel{46} & 47 & \cancel{48} & 49 & \cancel{50}
\end{array}
$$

If you proceed with the circling and crossing out, you will find all the primes less than 50:

$$2, 3, 5, 7, 11, 13, 17, 19, 23, 29, 31, 37, 41, 43, 47.$$

How Many Primes Are There?

Anyone who spends some time looking at primes soon notices that they get rarer as the numbers get larger. For example, there are fifteen primes between 1 and 50, but only ten between 51 and 100. Do they peter out entirely? This question was answered in Euclid's great book, *Elements*. His argument is based on a simple fact about primes.

Every integer greater than 1 is divisible by a prime.

We won't attempt a formal proof (which uses mathematical induction), but the idea is simple enough. Let n be an integer greater than 1. If n is prime, then it is divisible by a prime, namely, itself. If it is not a prime, then an integer other than 1 or n divides it, say m_1, with $1 < m_1 < n$. Now repeat this argument with m_1. If it is prime, then we have a prime dividing n. If not, we can find m_2 dividing m_1 with $1 < m_2 < m_1$. Keep on repeating this argument. Since each m_i is smaller than the last, this cannot go on forever. So eventually we arrive at some m_i being a prime, and since it divides n, we are done.

We can use this last result for Euclid's theorem on primes.

Euclid's Theorem.   There are infinitely many primes.

Here is Euclid's argument. Suppose that there are only finitely many primes, say p_1, p_2, ..., p_k. Let n be one greater than the product of all these primes: $n = p_1 p_2 \cdots p_k + 1$. Now notice that p_1 divides $p_1 p_2 \cdots p_k$, so p_1 cannot divide n. (Otherwise it would divide 1.) Similarly, p_2 does not divide n, and so on. None of p_1, p_2, ..., p_k divides n. But we know that *some* prime divides n, so we have a contradiction to p_1, ..., p_k being the only primes.

Thus there is no end to the primes, but they do seem to get rarer as they get bigger. It turns out that there is a pattern to this, although it took more than 2000 years after Euclid until it was found. The pattern, called the *Prime Number Theorem*, was conjectured in 1792 by Carl Friedrich Gauss (when he was 15 years old) and, in a slightly different form, in 1796 by Adrien-Marie Legendre. This theorem gives a function for computing the number of primes less than or equal to a number x. The function is not precise, but gives a closer and closer approximation as x gets larger. (For those of you who have seen natural logs, the theorem tells us that the ratio of the number of primes to $x / \ln x$ approaches 1 as x gets large.)

The Prime Number Theorem attracted a lot of attention, but wasn't proved until 1896, independently by Jacques Hadamard and Charles Jean de la Vallée-Poussin.

EXERCISES

5.21   How many primes end in a 5?

5.22   Use the Sieve of Eratosthenes to find all the primes less than 100.

5.23   The only primes that are consecutive integers are 2 and 3. Why?

5.24   *Twin primes* are primes that are two apart, for example, 3 and 5. Find five other such pairs.

5.25   Twin primes are always consecutive *odd* integers. Why is that?

5.26   Suppose that we define *prime triplets* to be three consecutive odd integers which are all prime, for example, 3, 5, and 7. How many such triplets are there?

5.27   Define $n!$ to be the product of the first n positive integers. For example, $2! = 1 \cdot 2 = 2$ and $3! = 1 \cdot 2 \cdot 3 = 6$. Show that the 999 consecutive numbers

$$1000! + 2, \ 1000! + 3, \ 1000! + 4, \ \ldots, \ 1000! + 1000$$

are all composite.

5.28   Use the idea of the last exercise to show that, whatever n is, it is always possible to find n consecutive composite numbers.

5.6 Relatively Prime Integers

Bézout's Identity

We begin this section with a result that turns out to be remarkably useful.

Bézout's Identity. Let a and b be integers, not both 0. Then there are integers u and v such that $\gcd(a, b) = au + bv$.

   As an example, let $a = 144, b = 22$. You can check that $\gcd(144, 22) = 2$. In this case, we can choose $u = 2$ and $v = -13$. They satisfy Bézout's Identity, since $2 = 144 \cdot 2 + 22 \cdot (-13)$. (If you don't see where u and v came from, it is because we just pulled them out of a hat. We will see later how one can find them.)
   Bézout's Identity was named after a French mathematician, Étienne Bézout (1730–1783), although the result was already known to Claude Gaspard Bachet de Méziriac (1581–1638) more than a century earlier. Bézout actually proved a generalization of it.
   We first apply Bézout's Identity to proving a nice theorem about common divisors.

The common divisors of a and b are the divisors of $\gcd(a, b)$.

Proof. Let $\gcd(a, b) = d$. First we show that every common divisor of a and b divides d. Let c be a divisor of a and b. Using Bézout, we can write $d = au + bv$, for some integers u and v. Now we use our properties of divisibility. Since $c \mid a$, we have $c \mid au$. Since $c \mid b$, we have $c \mid bv$. Since $c \mid au$ and $c \mid bv$, we have $c \mid (au + bv)$. So $c \mid d$.
   The other half is pretty easy to prove. Let e be a divisor of d, say $d = er$. Now d divides a and b, so we can write $a = ds$ and $b = dt$ for some integers s and t. Then

$a = ds = (er)s = e(rs)$ and $b = dt = (er)t = e(st)$, so we have e dividing both a and b, which is what we needed to show. ∎

Think about this. We have just reduced the problem of finding the common divisors of *two* integers to finding the divisors of *one* integer. Now that's progress.

▣ EXAMPLE 5.13

Find the common divisors of 84 and 192.

Solution: First we use the Euclidean Algorithm to find $\gcd(192, 84)$.

$$192 = 84 \cdot 2 + 24$$
$$84 = 24 \cdot 3 + 12$$
$$24 = 12 \cdot 2 + 0$$

So $\gcd(192, 84) = 12$. By the theorem, the common divisors of 84 and 192 are the divisors of 12, namely, $\pm 1, \pm 2, \pm 3, \pm 4, \pm 6, \pm 12$. ∎

Relatively Prime Integers

Consider the following two statements.

i. If $a \mid bc$ and $a \nmid b$, then $a \mid c$.

ii. If $a \mid c$ and $b \mid c$, then $ab \mid c$.

They may seem reasonable, but are they always true? Unfortunately, no. Here are counterexamples.

i. Let $a = 6$, $b = 9$, $c = 8$. Then $a \mid bc$ and $a \nmid b$, but $a \nmid c$.

ii. Let $a = 15$, $b = 5$, $c = 60$. Then $a \mid c$ and $b \mid c$, but $ab \nmid c$.

Is that the end of the story? No, if we are careful, we can find somewhat more restricted statements similar to the ones above, that are true. To formulate these, we need a new notion.

Definition. Let a and b be integers. We say that a and b are *relatively prime*, or *coprime*, if $\gcd(a, b) = 1$. In other words, relatively prime integers are those whose common divisors are only 1 and -1.

🖥 **EXAMPLE 5.14**

(a) 12 and 5 are relatively prime, since $\gcd(12, 5) = 1$.

(b) 27 and 96 are not relatively prime, since 3 divides both 27 and 96.

   The next two results are versions of the two statements with which we started this section.

Let a divide bc, and a and b be relatively prime. Then a divides c.

Proof. Since $\gcd(a, b) = 1$, Bézout lets us write $1 = au + bv$ for some integers u and v. Multiply this equation by c to get $c = acu + bcv$. Now look at the right-hand side of this new equation. Certainly $a \mid acu$. Also, since $a \mid bc$, we have $a \mid bcv$. Since a divides both terms on the right, it must divide the whole thing. Hence $a \mid c$.   ∎

Suppose that integers a and b both divide c, and that a and b are relatively prime. Then ab divides c.

Proof. Since $a \mid c$, there is some integer s such that $c = as$. Since $b \mid c$, there is some integer t such that $c = bt$. Since $\gcd(a, b) = 1$, we have integers u and v such that $1 = au + bv$. Multiply the last equation by c to get $c = acu + bcv$. Now plug in $c = bt$ for the first c, and $c = as$ for the second:

$$c = a(bt)u + b(as)v = ab(tu + sv).$$

Thus c is a multiple of ab, so $ab \mid c$.   ∎

🖥 **EXAMPLE 5.15**

(a) Does 7 divide 710,000? No. Apply the first result, with $a = 7$, $b = 10{,}000$, and $c = 71$. Because $10{,}000 = 10^4 = 2^4 \cdot 5^4$, we can see that 7 and 10,000 have no factors in common, so 7 and 10,000 are relatively prime, i.e., $\gcd(7, 10{,}000) = 1$. Therefore, if 7 were to divide 710,000, the theorem would tell us that 7 divides 71, which it does not.

(b) Suppose that a number x is even and is divisible by 5. Then it is divisible by 10. This is true by the second result, with $a = 2$, $b = 5$, and $c = x$.

Computing Bézout

How do we find the integers u and v of Bézout's Identity? It is possible using the Euclidean Algorithm!

Let's do an example, with $a = 69$ and $b = 31$. We will start by running the Euclidean Algorithm.

$$69 = 31 \cdot 2 + 7$$
$$31 = 7 \cdot 4 + 3$$
$$7 = 3 \cdot 2 + 1$$
$$3 = 1 \cdot 3 + 0$$

So $\gcd(69, 31) = 1$. Now we will find u and v such that $69u + 31v = 1$.

The method we use will express each of the remainders r_i above, that is, each of 7, 3, and 1, in the form $r_i = 69x_i + 31y_i$.

Starting with the first equation $69 = 31 \cdot 2 + 7$, we can write

$$7 = 69 \cdot 1 + 31 \cdot (-2).$$

Then using this and $31 = 7 \cdot 4 + 3$, we have

$$\begin{aligned}
3 &= 31 \cdot 1 - 7 \cdot 4 \\
&= 31 \cdot 1 - [69 \cdot 1 + 31 \cdot (-2)] \cdot 4 \\
&= 69 \cdot (-4) + 31 \cdot 9.
\end{aligned}$$

Finally, from $7 = 3 \cdot 2 + 1$ we have

$$\begin{aligned}
1 &= 7 - 3 \cdot 2 \\
&= [69 \cdot 1 + 31 \cdot (-2)] - [69 \cdot (-4) + 31 \cdot 9] \cdot 2 \\
&= [69 \cdot 1 + 31 \cdot (-2)] + [69 \cdot (8) + 31 \cdot (-18)] \\
&= 69 \cdot 9 + 31 \cdot (-20).
\end{aligned}$$

Thus $u = 9$ and $v = -20$ work.

EXERCISES

5.29 Find u and v such that $7u + 3v = \gcd(7, 3)$.

5.30 Find u and v such that $84u + 351v = \gcd(84, 351)$.

5.31 Find the common divisors of 84 and 351.

5.32 Does 11 divide 23,000?

5.33 Consider the set of all numbers of the form $6u + 8v$. We know from Bézout that 2, the gcd of 6 and 8, is in this set. What else is in the set? (Try some values of u and v, and look for a pattern.)

5.34 Repeat the last problem for the set of all numbers of the form $9u + 12v$.

5.35 Repeat the last problem for the set of all numbers of the form $au + bv$, where a and b are any integers.

5.7 Mersenne and Fermat Primes

One of the oldest problems in number theory is to find primes. Is there a formula for primes? Or, failing that, is there a formula for some of the primes?

Mersenne Primes

Consider numbers of the form $2^p - 1$, where p is prime. The first few numbers of this form are

$$2^2 - 1 = 3$$
$$2^3 - 1 = 7$$
$$2^5 - 1 = 31.$$

Note that 3, 7, and 31 are all primes. When a number of the form $2^p - 1$ is a prime, it is called a *Mersenne prime*. (Mersenne was a 17th century mathematician who studied such primes.) Could this then be a formula for obtaining primes? Unfortunately no, since not all numbers of this form are in fact primes. (See the exercises.)

Still, some numbers of this form *are* prime, and some are still being discovered. In 1996 the Great Internet Mersenne Prime Search (GIMPS) was launched (see http://www.mersenne.org/prime.htm). This is a distributed computing project, a collaborative project in which thousands of users around the world run free software as "screen savers" during their computers' idle time. Since its start, this project has steadily, albeit slowly, been finding monster primes. The biggest at the time of this writing is $2^{43112609} - 1$, found in 2008. It has 12,978,189 digits.

Euclid knew about primes of this type, but they are named after a French friar, Marin Mersenne (1588–1648), who checked the primality of all the numbers of the form $2^p - 1$, up to $p = 257$. He was actually mistaken in five cases. It isn't known whether there are infinitely many Mersenne primes.

Mersenne primes have a strong connection with perfect numbers. A positive integer n is a *perfect number* if it is the sum of all its divisors, excluding n itself. For example, 6 is a perfect number, since $6 = 1 + 2 + 3$. Another example is 28, since $28 = 1 + 2 + 4 + 7 + 14$. Mathematicians have been interested in finding perfect numbers since the ancient Greeks. Perfect numbers are rare; the next one after 28 is 496. Euclid proved that if $2^p - 1$ is a Mersenne prime, then $2^{p-1}(2^p - 1)$ is a perfect number. So, for example, if $p = 3$, the Mersenne prime is $2^3 - 1 = 7$, and $2^{3-1}(2^3 - 1) = 28$ is perfect. Conversely, Leonhard Euler in the 18th century proved that every even perfect number is of the form $2^{p-1}(2^p - 1)$, where $2^p - 1$ is a Mersenne prime. What about odd perfect numbers? Nobody knows. No one has found one, but nobody has proven that they don't exist.

Fermat Primes

Another attempt to find a formula for primes was initiated by the great Pierre de Fermat in the 17th century. He noticed that all these numbers are prime.

$$2^{2^0} + 1 = 3$$
$$2^{2^1} + 1 = 5$$
$$2^{2^2} + 1 = 17$$
$$2^{2^3} + 1 = 257$$
$$2^{2^4} + 1 = 65{,}537$$

This led him to conjecture, in 1650, that all numbers of the form $2^{2^k} + 1$ were prime. Primes of this form are called *Fermat primes*.

Fermat primes have a surprising connection to a very old problem, a connection discovered by Carl Friedrich Gauss some 150 years or so after Fermat.

The problem is from the ancient Greek mathematicians. They were interested in which plane figures could be drawn using only a straightedge (to make straight lines) and a compass (to make circles). The question that Gauss answered was this: for which values of n can one construct a regular n-gon, using only compass and straightedge? (A regular n-gon is a symmetric polygon with n sides, for example, an equilateral triangle or a square.) The answer is that a regular n-gon can be so constructed if and only if n is either a power of 2 or $n = 2^m p_1 p_2 \cdots p_r$, where $m \geq 0$ and $p_1, p_2, \ldots, p_r$ are distinct Fermat primes.

How has Fermat's conjecture fared? He showed that $2^{2^k} + 1$ is prime for $k = 0$, 1, 2, 3, 4. He tried to prove that $2^{2^k} + 1$ is always prime, but failed. The reason for his failure was discovered by Leonhard Euler in 1732: $2^{2^5} + 1 = 4{,}294{,}967{,}297$ is not prime. Some dozens of other cases have been settled. As for Mersenne primes, there also is a distributed computing project aimed at Fermat primes. So far, the *only* primes of this form are the ones that Fermat knew.

Nobody yet has been able to find a formula that generates an infinite number of primes and no composite numbers.

EXERCISES

5.36   Find the first five Mersenne primes and their associated perfect numbers.

5.37   What is the first number of the form $2^p - 1$, with p a prime, that is *not* a prime?

5.38   Use Euclid's theorem on perfect numbers to find a perfect number with four digits.

5.39   What is the biggest Mersenne prime known? (Check GIMPS.)

5.40   What is the largest prime of any type currently known? (See, for example, `http://primes.utm.edu/`.)

5.41   Show that $641 \mid (2^{2^5} + 1)$.

5.42   Has anyone found a Fermat prime that Fermat didn't already know? (Check the Web.)

5.43   Is a regular 100-gon constructible using only straightedge and compass? What about a regular 101-gon? A regular 102-gon?

5.8   The Fundamental Theorem of Arithmetic

The Theorem: Prime Factorization

The Fundamental Theorem of Arithmetic starts with an observation we recorded in Section 5.5—that every integer greater than 1 is divisible by a prime. So let n be an integer, with $n > 1$. Then there is some prime p such that $p \mid n$. Now look at n/p. It is possible that this is 1, so that $n = p$. If not, we can find an integer q such that q divides n/p. We can then repeat this process on $n/(pq)$; if $n/(pq) = 1$ then $n = pq$, else there is a prime r such that r divides $n/(pq)$, and so on. Eventually we will have expressed n as the product of primes.

The Fundamental Theorem of Arithmetic. Let n be any integer greater than 1. Then n can be expressed as a product of primes (with perhaps only one prime). Furthermore, this expression is unique, up to the order of the primes.

The product from the theorem is called the *prime factorization* of n.

The history of the Fundamental Theorem of Arithmetic is somewhat murky. Elements of the theorem can be found in the works of Euclid (c. 330–270 BCE), the Persian Kamal al-Din al-Farisi (1267–1319 CE), and others, but the first time it was clearly stated in its entirety, and proved, was in 1801 by Carl Friedrich Gauss (1777–1855). A formal proof is beyond the scope of this text.

◼ EXAMPLE 5.16

Find the prime factorization of 24.

Solution: First, we notice that $2 \mid 24$, so divide 24 by 2:

$$24 = 2 \cdot 12.$$

Next, we need to look at 12. Since 2 also divides 12,

$$24 = 2 \cdot (2 \cdot 6) = 2^2 \cdot 6.$$

Again, $2 \mid 6$, so

$$24 = 2^2 \cdot (2 \cdot 3) = 2^3 \cdot 3.$$

Now 3 is prime, so we are done. The prime factorization is $24 = 2^3 \cdot 3$. ■

■ EXAMPLE 5.17

Factor (i.e., find the prime factorization of) 1001.

Solution: It may not be obvious what, if any, prime divides 1001. So let us start with the smallest, then systematically work our way up. Since 1001 ends in an odd digit, 2 is not a divisor. The next prime is 3. Here is a trick for determining divisibility by 3; if 3 divides a number, it must divide the sum of the digits of that number. Here the sum of the digits is $1 + 0 + 0 + 1 = 2$, which 3 does not divide, so 3 is not a divisor of 1001. Next is 5, which does not divide 1001, since 1001 does not end in 0 or 5. The next prime is 7. We are in luck here: $1001 = 7 \cdot 143$. (Check!). So now we need only work with 143.

To factor 143, we do not need to check the primes 2, 3, and 5 again. If any of them divided 143, then they would divide its multiple 1001, which we know is not the case. So let's start with 7. You can check that $7 \nmid 143$. So we go to the next prime, 11. Here is success: $143 = 11 \cdot 13$. Finally, 13 is itself prime, so we have our prime factorization:
$$1001 = 7 \cdot 11 \cdot 13.$$

■

■ EXAMPLE 5.18

Factor 101.

Solution: Again, we start with the smallest prime and work our way up. You can check that none of 2, 3, 5, or 7 divides 101. Fortunately, we do not have to check any more primes. Why not? The next prime to check is 11. Suppose that $11 \mid 101$, say $101 = 11x$. We know that x cannot be less than 11, since we have checked all the primes less than 11. So $x \geq 11$. But then $101 = 11x \geq 11 \cdot 11 = 121$, but of course 101 is less than 121. By a similar argument, we know that no prime greater than 11 divides 101. We are done; 101 is prime.

The general principle is this: in checking primes for divisors, we need only check primes up to the square root of our number. To see this, suppose that we are factoring n, and checking p, having already verified that no number less than p divides n. If $p \mid n$, then $n = px$, for some integer x. Since we have checked all smaller numbers, we know that $x \geq p$. Then $n = px \geq p^2$, so $\sqrt{n} \geq p$, i.e., p is no larger than the square root of n. ■

If you find factoring numbers difficult, you are not alone. Computers find it hard as well, much harder than multiplying numbers. In fact, one of the most famous and widely used methods of encrypting data, called *RSA encryption* (for inventors

Ronald L. Rivest, Adi Shamir, and Leonard Adleman), is based on the difficulty of factorization.

Using Prime Factorization

You may not be surprised to learn that something called the Fundamental Theorem of Arithmetic is useful. There are many questions about an integer that can be reduced to questions about its prime factors. We will look at three applications of the theorem.

For our first application, we will look at the divisors of a number. Suppose that n is some positive integer, with prime factorization $n = p_1 p_2^2$ and k is a positive divisor of n. From the uniqueness of the prime factorization, we know that the only primes that can divide k are p_1 and p_2. So we can write $k = p_1^s p_2^t$ where either of s or t can be 0. (Recall that $x^0 = 1$ for any $x \neq 0$, so if s, say, is 0, that means that p_1 is not a factor of k.) Furthermore, p_1^2 doesn't divide n, so $s \leq 1$. Similarly, $t \leq 2$. Therefore, the complete list of divisors of n is

$$p_1^0 p_2^0, \; p_1^0 p_2^1, \; p_1^0 p_2^2, \; p_1^1 p_2^0, \; p_1^1 p_2^1, \; p_1^1 p_2^2$$

or, rewritten,

$$1, p_2, \; p_2^2, \; p_1, \; p_1 p_2, \; p_1 p_2^2.$$

Prime Factorization and Divisors.   Let n be a positive integer, with prime factorization $n = p_1^{r_1} p_2^{r_2} \cdots p_m^{r_m}$. Then the positive divisors of n are the integers $p_1^{s_1} p_2^{s_2} \cdots p_m^{s_m}$ with $0 \leq s_i \leq r_i$, for $i = 1, 2, \ldots, m$.

We can use this to understand more about greatest common divisors. Let's do an example.

◼ EXAMPLE 5.19

Find $\gcd(136, 52)$.

Solution: First, you can check that the prime factorizations of our numbers are

$$136 = 2^3 \cdot 17$$
$$52 = 2^2 \cdot 13.$$

It is useful to rewrite this so that both numbers have the same list of primes:

$$136 = 2^3 \cdot 13^0 \cdot 17^1$$
$$52 = 2^2 \cdot 13^1 \cdot 17^0,$$

where again we have used $p^0 = 1$ for any p. Now, by our result above, the divisors of 136 are

$$2^{s_1} \cdot 13^{s_2} \cdot 17^{s_3},$$

where s_1 is between 0 and 3, s_2 is 0, and s_3 is 0 or 1.

Similarly, the divisors of 52 are

$$2^{t_1} \cdot 13^{t_2} \cdot 17^{t_3},$$

where t_1 is between 0 and 2, t_2 is 0 or 1, and t_3 is 0.

A common divisor has to satisfy both of these, so the largest one is

$$2^2 \cdot 13^0 \cdot 17^0 = 4.$$

∎

Here is the general rule.

Prime Factorization and Greatest Common Divisors.   Let a and b be integers greater than 1. Let their prime factorizations be

$$a = p_1^{f_1} p_2^{f_2} \cdots p_r^{f_r}$$
$$b = p_1^{g_1} p_2^{g_2} \cdots p_r^{g_r},$$

where some of the f_i or g_i may be 0. Then the greatest common divisor of a and b is given by

$$\gcd(a, b) = p_1^{h_1} p_2^{h_2} \cdots p_r^{h_r}$$

where h_i is the minimum of f_i and g_i, for $i = 1, 2, \ldots, r$.

If you find this notation confusing, here is another way of thinking about it. To find the gcd of a and b, first find their prime factorizations. Then the prime factorization of their gcd contains just the primes that divide both a and b, and the power of each such prime is the lesser of the powers of that prime in the factorizations of a and b.

▣ EXAMPLE 5.20

Find the greatest common divisor of 990 and 4725.

Solution: First we factor the two numbers.

$$990 = 2 \cdot 3^2 \cdot 5 \cdot 11 = 2^1 \cdot 3^2 \cdot 5^1 \cdot 7^0 \cdot 11^1$$
$$4725 = 3^3 \cdot 5^2 \cdot 7 = 2^0 \cdot 3^3 \cdot 5^2 \cdot 7^1 \cdot 11^0$$

Then we read off the gcd, taking the minimum power of each factor.

$$\gcd(990, 4725) = 2^0 \cdot 3^2 \cdot 5^1 \cdot 7^0 \cdot 11^0 = 45$$

∎

Recall that the least common multiple of integers a and b, written $\operatorname{lcm}(a, b)$, is the smallest positive integer d such that $a \mid d$ and $b \mid d$. The Fundamental Theorem of Arithmetic can also be used to understand least common multiples.

Prime Factorization and Least Common Multiples.   Let a and b be integers greater than 1. Let their prime factorizations be

$$a = p_1^{f_1} p_2^{f_2} \cdots p_r^{f_r}$$
$$b = p_1^{g_1} p_2^{g_2} \cdots p_r^{g_r},$$

where some of the f_i or g_i may be 0. Then the least common multiple of a and b is given by

$$\text{lcm}(a, b) = p_1^{h_1} p_2^{h_2} \cdots p_r^{h_r}$$

where h_i is the maximum of f_i and g_i, for $i = 1, 2, \ldots, r$.

■ EXAMPLE 5.21

Find the least common multiple of 990 and 4725.

Solution: This is easy. Using the factorizations from the last example, we take maximum powers, instead of the minimum. The least common multiple is

$$2^1 \cdot 3^3 \cdot 5^2 \cdot 7^1 \cdot 11^1 = 103{,}950.$$

■

EXERCISES

5.44   Find the prime factorization.
  a) 16
  b) 59
  c) 72
  d) 105
  e) 1633 (Hint: $3 \cdot 1633 = 4899 = 70^2 - 1$.)

5.45   List all the positive divisors of 675.

5.46   List all the divisors of 675.

5.47   How many positive divisors does 1,000 have?

5.48   Find the gcd of $2^2 \cdot 3^3 \cdot 11^2 \cdot 17$ and $2 \cdot 3^4 \cdot 17^6 \cdot 71$.

5.49   Find the gcd and lcm of 198 and 429.

5.50   Find the gcd and lcm of the *three* numbers 198, 429, and 156.

5.51   Express 1,000,000,000 as the product of two integers, neither of which contains a 0 digit.

5.52   A: How many kids do you have?

B: I have three.

A: What are their ages?

B: The product of their ages is 36. The sum of their ages is my house number, which of course you know.

A: Of course. But I need more information.

B: The oldest has red hair.

A: Thank you, now I know.

What are the ages of the kids?

5.9 Diophantine Equations

Definitions and Examples

An enclosure at the zoo holds ostriches and opossums. Between them, they have 30 eyes and 44 feet. How many of each type of animal is there?

At two eyes per animal, we have a total of $30/2 = 15$ animals. If all the possums were to lift two feet off the ground, so that every animal would have two feet on the ground, we would have $2 \cdot 15 = 30$ feet left on the ground. So $44 - 30 = 14$ feet were lifted. At 2 feet per opossum, there must be $14/2 = 7$ opossums. That leaves $15 - 7 = 8$ ostriches.

We can also solve this problem by writing down and solving equations. Let x be the number of ostriches and y the number of opossums. That there are 44 feet can then be written $2x + 4y = 44$. Let us make the problem harder here, and forget that there are 30 eyes, and just consider this equation. What are the possible solutions? In fact, there are an infinite number of solutions. For any number x, we have a solution where $y = (44 - 2x)/4$.

But if we think about the origin of this equation, many of the above solutions don't work. We can't, for example, have $x = 21$ and $y = 1/2$, without being cruel to an opossum. Letting $x = -2$ and $y = 12$ doesn't work either. We must, under the circumstances, insist that both x and y be nonnegative integers. You can check that, with this restriction, there are exactly twelve solutions.

In this section and the next, we will consider equations whose solutions are restricted in a particular way. A *Diophantine equation* is an equation whose solutions are required to be integers. They are named in honor of the Greek mathematician Diophantus, who lived in Alexandria in the 3rd century CE.

We are already familiar with some types of Diophantine equations. Bézout's Identity concerns the Diophantine equation $ax + by = \gcd(a, b)$. Another famous Diophantine equation comes from the Pythagorean Theorem: $x^2 + y^2 = z^2$. For example, one solution is $x = 3$, $y = 4$, $z = 5$, since $3^2 + 4^2 = 5^2$. This type of Diophantine equation had already been studied by the time of Euclid.

We will now consider the simplest type of Diophantine equation with two variables: linear equations.

Problem: Let a, b, and c be integers, with a and b nonzero. Find all pairs of integers (x, y) such that $ax + by = c$.

Some Geometry

Let us consider the Diophantine equation $2x + 3y = 17$. If we forget for a moment that our solutions must be integers, we can think of this as a regular equation and graph it. It is a line (Figure 5.3).

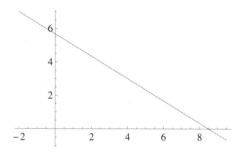

Figure 5.3   The graph of $2x + 3y = 17$.

What *is* the graph of this equation? It is the set of ordered pairs (x, y) such that $2x + 3y = 17$. So the solution to our problem consists of points on this line, but not just any points, since x and y have to be integers. If we mark all points (x, y) where both x and y are integers, the picture looks like Figure 5.4.

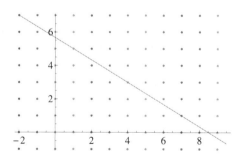

Figure 5.4   The graph of $2x + 3y = 17$ and the integer lattice.

The set of integer points in the plane is called the *integer lattice*. Solving our Diophantine equation means finding all the points that are on both the line and the integer lattice.

Now let's go back to the line, and put a point (x_0, y_0) on it (Figure 5.5).

Suppose that (x_0, y_0) is a solution to our Diophantine equation. The geometry of the line suggests another solution. Recall that the slope of the line $2x + 3y = 17$ can

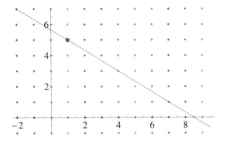

Figure 5.5   The graph of $2x + 3y = 17$ with one solution.

be obtained by putting the equation into slope-intercept form: $y = (-2/3)x + 17/3$, so the slope is $-2/3$. This means that, if we start from a point on the line, go 3 units to the right, and down 2, we get another point on the line (Figure 5.6).

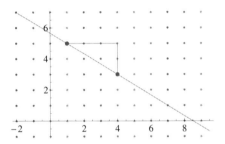

Figure 5.6   The graph of $2x + 3y = 17$ with two solutions.

To check this, you can plug $(x_0 + 3, y_0 - 2)$ into the equation and see that it works. And of course both $x_0 + 3$ and $y_0 - 2$ are also integers, so are solutions also of our Diophantine equation. In the same way, we can find more solutions (Figure 5.7).

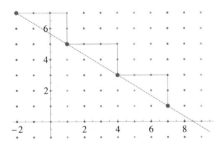

Figure 5.7   The graph of $2x + 3y = 17$ with multiple solutions.

We summarize this in the following theorem.

Theorem. Let (x_0, y_0) be a solution to the Diophantine equation $ax + by = c$, and t be any integer. Then $(x_0 + bt, y_0 - at)$ is also a solution.

Proof. We are given that $ax_0 + by_0 = c$. Using this,

$$a(x_0 + bt) + b(y_0 - at) = ax_0 + abt + by_0 - abt$$
$$= ax_0 + by_0$$
$$= c,$$

so $(x_0 + bt, y_0 - at)$ solves the equation. It is easy to see that $x_0 + bt$ and $y_0 - at$ are integers, so the theorem is proven. ∎

▣ EXAMPLE 5.22

Find four solutions to $5x + 6y = 17$.

Solution: By trial and error, we find that $(x, y) = (1, 2)$ works. Letting $t = -1, 1, 2$, we get three more solutions.

If $t = -1$, then we get the solution $(-5, 7)$.

If $t = 1$, then we get the solution $(7, -3)$.

If $t = 2$, then we get the solution $(13, -8)$.

■

EXERCISES

5.53 Find the 12 solutions to $2x + 4y = 44$, where x and y are nonnegative integers.

5.54 For each of the following Diophantine equations, find, if possible, three solutions.

 a) $4x + 6y = 26$
 b) $2x + 4y = 3$
 c) $6x - 3y = 15$

5.55 Alice bought a number of $3 pens and $7 notebooks. If she spent $37 altogether, what are the possibilities for the numbers of pens and notebooks?

5.56 A certain forest is made up of trees planted in a square pattern, with one foot between adjacent pairs of trees. A marksman is challenged to stand by one of the trees and shoot his rifle in such a way so as to miss every tree. It takes an exact aim to hit a tree. (The trees are very thin. In fact, the thickness of each trunk is so small that it can't be measured.)

The marksman, who is also a mathematician, thinks to himself: "Let this tree I'm standing by be the origin of a coordinate system. Then the other trees are located exactly at the points of the integer lattice. If I shoot my rifle, the bullet follows a straight line. Since I am at the origin, that line will be $y = mx$ for some slope m. All I have to do is pick m so that the line doesn't pass through any lattice point."

How can he pick such an m?

5.10   Linear Diophantine Equations

We will again address the problem we introduced in the last section.

Problem: Let a, b, and c be integers, with a and b nonzero. Find all pairs of integers (x, y) such that $ax + by = c$.

When Are There Any Solutions?

Consider the Diophantine equation $2x+6y = 17$. Suppose that (x_0, y_0) is a solution. Then $2x_0$ is even (divisible by 2). But, since 6 is even, so is $6y_0$. Therefore, $2x_0+6y_0$ is even. But 17 is odd, so $2x_0 + 6y_0$ cannot equal 17. In other words, there are *no* solutions to $2x + 6y = 17$. What makes this argument work? The fact that $2\,|\,2$ and $2\,|\,6$, but $2 \nmid 17$. This observation is the basis of the following theorem.

Theorem. The Diophantine equation $ax + by = c$ has a solution if, and only if, $\gcd(a, b)$ divides c.

Proof. Let $d = \gcd(a, b)$, and (x_0, y_0) be a solution of $ax + by = c$. Since $d\,|\,a$ and $d\,|\,b$, we have $d\,|\,ax_0$ and $d\,|\,by_0$. Hence $d\,|\,(ax_0 + by_0)$, so $d\,|\,c$. This gives us half of the theorem (only if).

Now suppose that $d\,|\,c$, say $c = de$. We need to show that $ax + by = c$ has a solution. We can use Bézout's Identity, which tells us that there are integers u and v such that $au + bv = d$. Multiply by e:

$$(au + bv)e = de = c.$$

So

$$a(ue) + b(ve) = (au + bv)e = c.$$

Thus (ue, ve) is a solution of the equation.   ∎

▣ EXAMPLE 5.23

Which of the following Diophantine equations are solvable?

(a) $14x + 35y = 119$

(b) $24x - 32y = 106$

(c) $7x - 17y = 10004$

Solution: (a) You can verify that $\gcd(14, 35) = 7$, and $7 \mid 119$ (since $119 = 7 \cdot 17$), so $14x + 35y = 119$ is solvable.

(b) The gcd of 24 and -32 is 8, but $8 \nmid 106$, so $24x - 32y = 106$ is not solvable.

(c) The gcd of 7 and -17 is 1, which divides 10004 since it divides everything. So $7x - 17y = 10004$ has solutions. ∎

We now know exactly which linear Diophantine equations have solutions. Next, we will see what the sets of solutions look like.

Complete Sets of Solutions

Let us look a bit closer at the first equation in the last example: $14x + 35y = 119$. We saw that 14, 35, and 119 are all divisible by 7, so let's divide the whole equation by 7, to get $2x + 5y = 17$. From elementary algebra, we know that the solutions of this new equation are *exactly the same* as the solutions to the old one. (If this doesn't sound familiar, take a moment to convince yourself—it isn't hard.) But the new equation is simpler. You can easily solve it. Try!

In general, suppose that $ax + by = c$ is a solvable Diophantine equation, i.e., that $\gcd(a, b)$ divides c. If $\gcd(a, b) > 1$, we can divide $ax + by = c$ by $\gcd(a, b)$ to get a new equation $a'x + b'y = c'$, which has the same solutions as the original one. The new equation is, however, simpler. In particular, a' and b' are relatively prime, since we have divided out the common factors.

So we have reduced the problem of finding solutions of linear Diophantine equations to that of finding solutions to equations where $\gcd(a, b) = 1$. The following theorem addresses that important case.

Theorem. Let $\gcd(a, b) = 1$ and (x_0, y_0) be a solution of $ax + by = c$. Then every solution is of the form $(x_0 + bt, y_0 - at)$ for some integer t.

Proof. Let (r, s) be a solution of our equation. We need to show that there is some integer t such that $r = x_0 + bt$ and $s = y_0 - at$. Let $t = (y_0 - s)/a$. This gives us $s = y_0 - at$. We need to show that t is an integer, and that $r = x_0 + bt$.

First we show that t is an integer, in other words, that a divides $y_0 - s$. Because (x_0, y_0) and (r, s) are solutions, we know the following.

$$ax_0 + by_0 = c \tag{5.1}$$

$$ar + bs = c \tag{5.2}$$

Subtracting equation (5.2) from equation (5.1) yields

$$(ax_0 + by_0) - (ar + bs) = c - c - 0.$$

Rearranging terms,

$$a(x_0 - r) + b(y_0 - s) = 0. \tag{5.3}$$

Now $a \mid a(x_0 - r)$ and $a \mid 0$, so $a \mid b(y_0 - s)$. We can apply a result from Section 5.6: since a divides the product $b(y_0 - s)$, and $\gcd(a, b) = 1$, it follows that a divides $y_0 - s$. This is what we needed for t to be an integer.

Finally, we show that $r = x_0 + bt$. From the definition of t, we have $y_0 - s = at$. Substituting at into equation (5.3) for $y_0 - s$ gives

$$a(x_0 - r) + b(at) = 0.$$

Since $a \neq 0$, we can divide this equation to get $x_0 - r + bt = 0$, which yields $r = x_0 + bt$. ∎

Putting this theorem together with the one from the last section gives us a satisfying result.

Theorem. Let $\gcd(a, b) = 1$ and (x_0, y_0) be any solution of the Diophantine equation $ax + by = c$. Then a pair of integers is a solution if, and only if, it is of the form $(x_0 + bt, y_0 - at)$ for some integer t.

◼ **EXAMPLE 5.24**

Find all solutions of the Diophantine equation $22x + 143y = -66$.

Solution: First we note that $\gcd(22, 143) = 11$. So divide the equation by 11 to get the equivalent $2x + 13y = -6$. One solution of this is $(-3, 0)$, so the solutions are pairs of the form $(-3 + 13t, \; 0 - 2t)$, or $(-3 + 13t, \; -2t)$.

It is always a good idea to check your answer. In this case it is easy; substitute $x = -3 + 13t$ and $y = -2t$ into the original equation.

$$22(-3 + 13t) + 143(-2t) = -66 + 286t - 286t = -66$$

◼

Computing Solutions

In the last section, we saw how to find entire solution sets to an equation, *provided* we had the first solution. How do we find that first solution? If the equation is simple, it may be easiest to eyeball a solution. For harder equations, or if we want to tell computers how to find solutions, we need a more organized way. There is a way, using Bézout's Identity.

We start with the theorem that we mentioned earlier: The equation $ax + by = c$ has a solution if, and only if, $\gcd(a, b)$ divides c. Suppose that we have such an equation, and let us call $\gcd(a, b) = d$. Then d divides c, say $c = ed$. Now we know from Bézout that we can find integers u and v such that $au + bv = d$. Then

$$a(eu) + b(ev) = e(au + bv) = ed = c.$$

In other words, (eu, ev) is a solution to our equation!

How to solve the linear Diophantine equation $ax + by = c$.

i. Run the Euclidean Algorithm to find $\gcd(a, b)$, say d.

ii. Find the pair (u, v) of Bézout's Identity.

iii. Multiply (u, v) by c/d to get a single solution of the original equation.

iv. Divide the equation by d to get an equivalent equation (one with the same solutions), $a'x + b'y = c'$, with $\gcd(a', b') = 1$.

v. Use the solution of (iii) and the last theorem to get the entire solution set.

■ EXAMPLE 5.25

Solve $145x + 160y = 35$.

Solution: (i) First we run the Euclidean Algorithm to find $\gcd(145, 160)$.

$$160 = 145 \cdot 1 + 15$$
$$145 = 15 \cdot 9 + 10$$
$$15 = 10 \cdot 1 + 5$$
$$10 = 5 \cdot 2 + 0$$

So $\gcd(145, 160) = 5$.

(ii) Next, we compute the integers u and v of Bézout's Identity. Proceeding as in Section 5.6, we express each of the remainders r_i above in the form $r_i = 145x_i + 160y_i$. First, express 15 that way,

$$15 = 145 \cdot (-1) + 160 \cdot (1),$$

then 10,

$$10 = 145 - 15 \cdot 9$$
$$= 145 - [145 \cdot (-1) + 160 \cdot (1)] \cdot 9$$
$$= +145 \cdot 10 + 160 \cdot (-9),$$

and finally 5:

$$5 = 15 - 10 \cdot 1$$
$$= [145 \cdot (-1) + 160 \cdot (1)] - [145 \cdot 10 + 160 \cdot (-9)] \cdot 1$$
$$= 145 \cdot (-11) + 160 \cdot 10.$$

Thus $u = -11$ and $v = 10$ work.

(iii) Since $35 = 7 \cdot 5$, we can multiply (u, v) by 7 to get our solution. In other words, multiply $(-11, 10)$ by 7, to get $(-77, 70)$. As always, we check our answer:

$$145(-77) + 160(70) = -11165 + 11200 = 35.$$

(iv) Divide our equation by 5 (the gcd) to get $29x + 32y = 7$.

(v) Finally, using the solution in (iii) and the equation in (iv), our entire set of solutions is

$$(-77 + 32t, \ 70 - 29t).$$

■

EXERCISES

5.57 Find all solutions to the following Diophantine equations.
  a) $3x + 4y = 5$
  b) $3x - 4y = 5$
  c) $10x + 22y = 15$
  d) $209x + 143y = 176$
  e) $209x + 143y = 99$

5.58 Cuthbert purchased a number of bottles of soft drinks, namely, Passionate Peach at \$2.40 per bottle, and Marvelous Mango at \$2.10 a bottle. He spent \$46.50 all together. How much of each type could he have bought?

5.59 A simplified version of American football allows teams to score 7 points on a touchdown, or 3 points on a field goal. What scores are possible?

5.11 Pythagorean Triples

Recall the Pythagorean Theorem: If a right triangle has legs of lengths a and b, and hypotenuse of length c, then $a^2 + b^2 = c^2$. As far back as the old Babylonian period, almost 4000 years ago, mathematicians were looking for triples of *integers* that satisfied this equation. In other words, they were looking at the Diophantine equation $x^2 + y^2 = z^2$. Solutions to the Diophantine equation are called *Pythagorean triples*. A couple of examples are the 3–4–5 triangle and the 5–12–13 triangle. You can check any of these by summing squares. For example,

$$5^2 + 12^2 = 25 + 144 = 169 = 13^2.$$

Is there a formula to give all Pythagorean triples?

Given any Pythagorean triple, we can multiply all three numbers by any positive integer and obtain another Pythagorean triple. For example, the 6–8–10 triangle is created by doubling the 3–4–5 triangle. This amounts to enlarging the triangle but keeping its shape. We say that a Pythagorean triple is *primitive* if the three side lengths have no common factor.

There is a simple method, called *Euclid's formula*, for generating all primitive Pythagorean triples.

Euclid's Formula. All primitive Pythagorean triples are given by $a = m^2 - n^2$, $b = 2mn$, and $c = m^2 + n^2$, where m and n are positive integers with no common factor, $m > n$, and one of m and n is even and the other is odd.

Let's do an example of the formula. Let $m = 5$ and $n = 2$. These numbers have no common factor, and one is odd and the other even. According to Euclid's formula, we obtain the Pythagorean triple 21, 20, and 29. Let's check:

$$21^2 + 20^2 = 441 + 400 = 841 = 29^2.$$

And it is easy to check that the numbers 21, 20, and 29 have no common factor.

Proof. It is a simple matter to prove that Euclid's formula always gives us a Pythagorean triple. Just sum the squares:

$$
\begin{aligned}
a^2 + b^2 &= (m^2 - n^2)^2 + (2mn)^2 \\
&= m^4 - 2m^2n^2 + n^4 + 4m^2n^2 \\
&= m^4 + 2m^2n^2 + n^4 \\
&= (m^2 + n^2)^2 \\
&= c^2.
\end{aligned}
$$

Now we will show that the triple a, b, c is primitive (the numbers have no common factor). We will give a proof by contradiction, assuming that our result is false and showing that this leads to a contradiction.

Suppose that a, b, and c have a common factor. Then c has a common factor with a. Such a factor may be a prime number or it may be a composite number (a product of primes). However, if the common factor is a composite number, then that number's prime factors also divide a and c. Hence, we may assume that a and c have a prime factor in common.

Let's call this common prime factor p. Notice that p cannot equal 2, since c is an odd number (why?). We will show that p is also a prime common factor of m and n, which is impossible because m and n have no common factor; this contradiction will show that the assumption that a, b, and c have a common factor is false.

Since p divides $m^2 + n^2$ and $m^2 - n^2$, it follows that p divides their sum, $2m^2$, and their difference, $2n^2$. Thus p divides m^2, and therefore p divides m. By the same reasoning, p divides n.

This is what we wanted to show. Thus, p is a common factor of m and n, which isn't possible (since m and n have no common factor). Therefore, our original assumption that a, b, and c have a common factor is false, and we conclude that (a, b, c) is a *primitive* Pythagorean triple.

We have proven that Euclid's formula always generates a primitive Pythagorean triple, but we haven't proven that all primitive Pythagorean triples can be obtained in this way. We now do this. Suppose that $a^2 + b^2 = c^2$, where a, b, and c are positive integers with no common factor.

First we note that a and b cannot both be even, since a, b, and c do not share the common factor 2. Suppose that a and b are both odd, say $a = 2k + 1$ and $b = 2l + 1$. Then

$$c^2 = a^2 + b^2$$
$$= (2k + 1)^2 + (2l + 1)^2$$
$$= (4k^2 + 4k + 1) + (4l^2 + 4l + 1)$$
$$= 4(k^2 + l^2 + k + l) + 2.$$

Note that $4(k^2 + l^2 + k + l) + 2$ is even, but not divisible by 4 (else 2 would be divisible by 4). This is a problem, however, since this number is also c^2, and every even square must be divisible by 4. We conclude, then, that a and b cannot both be odd. Since we already showed that they could not both be even, one must be even, and one odd.

Suppose, without loss of generality, that a is even and b is odd. Then we have

$$a^2 = c^2 - b^2 = (c + b)(c - b).$$

Since b and c are odd (check!), both $c + b$ and $c - b$ are even, so we can write

$$\left(\frac{a}{2}\right)^2 = \left(\frac{c + b}{2}\right)\left(\frac{c - b}{2}\right),$$

where the terms in parentheses are integers.

Next, we see that $(c + b)/2$ and $(c - b)/2$ are relatively prime. Suppose that some number d divides both. Then d divides $(c + b)/2 + (c - b)/2 = c$ and d divides $(c + b)/2 - (c - b)/2 = b$, so that d would be a common factor of a, b, and c, which we assumed cannot happen.

So we have a product of two relatively prime integers equal to a perfect square, $(a/2)^2$. It follows that both integers, that is, $(c + b)/2$ and $(c - b)/2$ are perfect squares (why?). Let's set

$$\frac{c + b}{2} = m^2, \quad \frac{c - b}{2} = n^2.$$

Solving for c and b, and then a, we have

$$a = 2mn, \quad b = m^2 - n^2, \quad c = m^2 + n^2.$$

Clearly, $m > n$. And m and n can have no common factor, or else a, b, and c would have a common factor (and we have assumed that they do not). Finally, we note that one of m, n is even and the other odd. (Details of this are left to the exercises.) This establishes the formula for Pythagorean triples. ∎

EXERCISES

5.60 Use Euclid's formula to find the primitive Pythagorean triangle corresponding to the choice $m = 10$, $n = 7$.

5.61 What choice of m and n in Euclid's formula gives the 3–4–5 right triangle?

5.62 Find all the primitive Pythagorean triples where the hypotenuse is at most 100.

5.63 Show that, if we define m and n as in the proof,

$$\frac{c + b}{2} = m^2, \quad \frac{c - b}{2} = n^2,$$

one of them is even and the other is odd. (Hint: since b and c are both odd, we can write $b = 2k + 1$ and $c = 2l + 1$ for integers k and l. See how the parity of m and n depends on the parity of k and l.)

5.12   An Introduction to Modular Arithmetic

Odds and Evens

The notion of the *parity* of a number, whether it is odd or even, is quite useful.

A Magic Trick. The magician takes a half-dozen or so coins, throws them on a table. She asks for a volunteer. Then she turns her back on the table and asks the volunteer to turn over two of the coins, any two. Then to turn two more, then two more. Then the volunteer should hide one coin with his hand. The magician turns back around, looks at the table, and announces whether the hidden coin is heads or tails. How does she do it?

The magician counts the number of heads before she looks away, and remembers whether it is odd or even. When the volunteer turns two coins, one of three things can happen. The two coins are both heads, in which case the number of heads increases by two; both tails, in which case the number of heads decreases by two; or one heads and one tails, in which case the number of heads stays the same. In all three cases, the parity of the number of heads is unchanged. So at the end, the magician still knows whether the number of heads is odd or even. She can then tell whether the hidden coin is heads or tails, to make the parity of total number of heads right.

Next is an 18th century classic.

The Bridges of Königsberg. In 1735 Leonhard Euler came across a puzzle based on the layout of the city of Königsberg in Prussia (now called Kaliningrad, in Russia). The city was located on the Pregel River, which had two islands. The two sides of the river and the islands were connected by bridges as in Figure 5.8.

The puzzle was this: design a tour, starting anywhere, that takes you across each bridge once and only once, and ends at the starting point.

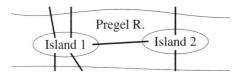

Figure 5.8 The seven bridges of Königsberg.

Euler, being a mathematician, abstracted out the important features of the problem, ignoring others (such as the lengths of the bridges) that wouldn't affect any solution. Using modern terms, he defined a *graph*, pictured in Figure 5.9. Each land mass—the two islands and the two sides of the river—is represented by a dot, called a *vertex* (plural vertices). Each bridge is represented by a line, called an *edge*, connecting its two land masses.

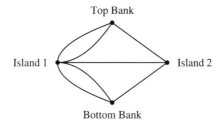

Figure 5.9 The bridges of Königsberg graph.

The puzzle then becomes this: starting at any vertex, traverse each edge once and only once, ending at the starting point. Of course, Euler didn't just rephrase the problem, but solved it. In fact, he solved a more general problem, for any graph. He argued as follows. Suppose we start at some vertex A. When we leave A, we cross one of its edges. The first time we return to A, we have used two of its edges. If we return again, we will have used two more edges, for a total of four. In general, each time we return, we have used two more edges, so the total number of its edges used is even. Since we have to return at the end, we will then have used an even number of edges. But each vertex in our graph has an *odd* number of edges, so this is not possible. The puzzle has no solution.

Euler's solution was even more general. Consider a vertex other than the starting vertex. Each time we enter and leave it, we use two edges. So at this vertex as well, we must use a total number of edges that is even. In other words, for any graph, in order for such a tour to exist, *every* vertex must have an even number of edges.

Euler published his solution in 1735. The general theorem is this: A graph has a solution if, and only if, every vertex has an even number of edges, and the graph is connected (you can get from any vertex to any other). Such a solution is now called an *Eulerian path*, in his honor. In fact, this theorem is considered the first in the area of *graph theory*, which has become a large and important topic of mathematical research and application in the last 50 years.

The Envelope Puzzle. Here is another classic puzzle, which you may have seen before. Starting anywhere, trace the shape in Figure 5.10, without repeating any line.

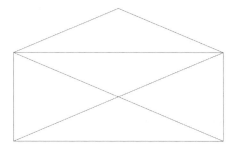

Figure 5.10   The envelope puzzle.

Do you see the connection to the last problem? Consider the graph in Figure 5.11.

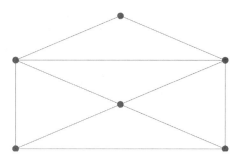

Figure 5.11   The envelope graph.

This problem is a variation of the last one. We need to traverse every edge exactly once, but we aren't obligated to start and end at the same vertex. By the argument we used above, every vertex *except* maybe the starting and ending vertices, must have an even number of edges. If you check the graph, you can see that exactly two vertices have an odd number of edges, the two at the bottom. So these two must be our starting and ending points. A solution is shown in Figure 5.12. The edges are traversed in the order given.

Opening and Closing Lockers. An army base has a cavernous room with a row of 1000 lockers, numbered 1 through 1000. A soldier walks down the row, opening every locker. A second soldier comes by and closes every second locker. Another either opens or closes every third locker, depending on whether it was closed or open before, and so on. The 1000th soldier either opens or closes the 1000th locker. Which lockers are now open?

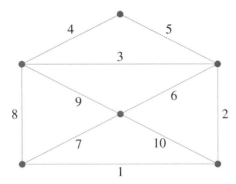

Figure 5.12   A solution to the envelope puzzle.

It is easy to figure out which of the lockers 1–10 are open at the end. Only the first soldier touches the first locker, so it is open. The second locker is opened by the first soldier, closed by the second, and bypassed by the others, so it is closed. After you figure out the first 10 lockers, you may notice a pattern.

How do we solve this problem? (If you enjoy puzzles, you may want to try this one before reading our solution.) The first question we need to answer is this: how do we tell if soldier m changes (opens or closes) locker n? Notice that soldier 2 changes lockers 2, 4, 6, 8, ... Soldier 3 changes lockers 3, 6, 9, ... In general, soldier m changes locker n if m divides n. In other words, the number of times a locker is changed is the number of divisors it has.

A locker ends up closed if it is changed an even number of times, open if it is changed an odd number of times. So, locker n is open at the end if n has an odd number of divisors. Therefore, we can answer the question if we can tell which numbers have an odd number of divisors.

Consider the divisors of 12, for example: they are 1, 2, 3, 4, 6, and 12. Another way to list them: 1 and 12, 2 and 6, 3 and 4. The divisors are paired off, each pair multiplying to 12. Because of this pairing, it is easy to see that the number of divisors of 12 is even. If we try this with 9, however, we get 1 and 9, 3 and 3. Because 9 is a square, 3 is matched with itself, so there are really an odd number of divisors of 9. Do you see the pattern? If n is not a square, its divisors pair up, so there are an even number of divisors. If n is a square, one of the divisors pairs with itself, so there are an odd number of divisors. Combining this with the observation of the last paragraph, we have our solution: locker n is open if, and only if, n is a square.

Modular Arithmetic

Let's look at odds and evens a bit more closely. Recall that every even number m can be written in the form $m = 2k$, and every odd number n can be written in the form $n = 2k + 1$.

How is parity affected when we do arithmetic on numbers? Suppose we add two even numbers m and n. We can write them as $m = 2k$ and $n = 2l$. Then

$$m + n = 2k + 2l = 2(k + l).$$

Thus the sum of two even numbers is always even. How about two odds? Using our standard form for odds,

$$(2k + 1) + (2l + 1) = 2k + 2l + 2 = 2(k + l + 1),$$

so the sum of two odds is even. As a final example, let us multiply an even and an odd number: $(2k)(2l + 1) = 2[k(2l + 1)]$, so the product is even. One can go through all of the other cases; the general principle is this: if we know the parity of two numbers, we can determine the parity of their sum, difference, and product.

If we are concerned only about the parity of numbers, we can represent a number by a simple one that has the same parity. We can use 0 and 1. For instance, if we don't remember the parity of the sum of two odd numbers, we simply consider $1 + 1$. It is even, so every sum of two odds must be even.

This arithmetic, taking into account only the parity, is an example of *modular arithmetic*, with 2, in this case, being the *modulus*. (The formal definitions are in the next section.) We can summarize modular arithmetic with modulus 2 by the following tables, writing $[0]$ to represent even numbers, and and $[1]$ to represent odd numbers.

+	[0]	[1]
[0]	[0]	[1]
[1]	[1]	[0]

−	[0]	[1]
[0]	[0]	[1]
[1]	[1]	[0]

×	[0]	[1]
[0]	[0]	[0]
[1]	[0]	[1]

For example, the first table tells us that $[1] + [0] = [1]$, in other words, that the sum of an odd number and an even number is odd.

There is another way to look at this arithmetic, using remainders upon division by 2. For example, let us look at $[1] + [1]$. According to our table, $[1] + [1] = [0]$, which may strike you as strange. But if we think of the 1s as numbers, $1 + 1 = 2$, and 2 has *remainder* 0. You can check that this works for everything in the tables; if you do normal arithmetic, then take the remainder on division by 2, you get the results in the table.

In the following sections, we will pursue these ideas in a more general setting.

EXERCISES

5.64 Here is a variation on the magic trick at the start of the section. Have the volunteer pick a number from 1 to the number of coins, tell you the number, then turn that many coins. As before, one coin is covered. How would you figure out whether the hidden coin was heads or tails? (Hint: suppose that the volunteer turned three coins. What would happen? How about four coins?)

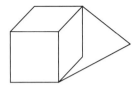

5.65   Starting anywhere, trace the figure below, without repeating any line.

5.66   What four odd integers add up to 19?

5.67   Produce tables for addition, subtraction, and multiplication, like those in this section, for modulus 3. (Hint: the possible remainders upon division by 3 are 0, 1, and 2.)

5.68   Modular arithmetic is sometimes called *clock arithmetic*, because twelve-hour clocks repeat in the same way. Construct the addition and multiplication tables for modular arithmetic with modulus 12.

5.13 Congruence

We will generalize the ideas introduced in the last section. This generalization is largely due to Leonhard Euler and Carl Friedrich Gauss. The notation is from an amazing book Gauss published in 1801, when he was 24 years old (and written three years before). It is called *Disquisitiones Arithmeticae*.

Definitions and Examples

Definition.   Let a and b be integers, and m a positive integer. We say that a *is congruent to b modulo m* if $m \mid (a - b)$. If this is the case, we write

$$a \equiv b \pmod{m}.$$

If a is not congruent to b modulo m, we write

$$a \not\equiv b \pmod{m}.$$

■ EXAMPLE 5.26

(a) $25 \equiv 17 \pmod 4$, because $4 \mid (25 - 17)$

(b) $6 \equiv -27 \pmod 3$, because $3 \mid [6 - (-27)]$

(c) $17 \not\equiv 117 \pmod 7$, because $7 \nmid (17 - 117)$

Basic Properties

$a \equiv b \pmod{m}$ if and only if there is some integer k such that $a = b + km$.

Proof. The following statements are equivalent, i.e., each is true if and only if the next is true.

  i. $a \equiv b \pmod{m}$

  ii. $m \mid (a - b)$

  iii. There is some integer k such that $a - b = km$.

  iv. There is some integer k such that $a = b + km$.

■

■ **EXAMPLE 5.27**

We saw in the last example that $6 \equiv -27 \pmod{3}$. Let $a = 6$, $b = -27$, $m = 3$. As promised in the theorem, there is a k such that $a = b + km$, namely, $k = 11$. How did we find k? We merely divided $6 - (-27)$ by 3.

Congruence shares many properties with equality. (This is the reason for the $\equiv$ notation.) Here are three.

Let a, b, and c be any integers, and m a positive integer. Then

  i. (Reflexive) $a \equiv a \pmod{m}$.

  ii. (Symmetric) If $a \equiv b \pmod{m}$ then $b \equiv a \pmod{m}$.

  iii. (Transitive) If $a \equiv b \pmod{m}$ and $b \equiv c \pmod{m}$, then $a \equiv c \pmod{m}$.

Proof. These are all pretty easy. Consider (iii). Suppose that $a \equiv b \pmod{m}$ and $b \equiv c \pmod{m}$. Then, by the last result, there are integers k and l such that

$$a = b + km$$
$$b = c + lm.$$

Substituting the second of these equations into the first,

$$a = (c + lm) + km = c + (k + l)m,$$

which, again by the last result, gives us $a \equiv c \pmod{m}$. ■

Here is a useful way of thinking about congruence.

Let a be any integer, and m be a positive integer. Then a is congruent to exactly one of $0, 1, \ldots, m - 1$, namely, the remainder when we divide a by m.

Proof. Divide a by m, say $a = qm + r$, with $0 \le r < m$. By our first basic property above, $a \equiv r \pmod{m}$. Hence a is congruent to one of $0, 1, \ldots, m - 1$.

Now suppose that a were congruent to more than one of $0, 1, \ldots, m - 1$, say, r and r'. Then we would have, for some k and k',

$$a = km + r, \quad 0 \le r < m$$
$$a = k'm + r, \quad 0 \le r' < m.$$

But this violates the uniqueness of the Division Theorem. Therefore a is congruent to only one of $0, 1, \ldots, m - 1$. ∎

▣ EXAMPLE 5.28

Let $a = 234$ and $m = 4$. The remainder when we divide 234 by 4 is 2 (check!), so $234 \equiv 2 \pmod{4}$.

We have to be careful with negative numbers here. We may be tempted to use the last theorem to say that $-9 \equiv 2 \pmod{7}$, because the remainder of 9 is 2 when we divide by 7. But 7 does not divide $-9 - 2$. Is the theorem wrong? No, because the remainder of -9 is *not* 2; it is 5. This is because $-9 = 7(-2) + 5$.

Remainders are also useful for determining when two integers are congruent. Suppose that a and b have the same remainder r on division by m. By the last result, $a \equiv r \pmod{m}$ and $b \equiv r \pmod{m}$. Using the symmetric property on the second of these congruences, $r \equiv b \pmod{m}$. Then using the transitive property on this and the first yields $a \equiv b \pmod{m}$.

The general result is this.

Let a and b be any integers, m a positive integer. Then $a \equiv b \pmod{m}$ if and only if they have the same remainder when divided by m.

Definition. The remainder of a upon division by m is called the *least residue of a modulo m*. The least residue of a is written $a \bmod m$.

▣ EXAMPLE 5.29

(a) The least residue of 1,000,009 modulo 10 is 9.

(b) $17 \bmod 5 = 2$

Congruence Classes

Definition. Let a be any integer. The set of all integers congruent to a modulo m is called the *congruence class of a modulo m*. The congruence class is written $[a]_m$.

■ **EXAMPLE 5.30**

The integers 1,000,009 and 109 belong to the same congruence class mod 10, because they have the same remainder. (Often, people use mod as shorthand for modulo.) The entire congruence class of 1,000,009 modulo 10 is

$$[1{,}000{,}009]_{10} = \{\ldots, -21, -11, -1, 9, 19, 29, \ldots\}.$$

■ **EXAMPLE 5.31**

What are the congruence classes modulo 2? There are two of them, corresponding to the remainders 0 and 1. An integer has remainder 0 if it is divisible by 2, i.e., even. A number has remainder 1 if it is odd. So the congruence classes are the sets of even and odd integers.

EXERCISES

5.69   Determine, for each a, b, and m, whether $a \equiv b \pmod{m}$. When the answer is yes, find a k such that $a = b + km$.

    a) $a = 53$, $b = 32$, $m = 7$
    b) $a = 17$, $b = 2334$, $m = 15$
    c) $a = -17$, $b = 37$, $m = 9$
    d) $a = 23$, $b = 0$, $m = 7$

5.70   For each value of a and m, find the least residue of a modulo m. Also find at least four other numbers in the same congruence class as a, at least two of which are negative.

    a) $a = 17$, $m = 8$
    b) $a = 87$, $m = 3$
    c) $a = 5$, $m = 17$
    d) $a = -5$, $m = 17$

5.71   A band director wants his band to march in rows of 7, but finds that, when he lines them up, he has only 5 people in the last row. So, after some careful figuring, he thinks that maybe rows of 15 are OK. But when he lines them up in rows of 15, the last row has only one person in it.

    a) Assuming that there are no more than 100 people in the band, how many people are there?
    b) What does this problem have to do with congruences?
    c) What if there might be more than 100 people?

5.14 Arithmetic with Congruences

One of the main reasons that congruence is useful is that most arithmetic on congruence classes is much like our familiar arithmetic.

Addition, Subtraction, and Multiplication Are Easy

Let a, b, c, and d be any integers, m a positive integer. Suppose that $a \equiv b \pmod{m}$ and $c \equiv d \pmod{m}$. Then

   i. $a + c \equiv b + d \pmod{m}$

  ii. $a - c \equiv b - d \pmod{m}$

 iii. $ac \equiv bd \pmod{m}$.

Proof. This is pretty straightforward. We will prove (i). Since $a \equiv b \pmod{m}$ and $c \equiv d \pmod{m}$, there are integers k and l such that $a = b + km$ and $c = d + lm$. Then

$$a + c = (b + km) + (d + lm) = (b + d) + (k + l)m.$$

Therefore $a + c \equiv b + d \pmod{m}$. ∎

EXAMPLE 5.32

If $a \equiv 5 \pmod{9}$, what is the remainder of $17a$ on division by 9?

Solution: We use part (iii) of the theorem. Note that we have $a \equiv 5 \pmod{9}$ and $17 \equiv 8 \pmod{9}$. So we have

$$17a \equiv 8 \cdot 5 \equiv 40 \equiv 4 \pmod{9}.$$

Definition: What does this theorem do for us? It means, for example, that the congruence class of the sum of two numbers depends only on the congruence classes of the two numbers, not on the particular numbers themselves. So it makes sense to talk of the sum of two congruence classes. Similarly, we can subtract and multiply classes. Using the notation we introduced in the last section, we can define

$$[a]_m + [c]_m = [a + c]_m$$
$$[a]_m - [c]_m = [a - c]_m$$
$$[a]_m \cdot [c]_m = [ac]_m.$$

In other words, we have arithmetic on *congruence classes*.

■ **EXAMPLE 5.33**

Revisiting the last example, we have

$$[17a]_9 = [17]_9[a]_9 = [8]_9[5]_9 = [40]_9 = [4]_9.$$

Notation: When the modulus is clear, we can skip the subscript, e.g., write $[17a]$ instead of $[17a]_9$.

■ **EXAMPLE 5.34**

Suppose that x is a number congruent to 3 mod 5. What is the congruence class of $2x^2 - 7x + 5$?

Solution: To solve this, first note that $[x^2] = [x][x]$. We will denote this by $([x])^2$. So

$$\begin{aligned}
[2x^2 - 7x + 5]_5 &= [2][x^2] - [7][x] + [5] \\
&= [2]([x])^2 - [7][x] + [5] \\
&= [2]([3])^2 - [7][3] + [5] \\
&= [2][9] - [21] + [0] \\
&= [18] - [1] \\
&= [3] - [1] \\
&= [2].
\end{aligned}$$

■

Using Congruence Arithmetic

■ **EXAMPLE 5.35**

Prove that no square has remainder 2 or 3 when divided by 4.

Solution: Suppose we have a square, say x^2. Let's consider all possible least residues (i.e., remainders) of x mod 4.

If $x \bmod 4 = 0$, then $[x^2]_4 = [x]_4[x]_4 = [0]_4[0]_4 = [0^2]_4 = [0]_4$.

If $x \bmod 4 = 1$, then $[x^2]_4 = [1^2]_4 = [1]_4$.

If $x \bmod 4 = 2$, then $[x^2]_4 = [2^2]_4 = [0]_4$.

If $x \bmod 4 = 3$, then $[x^2]_4 = [3^2]_4 = [1]_4$.

Thus the remainder is always 0 or 1. ■

Congruence Modulo 9: If you add up the digits of a positive integer, the sum is in the same congruence class as the original integer. For example,

$$512{,}834 \equiv (5+1+2+8+3+4) \equiv 23 \equiv (2+3) \equiv 5 \pmod 9.$$

Why does this work? Because $[10]_9 = [1]_9$. For our example,

$$
\begin{aligned}
[512{,}834]_9 &= [5 \cdot 100{,}000 + 1 \cdot 10000 + 2 \cdot 1000 + 8 \cdot 100 + 3 \cdot 10 + 4 \cdot 1] \\
&= [5 \cdot 100{,}000] + [1 \cdot 10000] + [2 \cdot 1000] + [8 \cdot 100] + [3 \cdot 10] + [4 \cdot 1] \\
&= [5 \cdot 10^5] + [1 \cdot 10^4] + [2 \cdot 10^3] + [8 \cdot 10^2] + [3 \cdot 10] + [4 \cdot 1] \\
&= [5]([10])^5 + [1]([10])^4 + [2]([10])^3 + [8]([10])^2 + [3][10] + [4][1] \\
&= [5]([1])^5 + [1]([1])^4 + [2]([1])^3 + [8]([1])^2 + [3][1] + [4][1] \\
&= [5][1] + [1][1] + [2][1] + [8][1] + [3][1] + [4][1] \\
&= [5] + [1] + [2] + [8] + [3] + [4] \\
&= [5+1+2+8+3+4].
\end{aligned}
$$

A Divisibility Rule for 9: Because the sum of the digits of a number n is congruent to n, we can easily determine whether a number is divisible by 9: n is divisible by 9 if and only if the sum of its digits is divisible by 9. For example, let $n = 12{,}345{,}623$. Adding the digits, gives

$$1+2+3+4+5+6+2+3 = 26.$$

Since 26 is not divisible by 9, neither is 12,345,623.

Casting Out Nines: If you can find a set of digits that add up to 9, you can throw them away (cast them out), without changing the congruence class. For example, in 512,834, the digits 1 and 8 add up to 9, so we can cast them out: $512{,}834 \equiv 5234$ (mod 9). This works because the sum of the digits in the new number differs only by 9, which is 0 mod 9, so it doesn't affect the congruence class of the sum of the digits. Thus it doesn't affect the congruence class of 512,834.

Continuing with our example, $2+3+4 = 9$, therefore we can cast these digits out: $5234 \equiv 5$ (mod 9), hence $512{,}834 \equiv 5$ (mod 9), which is the same answer we got above.

Checking Calculations By Casting Out Nines: An old way (pre-calculator) for checking arithmetic is based on casting out nines. As an example, consider this addition.

$$
\begin{array}{r}
203{,}495 \\
+121{,}209 \\
\hline
324{,}604
\end{array}
$$

Casting out nines, we see that $203{,}495 \equiv 5 \pmod 9$, that $121{,}209 \equiv 6 \pmod 9$, and $324{,}604 \equiv 1 \pmod 9$. But $[5]_9 + [6]_9 \neq [1]_9$, so the addition must be wrong. The correct sum is $324{,}\underline{7}04$.

Residues of Negative Integers: We saw in the last section that finding congruence classes of negative numbers could be tricky. But there is an easy way around the difficulty. Consider $[-8]_5$. We can write this as

$$[-1]_5[8]_5 = [-1]_5[3]_5 = [-3]_5 = [0]_5 + [-3]_5 = [5]_5 + [-3]_5 = [5-3]_5 = [2]_5.$$

To find the least residue of -8 mod 5, find the least residue of 8, and subtract it from 5.

Here is another example: $-102 \pmod 7 = 7 - 102 \pmod 7 = 7 - 4 = 3$.

EXERCISES

5.72   Find the congruence classes.
  a) $[31]_8 + [17]_8$
  b) $[2]_3[22]_3 - [1]_3$
  c) $[-9]_7 - [-2]_7$

5.73   What is the least residue of -26 modulo 17?

5.74   If $x \equiv 5 \pmod 8$, what is $[32x^3 - 45x^2 + 232x - 118]_8$?

5.75   What day of the week will it be exactly one year from today?

5.76   Find all integers x such that $x^2 + 2x + 1 \equiv 2 \pmod 7$.

5.77   The method of checking calculations by casting out nines is not perfect; there are cases in which the calculation is incorrect but the method does not reveal this fact. Find such an example.

5.78   Is there any number other than 9 for which something like casting out nines works?

5.79   How can you change a dollar into 26 coins? (Hint: what are the possible numbers of pennies?)

5.80   Recall that $n!$ is the product of the first n positive integers. For example, $2! = 1 \cdot 2 = 2$ and $3! = 1 \cdot 2 \cdot 3 = 6$. Find the remainder of $1! + 2! + 3! + \cdots + 100!$ upon division by 12.

5.81   Let a, b, c, and d be any integers, and m a positive integer. Suppose that $a \equiv b \pmod m$ and $c \equiv d \pmod m$. Show that the following congruences are always true.
  a) $a - c \equiv b - d \pmod m$
  b) $ac \equiv bd \pmod m$

5.15   Division with Congruences; Finite Fields

The Problem

You may have noticed the conspicuous absence of division from the congruence arithmetic so far. The reason for this is familiar: division presents special problems with integers.

What would the division equivalent of the basic result of Section 5.13 look like? Perhaps:

If $a \equiv b \pmod{m}$ and $c \equiv d \pmod{m}$,   then $a/c \equiv b/d \pmod{m}$.

But this only makes sense if a/c and b/d are integers. But the situation is more complicated even than this. For example, let $a = 4$, and b, c, d, and m all equal 2. Then $a \equiv b \pmod{m}$ and $c \equiv d \pmod{m}$, but although a/c and b/d are integers in this case, still $a/c \not\equiv b/d \pmod{m}$.

Here is another example to illustrate the difficulty. Let $a = 2$, $b = 9$, $c = 12$, and $m = 6$. Then $ab \equiv ac \pmod{m}$, but $b \not\equiv c \pmod{m}$. (Check!) So we can't cancel. Of course cancelling is equivalent to dividing, so this shouldn't surprise us. There is one important case, however, in which cancellation does work.

GCDs to the Rescue

Let $ab \equiv ac \pmod{m}$ and suppose that a and m are relatively prime. Then $b \equiv c \pmod{m}$.

Proof. Recall the theorem of Section 5.6: If $r \mid st$, and r and s are relatively prime, then $r \mid t$. We will use this result.

Now $ab \equiv ac \pmod{m}$, so $m \mid (ab - ac)$, i.e., $m \mid a(b - c)$. Since a and m are relatively prime, the theorem applies to give us $m \mid (b - c)$. But this tells us that $b \equiv c \pmod{m}$.   ∎

◨ EXAMPLE 5.36

Solve $2x \equiv 14 \pmod{25}$ for x.

Solution: Since 2 and 25 are relatively prime, we can divide by 2, to get $x \equiv 7 \pmod{25}$. So x is any number in $[7]_{25}$.   ∎

◨ EXAMPLE 5.37

Solve $4x \equiv 1 \pmod{6}$ for x.

Solution: Since 4 and 6 are *not* relatively prime, the theorem does not apply. In fact, there are no solutions. If x were to work, then we would have $6 \mid (4x - 1)$. But $4x - 1$ is odd, and 6 does not divide any odd number.   ∎

◼ **EXAMPLE 5.38**

Solve $6x \equiv 2 \pmod 8$ for x.

Solution: Since 6 and 8 are not relatively prime, the theorem doesn't apply. But in this case, there are solutions. Let's systematically consider each congruence class mod 8.

$$[6]_8[0]_8 = [0]_8$$
$$[6]_8[1]_8 = [6]_8$$
$$[6]_8[2]_8 = [4]_8$$
$$[6]_8[3]_8 = [2]_8$$
$$[6]_8[4]_8 = [0]_8$$
$$[6]_8[5]_8 = [6]_8$$
$$[6]_8[6]_8 = [4]_8$$
$$[6]_8[7]_8 = [2]_8$$

There are actually a lot of solutions: all the numbers in $[3]_8$ and all the numbers in $[7]_8$. ◼

The last set of examples suggests that a variety of things can happen when you try to divide. We will concentrate on a particular, very important case: when m is prime.

Let p be a prime, and consider division modulo p. First, consider the implications of the last theorem on canceling in the equation $ab \equiv ac \pmod p$. Whenever p does not divide a, because p is prime, a and p must be relatively prime, so $b \equiv c \pmod p$. Therefore we can cancel.

Finite Fields

Consider the equation $[a]_p[x]_p = [b]_p$, where p is prime, and a and b are fixed. Can we solve this for $[x]_p$? Thinking of regular integer arithmetic, we want to write $[x]_p = [b]_p/[a]_p$. But what in the world is $[b]_p/[a]_p$? It certainly is not $[b/a]_p$, since b/a may not be an integer.

Let us jettison the subscript p. So our problem is to solve the equation $[a][x] = [b]$.

◼ **EXAMPLE 5.39**

Solve the equation $[4][x] = [1]$, where $p = 5$.

Solution: Note that there are only five possible solutions, since there are only five congruence classes, corresponding to the remainders 0, 1, 2, 3, 4. So let's try them.

$$[4][0] = [0]$$
$$[4][1] = [4]$$
$$[4][2] = [3]$$
$$[4][3] = [2]$$
$$[4][4] = [1]$$

We have our solution: $[x]$ must be $[4]$.   ■

■ **EXAMPLE 5.40**

Solve the equation $[4][x] = [222]$, where $p = 7$.

Solution: First, since $222 \equiv 5 \pmod 7$, we can restate this as: solve $[4][x] = [5]$. Let us try each of the five possible solutions for $[x]$.

$$[4][0] = [0]$$
$$[4][1] = [4]$$
$$[4][2] = [1]$$
$$[4][3] = [5]$$
$$[4][4] = [2]$$
$$[4][5] = [6]$$
$$[4][6] = [3]$$

Our solution is $[x] = [3]$.   ■

If you do a few computations like the last two, a pattern emerges.

Suppose $[a] \neq [0]$. Then the products $[a][0], [a][1], \ldots, [a][p-1]$ are all distinct.

Proof. Suppose that two of the products were the same, say $[a][b] = [a][c]$, where b and c are in $\{0, 1, \ldots, p-1\}$. Then $ab \equiv ac \pmod p$. But since $[a] \neq [0]$, a and p are relatively prime. Hence we can cancel: $b \equiv c \pmod p$. But b and c are in $\{0, 1, \ldots, p-1\}$, and no two of those remainders are congruent. Therefore $b = c$. This proves the theorem.   ■

■ **EXAMPLE 5.41**

Let us look at the products in the case where $p = 5$ and $a = 2$.

$$[2][0] = [0]$$
$$[2][1] = [2]$$
$$[2][2] = [4]$$
$$[2][3] = [1]$$
$$[2][4] = [3]$$

Notice that all the products are distinct, as advertised. But notice something else. Since there are only five possible products (the five congruence classes), *every* congruence class must appear as one of the products. This means that the equation $[2][x] = [b]$ *always* has a solution, because one of the products has to be $[b]$. This argument generalizes to every prime, and we have the following satisfying result.

Let $[a] \neq [0]$. Then the equation $[a][x] = [b]$ has exactly one solution.

Definition. Let $[a] \neq [0]$. Then we define the *quotient* $[b]/[a]$ to be the solution to $[a][x] = [b]$. Note that we do not have a quotient when $[a] \neq [0]$. But this is a form of our familiar inability to divide by 0.

■ **EXAMPLE 5.42**

Find the quotients, where $p = 5$.

(a) $[3]/[2]$

(b) $[0]/[2]$

(c) $[1]/[2]$

Solution: We can read these off of our products in the last example.

(a) $[3]/[2] = [4]$ because $[2][4] = [3]$

(b) $[0]/[2] = [0]$ because $[2][0] = [0]$

(c) $[1]/[2] = [3]$ because $[2][3] = [1]$

■

■ **EXAMPLE 5.43**

Solve $[5][x] - [3] = [28]$, where $p = 11$.

Solution: If we add $[3]$ to both sides of the equation, we get

$$[5][x] = [31] = [9]$$

using $31 \bmod 11 = 9$. Now we can check all the classes $[0], [1], \ldots, [10]$ one by one, to solve $[5][x] = [9]$, until we find a solution: $[x] = [4]$. By our last result, we know that this is the only solution.

It is always a good idea to double-check our solution in the original equation:

$$[5][4] - [3] = [20] - [3] = [17] = [28],$$

using $17 \equiv 28 \pmod{11}$.   ■

Let us pause and admire the scenery. Consider how far we have come. When p is a prime, we can add, subtract, multiply, and divide just as we can with real numbers. And we can solve linear equations.

Mathematicians have a name for such a structure: it is called a *field*. The set of congruence classes modulo p, with the arithmetic we have defined, is an example of a field. It is called a *finite field* because there are a finite number of congruence classes. Finite fields have found some important applications in the last 100 years. For example, statisticians use finite fields in *design theory*, to design good statistical experiments.

Another application is *error-correcting codes*, which are based on finite fields. These are used to encode messages in possibly noisy channels, in such a way that if a small error is made in transmission, the error can be detected and corrected. For example, consider communicating with a spaceship orbiting Saturn. If a message received from the spaceship is slightly garbled, we can ask for the message to be sent again. But then we need to wait hours for a response, since Saturn is so far away. If, instead, the message was encoded using an error-correcting code, we can immediately correct the error and read the message. Similarly, if your computer mistakenly flips a bit (changes a 0 to 1 or vice versa), it is automatically detected and corrected. In fact, error-correcting codes were invented for this use, by Richard Hamming in the late 1940s.

As a final application, we consider a widely used *error-detecting code*, i.e., a code that detects errors, but cannot correct them.

International Standard Book Number (ISBN): Every book has an ISBN, which is used to identify the book uniquely. These numbers employ modular arithmetic to detect copying errors. Below is the system used until 2007. The current version is a slight modification of it.

An ISBN is a ten-digit number. The first nine digits identify a geographical grouping, publisher, and title. The tenth digit is a *check digit*. The check digit is chosen as follows. We start with the first nine digits $a_1a_2a_3a_4a_5a_6a_7a_8a_9$. Then we pick the tenth digit, a_{10}, so that

$$10a_1 + 9a_2 + 8a_3 + \cdots + 3a_8 + 2a_9 + 1a_{10} \equiv 0 \pmod{11}.$$

In other words, 11 divides the sum. One little twist: the check digit may have to be 10, in which case we write it as X.

To simplify our notation, let us write

$$A = a_1a_2a_3a_4a_5a_6a_7a_8a_9a_{10}$$

and

$$\text{chk}(A) = 10a_1 + 9a_2 + 8a_3 + \cdots + 3a_8 + 2a_9 + 1a_{10},$$

so that $\text{chk}(A) \equiv 0 \pmod{11}$. The number $\text{chk}(A)$ is called the *checksum* of A.

As an example, consider the ISBN number 0-672-51831-7, where 7 is the check digit. (The dashes in the ISBN help in identifying the grouping, publisher, and title. They are not important to us.) Then

$$\text{chk}(A) = 10 \cdot 0 + 9 \cdot 6 + 8 \cdot 7 + 7 \cdot 2 + 6 \cdot 5 + 5 \cdot 1 + 4 \cdot 8 + 3 \cdot 3 + 2 \cdot 1 + 1 \cdot 7$$
$$= 209,$$

and 209 is divisible by 11, since $209 = 11 \cdot 19$.

Here is another example. Suppose that the ISBN starts out 0-471-15408. What is the check digit? Let us call it y. Then

$$\text{chk}(A) = 10 \cdot 0 + 9 \cdot 4 + 8 \cdot 7 + 7 \cdot 1 + 6 \cdot 1 + 5 \cdot 5 + 4 \cdot 4 + 3 \cdot 0 + 2 \cdot 8 + 1 \cdot y$$
$$= 162 + y.$$

So $162 + y$ is a multiple of 11. It is easy to check that the next multiple of 11 above 162 is 165. Therefore, $y = 3$ and the ISBN is 0-471-15408-3.

What does the check digit do? It helps to detect the two most common types of copying errors: mis-typing a digit, or switching two digits. Let us see how that works.

First, suppose that we mistakenly write the wrong third digit. Let

$$A = a_1a_2a_3a_4a_5a_6a_7a_8a_9a_{10},$$

and suppose that we have mistakenly written, instead,

$$A' = a_1a_2a_3'a_4a_5a_6a_7a_8a_9a_{10},$$

where $a_3' \neq a_3$. The claim is that $\text{chk}(A')$ cannot be divisible by 11, in other words, that we can see that an error occurred. Suppose not, that $\text{chk}(A')$ is divisible by 11. We will show that this leads to a contradiction.

Since $\text{chk}(A)$ is also divisible by 11, we have

$$\text{chk}(A') \equiv \text{chk}(A) \pmod{11}$$

or

$$\text{chk}(A') - \text{chk}(A) \equiv 0 \pmod{11}.$$

Now $\text{chk}(A)$ and $\text{chk}(A')$ differ in only the third position, so all the other terms of $\text{chk}(A') - \text{chk}(A)$ cancel, and we are left with $\text{chk}(A') - \text{chk}(A) = 8a'_3 - 8a_3 = 8(a'_3 - a_3)$. Hence

$$8(a'_3 - a_3) \equiv 0 \pmod{11}.$$

In congruence notation, $[8][a'_3 - a_3] = [0]$, where $p = 11$. We know that the equation $[8][x] = [0]$ has only one solution, however, namely, $[x] = [0]$. Therefore $[a'_3 - a_3] = [0]$, so that $a'_3 - a_3$ must be divisible by 11. Both a_3 and a'_3 are between 0 and 10, so $a'_3 - a_3$ is between -10 and 10. Now the only multiple of 11 in this range is 0, so $a'_3 - a_3 = 0$, i.e., $a'_3 = a_3$. But we started out assuming that $a'_3 \neq a_3$, so we have a contradiction. Therefore, this error would be detected.

A similar argument can be used to see that the checksum detects the error when we switch two digits. For example, suppose that we have switched the first and third digits, say,

$$A = a_1 a_2 a_3 a_4 a_5 a_6 a_7 a_8 a_9 a_{10}$$

and

$$A' = a_3 a_2 a_1 a_4 a_5 a_6 a_7 a_8 a_9 a_{10}.$$

Arguing as above, if the error cannot be detected, then we have

$$\text{chk}(A') - \text{chk}(A) \equiv 0 \pmod{11},$$

which in this case gives us

$$(10a_3 + 8a_1) - (10a_1 + 8a_3) \equiv 0 \pmod{11}.$$

(Check!) Rewriting,

$$10(a_3 - a_1) - 8(a_3 - a_1) \equiv 0 \pmod{11},$$

or

$$2(a_3 - a_1) \equiv 0 \pmod{11}.$$

In congruence notation, $[2][a_3 - a_1] = [0]$, where $p = 11$. As before, this implies that $[a_3 - a_1] = [0]$, so $a_3 = a_1$. So what? Well, we started out assuming that we switched a_1 and a_3, and concluded that, if this was not detected, then $a_3 = a_1$. But if we switch two identical digits, nothing is changed. If there is a change, the checksum will detect it.

EXERCISES

5.82   Solve for x.

    a) $3x \equiv 12 \pmod{16}$

    b) $2x \equiv 1 \pmod{10}$

    c) $3x \equiv 6 \pmod{9}$

    d) $3x + 1 \equiv 2 \pmod{7}$

5.83   Solve for $[x]$.

    a) $[2]_3[x] = [1]_3$

    b) $[2]_6[x] = [1]_6$

    c) $[5]_{11}[x] = [3]_{11}$

    d) $[2]_7[x] - [5]_7 = [1]_7$

5.84   In Section 5.11 we wrote out addition, subtraction, and multiplication tables for arithmetic modulo 2. Write out the corresponding tables for arithmetic in the field of congruence classes modulo 7.

5.85   Since 7 is prime, we can divide its congruence classes. Write out a division table for arithmetic mod 7. There is a useful shortcut, using quotients of the form $[1]/[a]$. For example, you can see that $[1]/[5] = [3]$. We can use this to divide any other class by $[5]$. Here are two samples.

$$\frac{[2]}{[5]} = [2]\frac{[1]}{[5]} = [2][3] = [6]$$

$$\frac{[3]}{[5]} = [3]\frac{[1]}{[5]} = [3][3] = [2]$$

5.86   Use the tables from the last two exercises to solve for $[x]$.

    a) $[4][x] = [3]$

    b) $[x] - [5] = [2]$

    c) $[2][x] + [5] = [1]$

    d) $[x]^2 - [3][x] + [2] = [0]$

5.87   Compute the check digit y of this ISBN: 0-8053-0490-y.

5.88   Which of the following ISBNs contain an error?

    a) 8-0027-1331-9

    b) 0-1400-2547-2

    c) 0-321-38700-X (Remember that X stands for 10.)

5.16   Fermat's Last Theorem

We saw in Section 5.11 that the Diophantine equation $x^2 + y^2 = z^2$ has an infinite number of solutions, the Pythagorean triples. Given this, it is natural to ask about a similar Diophantine equation, $x^3 + y^3 = z^3$, or more generally,

$$x^n + y^n = z^n$$

where n is some fixed integer greater than 2. One type of solution is pretty easy: one of x or y is 0, and z equals the other. But this is too easy; let us call these trivial solutions. So the challenge is to find nontrivial solutions.

The great mathematician Pierre de Fermat studied these equations in the 17th century. After his death, his son discovered the following, written in the margin of his copy of Diophantus' book *Arithmetica*.

> I have discovered a truly marvelous proof that it is impossible to separate a cube into two cubes, or a fourth power into two fourth powers, or in general, any power higher than the second into two like powers. This margin is too narrow to contain it.

In other words, Fermat claimed that these Diophantine equations had *no* nontrivial solutions, for any $n > 2$. It was a remarkable assertion, given that there are lots of solutions with $n = 2$. Fermat did have a proof of the case $n = 4$, but no one could find anywhere in his papers the proof that couldn't fit into the margin.

Assertions of theorems without published proofs were not uncommon at this time. In fact, Fermat himself provided a number of them. Fermat's record of being correct was very good, however. Several of his theorems were first supplied with published proofs only 100 years later by the great Leonhard Euler. But even Euler couldn't prove this one. It became known as *Fermat's Last Theorem*.

Over the years, many mathematicians attempted to prove or disprove Fermat's Last Theorem. They succeeded in proving it for some values of n, starting with Euler, who gave a proof for $n = 3$ in 1770. In the early 19th century, one of the greatest women mathematicians, Sophie Germain (1776–1831), made a number of important advances, which inspired further work, although she did not succeed in proving the theorem. Later in the 19th century, Ernst Kummer (1810–1893) managed to prove many cases, in the process inventing a whole area of mathematics, the theory of ideals, that has proven useful in parts of mathematics far beyond its original inspiration. In the 20th century, computers were enlisted. By 1993 computers (with the assistance of mathematical theory) had verified the theorem for all n up to 4,000,000.

The fame of Fermat's Last Theorem spread beyond mathematics. In 1908 German industrialist Paul Wolfskehl offered 100,000 marks as a prize for a proof. This inspired many amateurs to pursue the proof. In addition to providing hours of instructive labor to the amateurs, this also wasted a fair bit of time for the professionals. Edmund Landau (1877–1938), who was for a time responsible for evaluating entries for the Wolfskehl prize, received so many incorrect proofs that he had cards printed that read:

> Dear ... ,
> Thank you for your manuscript on the proof of Fermat's Last Theorem.
> The first mistake is on: Page ... Line ...
> This invalidates the proof.
>
> Professor E. M. Landau

He assigned the task of filling in the page and line numbers to his students.

The theorem was finally proven in 1995. The final steps were provided by a number of mathematicians. The general approach was suggested by Gerhard Frey in 1984, based on work from earlier in the century. An important advance was made by Ken Ribet in 1986. After Andrew Wiles, a British mathematician working at Princeton University, learned of this work, he began trying to complete Frey's program. He worked in secrecy, not wanting anyone else to get the credit for the proof of Fermat's Last Theorem. In 1993 Wiles announced that he had succeeded. He had, however, made a mistake in his complicated proof. The following year, with the assistance of Richard Taylor, Wiles fixed the error and completed the proof, about 357 years after Fermat made his assertion.

The history of the search for this proof is quite revealing about the nature of mathematics. Perhaps most remarkable is the sustained effort over 350 years, in many countries. Also notable is the way that effort spent to solve this one problem, with no applications in sight, produced the theory of ideals, which has proven applicable. Finally, although the proof was completed by Wiles, it built upon the work of many earlier mathematicians. The proof has many authors, not just one. In the end, mathematics is a collective enterprise.

EXERCISES

5.89 In Fermat's equation, $x^n + y^n = z^n$, we can assume that n is a prime number. Why?

5.90 Watch the 1996 Nova film *The Proof*, about the proof of Fermat's Last Theorem. Who are some of the mathematicians interviewed in the film? Who are some of the mathematicians, throughout history, who took part in proving Fermat's Last Theorem? What are some of the key ideas involved in the proof?

5.17 Unfinished Business

Number theory has been one of the most studied of mathematical topics, but one must not get the impression that all the important problems have been solved. This area is famous for problems that are easy to state and to understand, but devilishly difficult to solve. (The Devil himself is recruited in the search for a proof of Fermat's Last Theorem in the short story "The Devil and Simon Flagg," by Arthur Porges.) Here are a few unsolved problems for your enjoyment.

The Twin Prime Conjecture. If n and $n + 1$ are consecutive integers, one of them must be even, i.e., divisible by 2. This means that, if both n and $n + 1$ are primes, the numbers must be 2 and 3. Hence every pair of primes other than 2 and 3 is at least two apart. *Twin primes* are primes that are exactly two apart, for example 3 and 5, or 11 and 13, or 41 and 43. Primes get sparser as the numbers get larger, but does the supply of twin primes run out? A famous conjecture, the twin prime conjecture, is that there are infinitely many twin primes.

There is currently no proof, or disproof, of the twin prime conjecture. As of this writing, the largest pair of twin primes known is $65516468355 \times 2333333 - 1$ and $65516468355 \times 2333333 + 1$, each of which has $100{,}355$ digits. These were found by a distributed computing project for twin primes.

Goldbach's Conjecture. Goldbach's conjecture is that every even number is the sum of two primes. So, for example, $8 = 3 + 5$, and $110 = 31 + 79$. This conjecture dates from correspondence in 1742 between Leonhard Euler and the Prussian mathematician Christian Goldbach (1690–1764). As with the twin prime conjecture, many people have worked on this one, and quite a few assertions of proofs have been made, but no actual proof has been discovered. As of this writing, the conjecture has been verified up to at least 10^{17}.

The Collatz Problem. This one is of more recent vintage, and has a different flavor. It was proposed in 1937 by Lothar Collatz (1910–1990).

Let n be an integer greater than 1. If n is even, divide it by 2. If n is odd, multiply by 3 then add 1. Repeat until you get 1. Here are a couple of examples, starting with 10 and 50.

$$10, 5, 16, 8, 4, 2, 1$$

$$50, 25, 76, 38, 19, 58, 29, 88, 44, 22, 11, 34, 17, 52, 26, 13, 40, 20, 10, 5, 16, 8, 4, 2, 1$$

Of course, there is no guarantee that this ends. Maybe you never get to 1. Maybe you get into an infinite loop, or keep getting larger and larger numbers. The conjecture is that, whatever your starting number, you *always* end up at 1.

The number of steps required to get to 1 may be fairly large; If you start with $2{,}362{,}741{,}986{,}945{,}773{,}554{,}503$, it takes $2{,}589$ steps to reach 1. There is a distributed computing project for this problem as well, at

```
http://boinc.thesonntags.com/collatz/+.
```

The Collatz conjecture has been verified by computer for many numbers. But no one has been able to prove it. The great number theorist Paul Erdős (1913–1996) said of this conjecture, "Mathematics is not yet ready for such problems."

EXERCISES

5.91   Find ten pairs of twin primes.

5.92   Verify Goldbach's conjecture for numbers up to 50.

5.93   Verify the conjecture of the Collatz problem for $n = 13$ and $n = 104$.

5.94   What happens in the Collatz problem if n is a power of 2?

5.95   Prove any of the conjectures in this section.

APPENDIX A

ANSWERS TO SELECTED EXERCISES

SOLUTIONS FOR CHAPTER 1

1.5.b 10,862

1.5.d $\frac{1}{36} = .02777\ldots$

1.6.c 2,1,5

1.6.e 1;20

1.7.b 1;19

1.9.b 1; 26

1.11 The squares have sides $1/2$ and $1/3$.

1.16.a 319,700 ft^3

1.19 7.5

1.21.a 20

Mathematics for the Liberal Arts.
By Donald Bindner, Martin Erickson, Joe Hemmeter Copyright © 2012 John Wiley & Sons, Inc.

1.21.b 31

1.21.c 213

1.21.d 101

  1.23 24 meters

  1.34 $k = \pi/4$

  1.41 $45°$

  1.47 6 feet

  1.50 $A = 16\pi$ ft^2, $V = 32\pi/3$ ft^3

1.54.a $a \approx 5.196$, $b = 3$

  1.59 $\sin 15° = \sin(60° - 45°) = \sin 60° \cos 45° - \cos 60° \sin 45° \approx .259$

  1.62 $90°$, $90°$, and $90°$

  1.65 $\Delta^\upsilon \gamma_\varsigma \beta$

SOLUTIONS FOR CHAPTER 2

  2.2 $x = 1, y = 2, z = -2$

2.6.a True

2.6.b False

2.6.c True

2.6.d True

  2.13 The sum of all the entries in the square is

$$1 + 2 + 3 + \cdots + 9 = \frac{9 \cdot 10}{2} = 45.$$

Since there are three rows, each row sum must be $45/3 = 15$.

  2.18 $C_5^{52} = \frac{52 \cdot 51 \cdot 50 \cdot 49 \cdot 48}{5 \cdot 4 \cdot 3 \cdot 2 \cdot 1} = 2{,}598{,}960$

  2.21 Using Brahmagupta's method on $(12, 17)$ and $(12, 17)$, we get $(408, 577)$.

  2.23 We have $s = \frac{1}{2}(8 + 12 + 6 + 16) = 21$, and hence

$$A = \sqrt{(21 - 8)(21 - 12)(21 - 6)(21 - 16)} = \sqrt{8775} = 15\sqrt{39} \approx 93.675.$$

2.27 Using the labeling in the image below, $\sin \alpha = a/c$ and $\cos \alpha = b/c$. By the Pythagorean Theorem, we have $a^2 + b^2 = c^2$. Putting these together,

$$\sin^2 \alpha + \cos^2 \alpha = \left(\frac{a}{c}\right)^2 + \left(\frac{b}{c}\right)^2 = \frac{a^2 + b^2}{c^2} = \frac{c^2}{c^2} = 1.$$

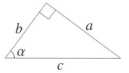

2.28.a $\pi/3$

2.28.b $\pi/4$

2.28.c $\pi/12$

2.28.d 4π

2.30 $\arctan 0 = 0$

2.34 Since $1 - z^2 = 2z$, or $z^2 + 2z - 1 = 0$, the quadratic formula gives $z = -1 \pm \sqrt{2}$. Abu Kamil would probably have thrown out the negative solution, so $z = -1 + \sqrt{2} = \sqrt{2} - 1$. Since $y^2 = z$, we have $y = \sqrt{\sqrt{2} - 1}$ (again ignoring negative solutions). Since $y = (10 - x)/x$, we have $(10 - x)/x = \sqrt{\sqrt{2} - 1}$. Solving this yields

$$x = \frac{10}{1 + \sqrt{\sqrt{2} - 1}} \approx 6.08.$$

(Note that setting $y = -\sqrt{\sqrt{2} - 1}$ gives another root. Even though y is negative, x is not. What is the approximate value of x?)

2.41

$$1^4 + 2^4 + \cdots + 10^4 = \left(\frac{10}{5} + \frac{1}{5}\right) \cdot 10 \cdot \left(10 + \frac{1}{2}\right)\left[(10 + 1) \cdot 10 - \frac{1}{3}\right] = 25333$$

2.42

$$C = 180° - 20° - 25° = 135°$$

$$a = 7 \frac{\sin 20°}{\sin 25°} \approx 5.67$$

$$c = 7 \frac{\sin 135°}{\sin 25°} \approx 11.71$$

2.45 $121/17$ and $167/17$ *denarii*

2.47.a 6

2.47.b 9

2.47.c 700

2.47.d 2900

2.53.a $\sqrt[3]{\sqrt{325} + 18} - \sqrt[3]{\sqrt{325} - 18}$

2.53.b 3

2.54

$$
\begin{aligned}
x^3 + bx^2 + cx + d &= (y - b/3)^3 + b(y - b/3)^2 + c(y - b/3) + d \\
&= [y^3 + 3y^2(-b/3) + 3y(-b/3)^2 + (-b/3)^3] \\
&\quad + b[y^2 + 2y(-b/3) + (-b/3)^2] + c(y - b/3) + d \\
&= y^3 + y^2(-b + b) + y[3(-b/3)^2 + 2b(-b/3) + c] \\
&\quad + [(-b/3)^3 + b(-b/3)^2 + c(-b/3) + d] \\
&= y^3 + y(c - b^2/3) + (2b^3/27 - bc/3 + d)
\end{aligned}
$$

SOLUTIONS FOR CHAPTER 3

3.2 -0.17609

3.5 302

3.8 $2/3$

3.10 $(x - 3)^2 + (y - 2)^2 = 25$

3.14 $x^5 + 5x^4y + 10x^3y^2 + 10x^2y^3 + 5xy^4 + y^5$

3.17.a $y - 2 = \dfrac{1}{4}(x - 4)$

3.22 $0, 1, i, -i$

3.24 $x^5 - x$

3.25 i

3.30 $\pi^2/12$

3.33 $e^i \approx 0.54 + 0.84i$

3.40 There are 24 $(= 4 \cdot 3 \cdot 2 \cdot 1)$ permutations.

3.52 $\overline{P + Q} = \overline{P} \cdot \overline{Q}$

3.61 No. The left side is divisible by 4 and the right side is not.

3.62 Number the coaches 1, 2, 3, ..., and number the people in each coach by their seat number, 1, 2, 3, ... Then give a card to each person, with the sum of the two numbers on the card. Assign people rooms in order of their card numbers. There will be multiple people with the same card number (except for the smallest card number). Serve people with the same card number in order of their coach number. This defines an ordering on all the arriving people, so they can all be given rooms.

3.64.d Note that we cannot divide by 0.

÷	0	1	2	3	4
0	-	0	0	0	0
1	-	1	3	2	4
2	-	2	1	4	3
3	-	3	4	1	2
4	-	4	2	3	1

3.73 The entropy of an information source with two symbols is

$$-p \log_2 p - (1 - p) \log_2(1 - p),$$

where the symbols occur with probabilities p and $1 - p$. Let this function be $f(p)$. We want to show that the maximum of f occurs where $p = 1/2$. We do this by taking the derivative of f and finding where it is equal to 0. To take a derivative, we may as well assume that the logarithms are natural logs, since the conversion to logs base 2 only involves a multiplicative constant (it won't affect where f' is 0). Using the product rule, we obtain

$$f'(p) = -\ln p - \frac{p}{p} - (-1)\ln(1 - p) - (1 - p)\frac{(-1)}{1 - p}$$
$$= -\ln p + \ln(1 - p).$$

Setting this equal to 0, we find that $p = 1/2$. Do we really have a maximum? Find the second derivative and show that it is negative.

SOLUTIONS FOR CHAPTER 4

4.1 We are told in the section that the height function for a falling pencil is $y = 4 - 16t^2$. We can verify that when $t = 0.25$, the height is $y = 4 - 16(0.25)^2 = 3$ ft.

Let h be a small positive number. The average velocity of the pencil over the time interval $[0.25, 0.25 + h]$ is

$$
\begin{aligned}
\frac{\text{distance}}{\text{time}} &= \frac{y(0.25 + h) - y(0.25)}{(0.25 + h) - 0.25} \\
&= \frac{[4 - 16(0.25 + h)^2] - [4 - 16(0.25)^2]}{h} \\
&= \frac{[4 - 16(0.25 + h)^2] - 3}{h} \\
&= \frac{[4 - 16(0.0625 + 0.5h + h^2)] - 3}{h} \\
&= \frac{[4 - (1 + 8h + 16h^2)] - 3}{h} \\
&= \frac{4 - 1 - 8h - 16h^2 - 3}{h} \\
&= \frac{-8h - 16h^2}{h} \\
&= -8 - 16h.
\end{aligned}
$$

If we let h go to zero, we see that the instantaneous velocity is -8 ft/s.

4.2.a To find the time t when the pencil strikes the ground, set

$$
\begin{aligned}
0 &= 16 - 16t^2 \\
16t^2 &= 16 \\
t^2 &= 1.
\end{aligned}
$$

This has solutions of $t = -1$ s and $t = 1$ s. Since the pencil strikes the ground *after* we drop it, we use $t = 1$ s.

Let h be a small positive number. Then the average velocity of the pencil over the time interval $[1 - h, 1]$ is

$$
\frac{\text{distance}}{\text{time}} = \frac{y(1) - y(1 - h)}{1 - (1 - h)}
$$

$$
\vdots
$$

$$
= -32 + 16h.
$$

Letting h go to zero, we get a velocity of -32 ft/s.

4.5.a Hint: find y when $t = 0$.

4.5.c The instantaneous velocity is approximately 0. The arrow reaches a height of slightly more than 627 ft.

4.6 Let $x = 3 + h$ be a value near the point of tangency $(3, 9)$. Since $y = x^2$, the second point on our secant is $(x, y) = (3 + h, (3 + h)^2)$. That makes the slope of the secant line

$$
\begin{aligned}
m &= \frac{\text{change in } y}{\text{change in } x} \\
&= \frac{(3 + h)^2 - 9}{(3 + h) - 3} \\
&= \frac{(9 + 6h + h^2) - 9}{h} \\
&= \frac{6h + h^2}{h} \\
&= 6 + h.
\end{aligned}
$$

By letting h approach 0, we find that the slope of the tangent line is $m = 6$. The equation for the tangent can then be obtained from the point-slope formula for a line, using the base point $(3, 9)$:

$$
\begin{aligned}
y - y_0 &= m(x - x_0) \\
y - 9 &= 6(x - 3).
\end{aligned}
$$

4.7 Hint: you should get a slope of $m = -8$ for the tangent line.

4.9

t (s)	v (ft/s)
0.0	4.0
0.1	3.84
0.2	3.36
0.3	2.56
0.4	1.44
0.5	0.0

4.12.d Pencils strike the floor faster on Earth and slower on the Moon.

4.13 By definition,

$$
\begin{aligned}
f'(x) &= \lim_{h \to 0} \frac{f(x + h) - f(x)}{h} \\
&= \lim_{h \to 0} \frac{((x + h)^2 + 3) - (x^2 + 3)}{h} \\
&= \lim_{h \to 0} \frac{(x^2 + 2xh + h^2 + 3) - (x^2 + 3)}{h} \\
&= \lim_{h \to 0} \frac{2xh + h^2}{h} \\
&= \lim_{h \to 0} (2x + h) \\
&= 2x.
\end{aligned}
$$

So $f'(x) = 2x$.

4.16 $f'(x) = 3$.

4.20.b $f'(x) = 2x + 1$

4.21.c $x'(t) = 7.5t^2 + 14t + 6.3$

4.22 Hint: find the derivative, set it equal to 0, and solve for x.

4.27.b Using the product rule,

$$
\begin{aligned}
g'(x) &= (2x^2 + 7x + 7)'(5x^2 + 4x + 5) + (2x^2 + 7x + 7)(5x^2 + 4x + 5)' \\
&= (4x + 7)(5x^2 + 4x + 5) + (2x^2 + 7x + 7)(10x + 4) \\
&= (20x^3 + 51x^2 + 48x + 35) + (20x^2 + 78x^2 + 98x + 28) \\
&= 40x^3 + 129x^2 + 146x + 63.
\end{aligned}
$$

If we expand g first, we get

$$g(x) = 10x^4 + 43x^3 + 73x^2 + 63x + 35,$$

which has the same derivative.

4.30 The fastest solution uses $f(x) = x(x + 1)(x + 2)$ and $g(x) = x + 3$, since you computed $f'(x)$ in the previous exercise.

Another reasonable choice is $f(x) = x(x + 1)$ and $g(x) = (x + 2)(x + 3)$.

For fun, you could try $f(x) = x(x + 2)$ and $g(x) = (x + 1)(x + 3)$. For this choice,

$$
\begin{aligned}
h'(x) &= f'(x)g(x) + f(x)g'(x) \\
&= \big(1(x + 2) + x(1)\big)(x + 1)(x + 3) + x(x + 2)\big(1(x + 3) + (x + 1)1\big) \\
&= (2x + 2)(x + 1)(x + 3) + x(x + 2)(2x + 4) \\
&= (2x^3 + 10x^2 + 14x + 6) + (2x^3 + 8x^2 + 8x) \\
&= 4x^3 + 18x^2 + 22x + 6.
\end{aligned}
$$

4.34.b Hint: although you *can* take the derivative of the bottom using the product rule, it will go a bit faster if you expand the denominator before you apply the quotient rule.

4.34.d The derivative is

$$-\frac{36x^3 + 9x^2 + 18x + 6}{(9x^4 + 3x^3 + 9x^2 + 6x + 6)^2}.$$

4.35.b From 4.34.d, the point of tangency is $(1, 1/33)$ and the slope is $m = -23/363$. Using the point-slope form of a line gives a tangent line of

$$y - y_0 = m(x - x_0)$$

$$y - \frac{1}{33} = -\frac{23}{363}(x - 1).$$

4.38.b Hint: rewrite the function as $f(x) = (x^2 + 7)^{1/3}$ and apply the chain rule.

4.40.c The derivative is

$$f'(x) = 3x^3 - 12x^2 + 12x$$
$$= 3x(x^2 - 4x + 4)$$
$$= 3x(x - 2)^2.$$

Critical points occur where $x = 0$ or $x = 2$. Checking values around 0,

x	-0.1	0.0	0.1
$f(x)$	0.64	0.0	0.056

indicates that $(0, 0)$ is a minimum. Near 2, the table

x	1.9	2.0	2.1
$f(x)$	3.998	4.0	4.002

shows that the function has a critical point at $(2, 4)$ that is not an extremum.

4.40.e Hint: the derivative is never zero, but don't forget to check where it is undefined.

4.41.c The derivative is $y'(t) = 2t - 5$, so the only critical point occurs when $t = 5/2$. If we realize that y is the formula for a parabola that opens upward, we conclude that $(5/2, -9/4)$ is a minimum. We can also check values directly:

t	2.4	2.5	2.6
$y(t)$	-2.24	-2.25	-2.24

4.42.c In the last exercise, we found that the only critical point happens when $x = 5/2$. We check that value and the ends of the interval:

t	0	2.5	4
$y(t)$	4	-2.25	0

The maximum occurs at $(0, 4)$, and the minimum occurs at $(2.5, -2.25)$.

4.46.b $y = 6/x$

4.46.c $z(x) = \sqrt{\left(\dfrac{6}{x} + 3\right)^2 + (x + 2)^2}$

4.46.e approximately 7.02 m

4.47.a Hint: use the Pythagorean Theorem.

4.47.c Hint: the area is hard to optimize since it has a square root in it. Use this trick instead. The area will be smallest at the same time that area^2 is smallest. Use algebra to show that area$^2 = 90x^3 - 10125x^2 + 364500x - 4100625$. Then use a derivative to solve for the values of x that make this smallest or largest.

4.48.a $I(1) = \dfrac{9}{100 + 0.15} \approx 0.090$ amps or 90 milliamps

4.48.b Use the quotient rule, and simplify selectively. $I'(n) = \dfrac{900 - 1.35n^2}{\left(100 + 0.15n^2\right)^2}$

4.48.c Hint: a fraction is 0 if the numerator is 0. Notice, by the way, that this fraction is never undefined. How do we know that?

4.48.d The maximum current is approximately 1.16 amps.

4.48.e Don't think too hard.

4.50 $\Delta S \approx \pm 90$ cm^2

4.55.c $\dfrac{80}{27}$

4.56.b

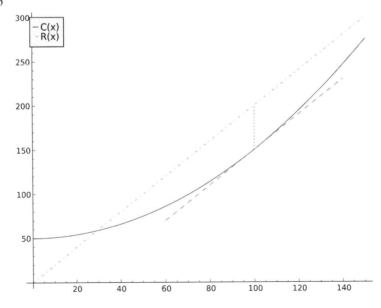

4.59.a 16,000

4.59.b $3,199,000

4.61 $1,680

4.64.a 11.896 years compared to $72/6 = 12$ years

4.65.c $f(t) = 1000e^{(\ln 1.05)t} \approx 1000e^{0.04879t}$

4.67.a Hint: use the product rule.

4.67.b There is more than one possible approach. Write $A(t) = 100/e^t$ using rules of exponents, and apply the quotient rule to get

$$A'(t) = \frac{0 - 100e^t}{(e^t)^2} = \frac{-100}{e^t}.$$

Another approach is to use the chain rule. The outside function is $g(t) = 100e^t$ and the inside function is $f(t) = -t$. Hence

$$A'(t) = f'(g(t))g'(t) = 100e^{-t}(-1) = -100e^{-t}.$$

4.67.g There is more than one possible approach. We can write $f(x) = \ln 3 + \ln x$ using the logarithm identities. Now, $\ln 3$ is a constant, so

$$f'(x) = 0 + \frac{1}{x} = \frac{1}{x}.$$

Another approach is to use the chain rule, and this also gives us

$$f'(x) = \frac{1}{3x}(3) = \frac{1}{x}.$$

4.68 The point of tangency is $(1, e^1) \approx (1, 2.71828)$. The slope is $m = y'(1) = e^1 \approx 2.71828$, since the exponential function is its own derivative. Using the point-slope form of a line, the tangent is

$$y - y_0 = m(x - x_0)$$
$$y - e^1 = e^1(x - 1)$$
$$y = 2.71828x.$$

4.70.a We have $A(138) \approx 5.5$ mg, which is half of the original 11 mg. The half-life of Polonium-210 is approximately 138 days.

4.70.b We have $A'(0) \approx -0.05522$ mg/day. The negative sign indicates that the amount of Polonium-210 is decreasing.

4.71.a $f'(x) = \dfrac{\sin x - x \cos x}{\sin^2 x}$

4.71.d Write $g(t) = (\sin(t/7))^{1/2}$ and apply the chain rule.

$$g'(t) = \frac{1}{2} (\sin(t/7))^{-1/2} \frac{d}{dx} (\sin(t/7))$$
$$= \frac{1}{2} (\sin(t/7))^{-1/2} \frac{1}{7} \cos(t/7)$$

4.73 Using the quotient rule,

$$\frac{d}{dx} \frac{\sin x}{\cos x} = \frac{(\cos x)(\cos x) - (\sin x)(-\sin x)}{\cos^2}$$
$$= \frac{\cos^2 x + \sin^2 x}{\cos^2 x}$$
$$= \frac{1}{\cos^2 x},$$

and most books use the identity $1/\cos x = \sec x$ to rewrite this as
$$= \sec^2 x.$$

4.77.c They must be equal.

4.81 By the Fundamental Theorem of Calculus,

$$\text{area} = F(1) - F(0)$$
$$= \left(\frac{1^3}{3} + 7 \right) - \left(\frac{0^3}{3} + 7 \right)$$
$$= \frac{1}{3}.$$

Notice the way the 7 cancels out.

4.83.a $F(x) = \dfrac{x^3}{3} + \dfrac{x^2}{2} + x$

4.83.e $F(x) = \dfrac{e^{3x+2}}{3}$

4.84.b $7 \sin x - \ln x + C$

4.84.e Hint: we don't have a specific rule for this type of function. You'll have to guess the answer and take a derivative to check that you are correct.

4.85.c Use $F(x) = (x + 1)^4$. The area is $F(1) - F(0) = 15$.

4.87.a Hint: is $v(0)$ positive or negative?

4.87.b When $t \approx 0.92$ s, because the professor has quit rising and is now falling.

4.87.c Approximately 1.94 m high.

4.87.f When $t \approx 1.84$ s

4.90.b 4.5 m/s^2

4.91.c By the chain rule,

$$\frac{d}{dt}\left(\ln(2-9t)\right) = \left(\frac{1}{2-9t}\right)\frac{d}{dt}(2-9t)$$
$$= \frac{1}{2-9t}(-9)$$
$$= \frac{-9}{2-9t}.$$

4.92.c Use the product and the chain rule to get $\ln(2-9t)$.

4.95 $20/9$

4.97.b The bartender would have run out of beer. Or maybe not. If the bar could hold an infinite number of mathematicians, ...

SOLUTIONS FOR CHAPTER 5

5.1 False

5.3.b The last two digits must be 00, 25, 50, or 75.

5.4.a $q = 11, r = 5$

5.4.b $q = 1501, r = 3$

5.4.c $q = -4, r = 3$

5.6.b True

5.6.e False

5.7 $\sqrt{4}$ and $117/3459753$

5.14.a 11

5.14.c 1

5.20 The last row is 360, upper left, and 11.

5.21 Just 5

5.26 There is only one such triplet.

5.27 Since $2 \mid 1000!$ and $2 \mid 2$, we have 2 dividing $1000!+2$, so $1000!+2$ is composite. The same argument works in general. For $2 \le n \le 1000$, since $n \mid 1000!$ and $n \mid n$, we have n dividing $1000! + n$, so $1000! + n$ is composite.

5.29 $u = 1, v = -2$

5.31 $\pm 1, \pm 3$

5.34 The set consists of the multiples of 3.

5.41 $2^{2^5} + 1 = 641 \times 6{,}700{,}417$

5.44.a $16 = 2^4$

5.44.b 59 is prime

5.44.c $72 = 2^3 \cdot 3^2$

5.44.d $105 = 3 \cdot 5 \cdot 7$

5.44.e $3 \cdot 1633 = 70^2 - 1 = (70 - 1)(70 + 1) = 69 \times 71$, so $1633 = 23 \times 71$

5.48 $2 \cdot 3^3 \cdot 17 = 918$

5.53 $x = 0, y = 11; x = 2, y = 10; x = 4, y = 9; \ldots; x = 22, y = 0$

5.55 10 pens and 1 notebook, or 3 pens and 4 notebooks

5.57.a $(3 + 4t, \ -1 - 3t)$

5.57.d $(-32 + 13t, \ 48 - 19t)$

5.60 51–140–149

5.61 $m = 2, n = 1$

5.64 If you turn an even number of coins, the parity of the number of heads is unchanged. If you turn an odd number of coins, the parity of the number of heads is changed.

5.67

$+$	$[0]$	$[1]$	$[2]$
$[0]$	$[0]$	$[1]$	$[2]$
$[1]$	$[1]$	$[2]$	$[0]$
$[2]$	$[2]$	$[0]$	$[1]$

$-$	$[0]$	$[1]$	$[2]$
$[0]$	$[0]$	$[2]$	$[1]$
$[1]$	$[1]$	$[0]$	$[2]$
$[2]$	$[2]$	$[1]$	$[0]$

$\times$	$[0]$	$[1]$	$[2]$
$[0]$	$[0]$	$[0]$	$[0]$
$[1]$	$[0]$	$[1]$	$[2]$
$[2]$	$[0]$	$[2]$	$[1]$

5.69.a Yes, $k = 3$

5.69.b No

5.69.c Yes, $k = -6$

5.69.d No

5.72.a $[0]_8$

5.72.b $[1]_3$

5.72.c $[0]_7$

5.74 $[5]_8$

5.78 3

5.80 Note that 12 divides $4! = 24$, and, since $n!$ is a multiple of $4!$ for every $n > 4$, we know that 12 divides all of them as well. So

$$1! + 2! + 3! + 4! + 5! + \cdots + 100! \equiv 1! + 2! + 3! + 0 + 0 + \cdots + 0 \equiv 9 \pmod{12}.$$

5.82.a $x \equiv 4 \pmod{16}$

5.82.b No solutions

5.82.c x is congruent to 2, 5, or 8 modulo 9

5.82.d $x \equiv 5 \pmod{7}$

5.83.a $[x] = [2]_3$

5.83.b No solutions

5.83.c $[x] = [5]_{11}$

5.83.d $[x] = [3]_7$

5.87 $y = 8$

5.89 Suppose that $x^n + y^n = z^n$, and let p be a prime divisor of n. Then $n = pk$ for some k, and we have a solution $(x^k)^p + (y^k)^p = (z^k)^p$.

5.95 This appendix is too small to contain it.

APPENDIX B

SUGGESTED READING

Chapters 1–3.

Jason Socrates Bardi, *The Calculus Wars: Newton, Leibniz, and the Greatest Mathematical Clash of All Time*, Thunder's Mouth Press, New York, 2006

Carl B. Boyer, *A History of Mathematics*, second edition, Wiley, New York, 1991

Ronald Calinger, *A Contextual History of Mathematics*, Prentice Hall, Upper Saddle River, NJ, 1999

Keith Devlin, *Mathematics: The Science of Patterns*, Scientific American Library, New York, 1997

Euclid, *The Thirteen Books of The Elements*, Vol. 1 (Books I and II), with introduction and commentary by Sir Thomas L. Heath, second edition, Dover, New York, 1956

By Donald Bindner, Martin Erickson, Joe Hemmeter Copyright © 2012 John Wiley & Sons, Inc.

Howard Eves, *Great Moments in Mathematics Before 1650*, The Mathematical Association of America, Washington, DC, 1983

Georges Ifrah, *From One to Zero: A Universal History of Numbers*, Penguin Books, New York, 1985

Ioan James, *Remarkable Mathematicians: from Euler to von Neumann*, Cambridge University Press, Cambridge, 2002

Victor J. Katz, *A History of Mathematics: An Introduction*, third edition, Addison-Wesley, New York, 2009

Nicholas Ostler, *Empires of the Word: A Language History of the World*, Harper Collins, New York, 2005

William Poundstone, *Prisoner's Dilemma*, Anchor, New York, 1993

Steven Schwarzman, *The Words of Mathematics: An Etymological Dictionary of Mathematical Terms Used in English*, The Mathematical Association of America, Washington, DC, 1994

Ian Stewart, *The Story of Mathematics: From Babylonian Numerals to Chaos Theory*, Quercus, London, 2007

Philip D. Straffin, *Game Theory and Strategy*, The Mathematical Association of America, Washington, DC, 1993

Chapter 4.

W. M. Priestley, *Calculus: A Liberal Art*, second edition, Springer-Verlag, New York, 1998

George F. Simmons, *Calculus Gems: Brief Lives and Memorable Mathematics*, McGraw-Hill, New York, 1992

Robert M. Young, *Excursions in Calculus: An Interplay of the Continuous and the Discrete*, Mathematical Association of America, Washington, DC, 1992

Chapter 5.

Martin Erickson, Anthony Vazzana, *Introduction to Number Theory*, Chapman & Hall/CRC, Boca Raton, FL, 2008

Joseph H. Silverman, *A Friendly Introduction to Number Theory*, Pearson/Prentice Hall, Upper Saddle River, NJ, 2006

Simon Singh, *Fermat's Enigma: The Epic Quest to Solve the World's Greatest Mathematical Problem*, Walker and Company, New York, 1997

Index